Models and Modeling in Science Education

Volume 12

More information about this series at http://www.springer.com/series/6931

Annette Upmeier zu Belzen • Dirk Krüger
Jan van Driel
Editors

Towards a Competence-Based View on Models and Modeling in Science Education

Springer

Editors
Annette Upmeier zu Belzen
Department of Biology, Biology Education
Humboldt-Universität zu Berlin
Berlin, Germany

Dirk Krüger
Department of Biology, Biology Education
Freie Universität Berlin
Berlin, Germany

Jan van Driel
Department of Education
Melbourne University
Melbourne, Australia

ISSN 1871-2983 ISSN 2213-2260 (electronic)
Models and Modeling in Science Education
ISBN 978-3-030-30254-2 ISBN 978-3-030-30255-9 (eBook)
https://doi.org/10.1007/978-3-030-30255-9

This Springer imprint is published by the registered company Springer Nature Switzerland AG.
The registered company address is: Gewerbestrasse 11, 6330 Cham, Switzerland

Preface

The book *Toward a Competence-Based View of Models and Modeling in Science Education* addresses the theoretical and empirical basis for a competence-based view of models and modeling in science learning and science education research.

The science education standards from many countries have unanimously demanded that students develop and apply models as a practice and understand the role of models in terms of scientific knowledge (KMK, 2005; NGSS Lead State, 2013).

However, the ways in which this unanimous demand is addressed in different countries sometimes differs with respect to theoretical foundations and empirical results from science education research.

Different approaches will be described and analyzed, and the extent to which they can be combined into a competence-based approach to form the concept of models as a research instrument and modeling as a scientific practice in science education will be discussed. In addition, the book is aimed at providing practical guidance by outlining evidence-based approaches to diagnosing and promoting modeling competence.

There are currently several theoretical frameworks for competences in the literature, some of which have been based on theoretical views, such as frameworks for modeling competence (e.g., Grünkorn, Upmeier zu Belzen, & Krüger, 2014; Nicolaou & Constantinou, 2014). Others describe the development of students' views (learning progressions, e.g., Schwarz, Reiser, Acher, Kenyon, & Fortus, 2012) or focus on the modeling process from a scientific perspective (e.g. Gilbert & Justi, 2016). The book explores, interprets, and discusses models and modeling through these different perspectives. It is aimed at initiating reflection on the relationships between these different perspectives and their relationships to the idea of competence. With this book, we want to convey a strong understanding of models and modeling for professions such as teacher educators, science education researchers, and scientists.

Finally, we discuss the findings presented in this book and identify research challenges for the future.

The book is divided into four sections. Section A deals with current thinking about the terms model and modeling, focusing on the development of modeling competence in science education and philosophical aspects, including perspectives on the nature of science. In Section B, different methods for the diagnosis and assessment of modeling competence are presented and discussed with regard to their potential and limitations. Section C provides evidence-based ideas about how teachers can be supported to teach with models and modeling from a competence-based perspective. Section D describes how students can develop modeling competence. This is done from a competence-based perspective that combines knowledge acquisition and performance expectations (Upmeier zu Belzen, Alonzo, Krell, & Krüger, 2019).

Berlin, Germany — Annette Upmeier zu Belzen
Melbourne, Australia — Jan van Driel
Berlin, Germany — Dirk Krüger

References

Gilbert, J. K., & Justi, R. (2016). *Modelling-based teaching in science education* (Vol. 9). Cham, Switzerland: Springer.

Grünkorn, J., Upmeier zu Belzen, A., & Krüger, D. (2014). Assessing students' understandings of biological models and their use in science to evaluate a theoretical framework. *International Journal of Science Education, 36*(10), 1651–1684.

KMK [Sekretariat der Ständigen Konferenz der Kultusminister der Länder in der BRD]. (2005). *Bildungsstandards im Fach (a) Biologie, (b) Chemie, (c) Physik für den Mittleren Schulabschluss [educational standards in (a) biology, (b) chemistry, (c) physics for middle school graduation]*. München/Neuwied, Germany: Wolters Kluwer.

NGSS Lead States (Ed.). (2013). *Next generation science standards: For states, by states*. Washington, DC: National Academies Press.

Nicolaou, C. T., & Constantinou, C. P. (2014). Assessment of the modeling competence: A systematic review and synthesis of empirical research. *Educational Research Review, 13*(3), 52–73.

Schwarz, C. V., Reiser, B., Acher, A., Kenyon, L., & Fortus, D. (2012). MoDeLS: Challenges in defining a learning progression for scientific modeling. In A. Alonzo & A. Gotwals (Eds.), *Learning progressions in science (LeaPS)* (pp. 101–137). Boston: Sense Publishers.

Upmeier zu Belzen, A., Alonzo, A. C., Krell, M., & Krüger, D. (2019). Learning progressions and competence models – A comparative illustration through models and modeling in science education. In E. McLoughlin, O. Finlayson, S. Erduran, & P. Childs (Eds.), *Bridging research and practice in science education (Selected papers from the ESERA 2017 Conference)*. Chapter 16. Springer.

Abstract

The chapters of the book cover two perspectives in science education and science education research.

The first perspective deals with the development of the term model in science and science education and the practice of modeling in the context of inquiry processes. This development has been influenced by philosophy of science, science itself, science education, and aspects of teaching and learning.

The so-called practice turn can be described as a shift from the conceptualization of models as ontological entities to the perception of models as functional tools that can be used for scientific inquiry (i.e., as epistemological entities). The underlying idea is not to simply do science and think about how experts conduct their research to improve authentic student learning. Rather, students should practice and discover their pathways of inquiry in ways that are appropriate for their individual learning environment and situation. Within this development, modeling has become a core practice in science teaching and learning and in science education.

The second perspective reflects the idea of competence-oriented teaching and learning. At least in German-speaking countries, the invention of this competence-oriented approach was linked to the implementation of standards as a result of the so-called PISA shock. Learning outcomes in terms of observable performances, as they are diagnosed according to given standards, provide insight into the underlying latent disposition: the individual competence. Competences are understood as domain-specific and learnable. By definition, they include cognitive, motivational, and volitional aspects. The theoretical foundations of cognitive aspects of a competence are derived a priori from the literature and structured in competence models. These models for student learning provide the foundation for teaching, learning, assessing, and diagnosing the competence.

In this book, the authors discuss different approaches by which to promote students' learning about models and modeling in science education, taking into account the two perspectives presented above. The preface explains the reason and main idea for creating this book, and the summaries of the chapters provide a first glance into the chapters. In the introductory part (Chap. 1), the main concepts of the book such as the term model, the practice of modeling, and the term competence as well as competence-based teaching and learning are defined with respect to the purpose of the book. After this first chapter, the book contains 17 subsequent chapters divided into four sections: views on models and modeling, assessing and diagnosing competences with respect to models and modeling, teacher training for competence-based teaching about models and modeling, and development of students' competences with respect to models and modeling. The book ends with a final chapter in which the relationships between the sections and the various chapters of the book are examined before further discoveries and challenges for research on models and modeling in science education are presented.

However, this book shows the current state of thinking about models and modeling in science education and science education research. The authors discuss the relevance of modeling competences, the various forms of knowledge required to fulfill these competences, and the possible ways of reasoning while using models for a specific purpose that is appropriate for the particular situation. In summary, the book will raise new questions that need to be considered in the further development of our field through research and by bridging multidisciplinary considerations in order to contribute to the development of science education that meets the challenges of the twenty-first century.

Contents

Abbreviations

AOI	Area of Interest
CODAP	Common Online Data Analysis Platform
CTA	Concurrent Think-Aloud
ECD	Evidence-Centered Design
EIP	Epistemologies in Practice
EMME	Eye Movement Modeling Examples
FMC	Framework for Modeling Competence
MBL	Model-Based Learning; Modeling-Based Learning
MLF	Modeling-Based Learning Framework
MMI	Multi-Method Interview
NOS	Nature of Science
RTA	Retrospective Think-Aloud

Part I
Theoretical Approaches to Models and Modeling

Chapter 1
Introducing a Framework for Modeling Competence

Annette Upmeier zu Belzen, Jan van Driel, and Dirk Krüger

1.1 Introduction

The purpose of this chapter is to locate the book's idea within a larger context regarding the definitions of models, modeling, and competence, beginning by describing the increasing relevance of models over time and resulting in the presentation of a competence-based approach for structuring different aspects and levels of modeling competence.

The "career" of the term "scientific models" began in the 1980s[1] and was related to shifts "[…] *from disregard to popularity, from formal accounts to a functional characterization of models, from the role of models in science to their role in human cognition*" by Bailer-Jones (1999, p. 24).

The disregard is related to the substitutive role of models as appendices to theory without their own relevance to scientific thinking. The gain in popularity began when models were considered relevant for scientific discovery and thus theory change, which led to an increasing focus on the functions of models in research

[1] A database research study in Scopus revealed an increasing number of publications on the terms "scientific models," "models," and "modeling" OR "modelling" between 1980 and 1990 at the time of this writing (query dated 10.01.2019).

A. Upmeier zu Belzen (✉)
Department of Biology, Biology Education, Humboldt-Universität zu Berlin, Berlin, Germany
e-mail: annette.upmeier@biologie.hu-berlin.de

J. van Driel
Department of Education, Melbourne University, Melbourne, Australia

D. Krüger
Department of Biology, Biology Education, Freie Universität Berlin, Berlin, Germany

A. Upmeier zu Belzen et al. (eds.), *Towards a Competence-Based View on Models and Modeling in Science Education*, Models and Modeling in Science Education 12, https://doi.org/10.1007/978-3-030-30255-9_1

processes. In this, a universal conception relying on formal accounts of models was displaced, and a characterization open to a diversity of conceptions corresponding to different functions arose. But still, understanding the different roles of models in science means understanding the epistemology *of* science in the sense of a property of science. However, from a rather constructivist perspective, research is more about sensemaking and figuring out personal epistemologies *for* science. In the latter, the function of models is considered not just within science but also in human cognition so that models are now also viewed as tools of actual scientific thinking (Bailer-Jones, 1999; Russ, 2014).

Nevertheless, the roles of models remain twofold. Models as media are needed to communicate about scientific research and to convey content learning in schools. Using models in the sense of a modeling practice means using models and modeling as research tools for inquiry purposes to gain insights into previously unknown aspects of a phenomenon (Krüger, Kauertz, & Upmeier zu Belzen, 2018; Upmeier zu Belzen & Krüger, 2010).

1.2 Theoretical Background

1.2.1 Relationships Between Modeling and Other Inquiry Practices

The process of modeling can be considered a concrete inquiry practice in which hypotheses about a phenomenon are derived from an initial model with a certain theoretical focus about a structure, process, or system (Upmeier zu Belzen & Krüger, 2010). In science education, basic inquiry practices consist of observing, comparing, and classifying as well as experimenting (Nowak, Nehring, Tiemann, & Upmeier zu Belzen, 2013). Against this background, the question of the position of modeling arises and can be answered from either an explicit or implicit position. Scientific modeling might be seen explicitly as one method alongside others (Fig. 1.1). Alternatively, modeling steps might be considered integral within all inquiry processes (cf. Hartmann, Upmeier zu Belzen, Krüger, & Pant, 2015; cf. Lehrer & Schauble, 2015, p. 707). In the latter, inquiry practices in general are considered modeling activities, each with specific characteristics of scientific thinking that depend on the content and scientific question and result in specific hypotheses and investigations (Fig. 1.1). Both trains of thought are allowed, each with a different main emphasis. Treating a modeling process as a practice of inquiry is helpful for teaching and learning because models and modeling can be experienced in their epistemological functioning as research tools for human cognition. From a philosophical point of view, the two ideas focus on the reasoning behind the modeling activities and, at the same time, the semantic view of models and modeling (Sect. 1.2.3).

Scientific thinking / Inquiry practice	Observing	Comparing/Classifying	Experimenting	Modeling
Research question on …	… functional relationships between systems (e.g. structure and function)	… criteria, categories, orders	… causes and effects	… modeling variables of phenomena
Hypothesis on …	… relations	… differences	… causes and effects	… all kinds of hypotheses
Investigation as …	… systematic observations of features with feature types or over time	… comparisons, categorizations, systematizations	.. controlled experiments	… all kinds of investigations
Results in terms of …	… correlations between variables, descriptions of features, structures, and systems, possibly over time	… categories and category systems, possibly hierarchical, matrices of objects and superior comparison criteria	… relations between causes and effects, causal explanations	… findings about a phenomenon or changing of the model

Fig. 1.1 Matrix of scientific thinking and inquiry practice

More recently, the idea of using models and modeling as personal cognitive tools for inquiry has become increasingly prominent. Therefore, different types of inquiry and reasoning have been discussed, such as modeling pedagogies (Campbell & Oh, 2015), reasoning styles (Osborne, 2018), and modeling frames (Louca, Zacharia, & Constantinou, 2011). These alternative structures of cognitive strategies during inquiry processes help in both school practice and science education research and can be applied to models and modeling as well as to other methods of inquiry.

1.2.2 Teaching and Learning with Models and Modeling for Inquiry and Thinking

To use models and modeling for scientific thinking and inquiry practices in schools, it is necessary to consider the perspective of learners and learning. Science education curricula (e.g. KMK, 2005; NGSS Lead States, 2013; NRC, 2012) entail standards to bring knowledge into action in terms of skills, performances, or competences. Research findings have suggested that models are used as media to describe and understand content rather than as research tools to gain new knowledge and to understand the role of scientific inquiry (Grünkorn, Upmeier zu Belzen, & Krüger, 2014). When using models and modeling for inquiry, expressive and exploratory modeling are the most commonly used pedagogies in science education, whereas cyclic modeling is used the least (Campbell & Oh, 2015; Krell & Krüger, 2016).

Against this background, one reason to publish this volume is to strengthen the systematic application of models and modeling in science education to go beyond their use as media. Their use as media will always remain important for teaching and learning content knowledge, but this use of models is not sufficient and must be complemented with the use of models as research tools when the goal is to acquire competence in scientific thinking and inquiry practices (Fig. 1.1).

1.2.3 Modeling Student Learning in Competence Models

At this point, models of student learning must come into play. For example, student learning can be modeled with competence models or learning progressions, two prominent examples from different cultural backgrounds (Upmeier zu Belzen, Alonzo, Krell, & Krüger, 2019). They have in common that they model a skill or a competence to be acquired. This book broadly discusses the competence-based approach to models and modeling. Along with Koeppen, Hartig, Klieme, and Leutner (2008, p. 68), Rychen and Salganik (2003, p. 43) defined the construct of competence coherently as “domain-specific cognitive dispositions that are required to successfully cope with certain situations or tasks, and that are acquired by learning processes.” An essential element of this definition is the contextual specificity and learnability of competence, as it has been introduced as an alternative to the focus on context-independent cognitive dispositions that are limited in learning (e.g. McClelland, 1973; “Testing for competence rather than for ‘intelligence’”). “In contrast, competences reflect a person’s potential to meet cognitive demands in specific areas of learning and behavior” (Koeppen et al., 2008, p. 62) in order to successfully solve problems in various situations (Klieme, Hartig, & Rauch, 2008). Competences “are, thus, more closely related to ‘real life.’” Connell, Sheridan, and Gardner (2003, p. 142) concisely characterized competences as ‘realized abilities’” (Koeppen et al., 2008, p. 62). In other words, competences are latent and complex constructs that encompass both the knowledge and skills that manifest during performance. However, according to Ropohl, Nielsen, Olley, Rönnebeck, and Stables (2018), the concept of competence is still under discussion due to its many components. Whereas cognitive aspects are always considered part of competence and therefore included in competence models, volitional components are often not considered (Koeppen et al., 2008).

Competences or performance expectations describe current goals for education rather than content lists students should learn (Koeppen et al., 2008). Models of student learning provide information about educational goals, curricula, teaching, and assessment (e.g., Gotwals, 2012; Reusser, 2014). As such, they mediate between standards, educational goals, teaching activities, and student learning. Thus, they can support lessons tailored to students’ learning needs (e.g., Alonzo, 2011).

Competences in terms of an expected outcome of learning processes are empirically investigated, and competence characteristics are diagnosed as clearly as possible using test procedures. With a focus on models and modeling in science education, modeling competence has been defined and structured in a framework for modeling competence (FMC) that incorporates both, models as media and models as research tools. Empirical studies have shown that the assumed structure is predominantly supported and can thus be used as a basis for the evidence-based promotion of modeling competence (Krell, Upmeier zu Belzen, & Krüger, 2016). As models and modeling are the central constructs of the FMC, we offer a theoretical clarification of them in the following.

1.2.4 The Term "Model"

Models are the central tools and resources of science. Models are used as tools to gain new insights and as media to communicate already known facts (Gilbert & Justi, 2016; Giere, Bickle, & Mauldin, 2006; Gouvea & Passmore, 2017; Passmore, Gouvea, & Giere, 2014). The scientific importance of models also explains their use in the science education curricula of schools around the world (e.g., in Germany (KMK, 2005); in the U.S. (NRC, 2012; NGSS Lead States, 2013)).

Given the importance of models, it may be surprising that in interdisciplinary discourses, no general classification systems are available for models (Mittelstraß, 2005), and even within the sciences, different classifications of models have been proposed in education research (e.g. Crawford & Cullin, 2005 for Biology; Justi & Gilbert, 2002 for Chemistry; Kircher, 2015 for Physics, pp. 804ff). These categorizations, which are phenomenologically oriented (Ritchey, 2012) or ontologically oriented (Oh & Oh, 2011), do not produce a satisfactory result because they provide only a one-criterion-based system without demonstrating insights into the functions and epistemologies of these models. However, what these models do have in common is that they are all connected by subject, purpose, and time (Giere, 2010; Stachowiak, 1973). People can therefore judge and interpret models as representations of original objects, phenomena, or systems of the experiential world. These representations depict the experiential world and also allow a person to derive and test hypotheses for a particular purpose and for a limited period of time. From the need to optimize models when needed, it follows that no one can claim that there is only one valid model. Because of this, models have focused meaning and a limited scope, that is, a special theoretical focus.

Despite the recognition of the scientific importance of models and modeling, there is no unified definition of the concept of the model in science and science education (Agassi, 1995; Gilbert & Justi, 2016; Sect. 1.2.2) nor is there a unified modeling theory (Ritchey, 2012). Mittelstraß (2005, p. 65) provided a general framework: "Models are replicas of a real or imaginary object with the aim of learning something about it or learning something from it." Mittelstraß pointed to both the descriptive and the research function of modeling. Special approaches to the concept of models have been presented by scientists in Cognitive Psychology (e.g. Nersessian, 2008), Philosophy (e.g. Bailer-Jones, 2003; Giere, 2010), Computer Science (e.g. Mahr, 2012), and subject-related Education Research (e.g. Gilbert & Justi, 2016; Gouvea & Passmore, 2017; Upmeier zu Belzen, 2013). Different terms have been used in their definitions of models: e.g. mental model (Nersessian, 2008), representation (Giere, 2004), theoretical model (Upmeier zu Belzen, 2013), abstractor, or analogy (Oh & Oh, 2011). Nersessian (2008, p. 93) defined a mental model as a "structural, behavioral, or functional analog representation of a real-world or imaginary situation, event, or process." In Giere's (2004) early view, models were described as representations of natural objects, processes, or phenomena that have been developed for a particular purpose and have a similarity to what they represent. When such models relate to the world, hypotheses arise with regard to adapting a model to a particular section of the world. In the most recent discussions (Mahr, 2015), this ontological definition of the concept of models is based on the existence of the represented steps of the model object in the background, and an epistemological position is taken in which models are used to understand the experiential world. For example, Giere (2010, p. 269) wrote: "Agents intend to use model M to represent a part of the world W for some purpose P." In this definition, the role of the modeler is significant when an object is used as a model. From this point of view, there are two conclusions that can be drawn: Depending on the purpose, there can be several models for a phenomenon that allow different applications so that one and the same representation can be used for different purposes (Gilbert & Justi, 2016, p. 21). To account for this epistemological function of models, models should not solely be considered as representations that are judged by how well they fit the particular phenomenon (Gilbert & Justi, 2016; Gouvea & Passmore, 2017). Rather, the nature of models as cognitive tools should be emphasized. Gouvea and Passmore (2017) suggested talking about *models for* (as method) instead of *models of* (as media). According to Gilbert and Justi (2016, p. 21), it helps to conceive of models as substitute systems (see Mäki, 2005) or to describe models as epistemic tools (Ritchey, 2012). This opens the tool-like character of the models for exploration (Gouvea & Passmore, 2017; Passmore et al., 2014).

1.2.5 The Idea of Model-Being

More and more authors are approaching the concept of models from an epistemological point of view (Gilbert & Justi, 2016; Mahr, 2012, 2015; Passmore et al., 2014). In this case, something becomes a model when it is used (Giere, 2010), developed (Ritchey, 2012), or conceived (Mahr, 2015) by a subject as a model because the subject made a judgment about model-being. A consistent epistemological perspective is presented through Mahr's (2015) model of model-being in the following approach. It can be used as a basis for theoretical justifications in the levels of modeling competence (Upmeier zu Belzen & Krüger, 2010).

Because the term "model" is a homonym with different meanings (e.g. for people from the fashion industry or art, true-to-scale organ sculptures, mathematical systems of equations, architectural designs, or map drawings), Mahr (2015) refrained from investigating and defining the ontological properties of models. Rather, he tried to epistemologically elucidate why an object is conceived as a model. He distinguished between an imagined (mental) model (e.g. climate change) and a model object that represents the model in the broadest sense (computer simulation of climate change). According to Mahr (2015), the mental model is thus represented by the model object, where it has two relationships to the perception of a subject: it is both a model *of* something and a model *for* something (cf. Gouvea & Passmore, 2017). These constructive relationships, being a model *of* something (perspective of construction; Fig. 1.2) and being a model *for* something (perspective of application; Fig. 1.2), justify the judgment of model-being (Mahr, 2015; Passmore et al., 2014).

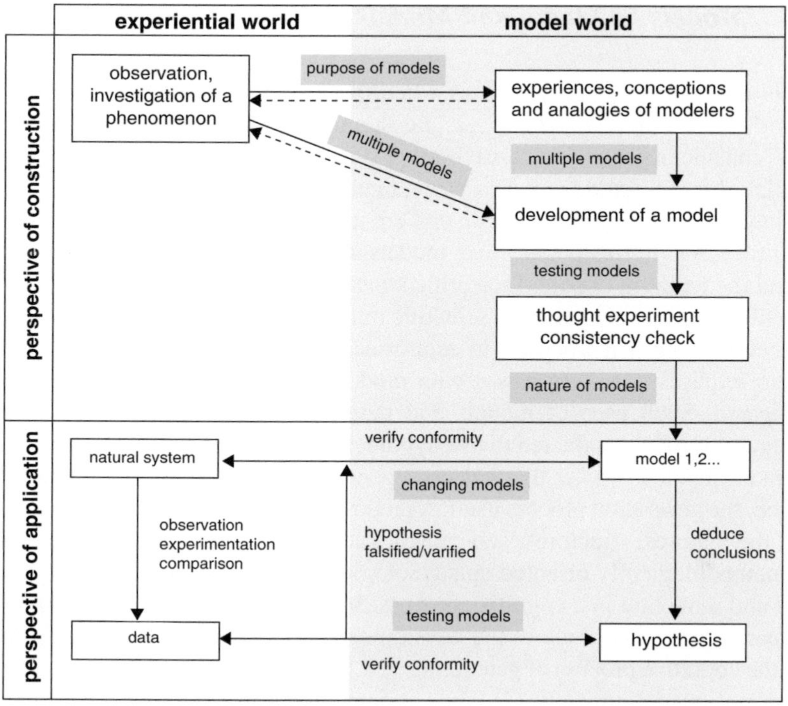

Fig. 1.2 Framework for the modeling process (Krell et al., 2016). Gray boxes indicate aspects of the FMC (Fig. 1.3) when going through inquiry processes

With this definition, Mahr (2015) provided conditions that, separated from inherent properties permanently associated with the model object, conceived of an object as a model that can be used by a person or group as a model *of* something and *for* something for a given time. The described aspects, that is, the model object that is a model *of* something (representative of a phenomenon) and, from an application perspective, a model *for* something (a medium in a mediation situation or a tool in the process of knowledge realization), can be used to think about how models are used to describe levels of competence. The relationship between a person who makes or uses the model and the model itself plays a central role. Giere (2010) described the subject as an *agent*, the person who decides both the focus of the similarities (intent) and the goal of that focus (purpose). In his approach to model-being, Mahr (2012) also consistently thinks with the subject when he distinguishes between the mental model that is modeled by the subject and the model object (i.e., the mental model externalized by the subject) as well as the creation and application of the model object. The perspectives of model-being on the model object, on the construction and application of the model provide descriptions for levels of competence. They are based on the fact that these perspectives can be considered in problem-solving situations when creating and using models (Upmeier zu Belzen & Krüger, 2010).

1.2.6 Models as Media and Models as Research Tools

Models are used in science classes when, for example, structures, processes, or systems are not immediately accessible due to their size, speed, abstractness, or ethical considerations. Models are both physical and theoretical representations of an initial object filtered through a theoretical perspective for a particular purpose. They have a representative function and are used to describe structures or to understand processes. In this perspective, models act as media that support communication and the learning of sound scientific knowledge.

Gaining new insights through scientific thinking with modeling not only explains retrospectively known processes to uninformed people by using models or modeling but also requires new hypotheses with models of phenomena that have yet to be investigated. Such uses of models and modeling allow theoretical or empirical investigations to be conducted to test hypotheses. At the center of this process is not the content-related answer that has been generated, but in order to understand and question, the generation process itself. A predominantly medially oriented ontological use of the features, structures, and categorizations of a model is therefore extended by a methodologically oriented epistemological perspective on the function of the model and modeling in a cognitive process. Models that are used methodologically as research tools contribute to the development of competences in three ways: (1) when the cognitive process of generating new knowledge is reflected through models and through modeling itself, (2) when an understanding of the nature of science is developed, and (3) when content knowledge is gained.

1.2.7 Modeling as a Process

Thereby, the particular importance of modeling becomes clear when it is recognized that modeling can be linked to scientific practices (e.g. observing, comparing, classifying, or experimenting; Mäki, 2005; Morgan, 2005; Fig. 1.1). Whereas in the case of an experiment, the isolation and manipulation of variables in the modeling process is done on a theoretical level in the model world (Clement, 2009), the isolation and manipulation of selected variables follows a material transformation in the experiential world (Mäki, 2005). This allows an empirical examination of predictions in the experiential world, predictions that have been derived from modeling (Giere et al., 2006; Fig. 1.2).

If the hypotheses derived from the model contradict the data, then the conclusion that the model and the system are mismatched can be drawn. In this case, the model has to be optimized or the concept of the modeled phenomenon has to be changed. This requires a new test and demonstrates the cyclical character of the modeling process. In the hypothesis-based comparison of the theoretical model world and the experiential world (Giere et al., 2006), the functions of modeling are recognized as a method of finding scientific knowledge.

Modeling is a complex process that is fundamental to both scientific knowledge generation and people's problem-solving skills (e.g. Nersessian, 2008). Various authors have described the purpose-oriented design as well as the corresponding testing and modification of (mental) models from different perspectives, for example, cognitive-psychological (e.g. Nersessian, 2013) or the school context (e.g. Clement, 1989; Fleige, Seegers, Upmeier zu Belzen, & Krüger, 2012; Gilbert & Justi, 2016).

In principle, the process of modeling dispenses with a strict procedural description and the definition of certain rules because modeling can be seen as an art with creative elements (Morrison & Morgan, 1999). Therefore, it is not only theory or data that determine modeling, but modeling also depends on the intuition and experience of the modeler, as in the case of the hypothetical-deductive approach to knowledge generation (Clement, 1989). Nevertheless, it is possible to identify recurring elements arranged according to the ideal type of research logic (Popper, 2005), which can also be used in other practices, for example, in a scientific observation or an experiment. The starting point of the modeling process is an observed phenomenon, which, taking into account the purpose of the model and the prior knowledge and experience of the person doing the modeling, leads to a first draft of a model that presents the relevant variables of the phenomenon. This step is referred to as the construction of a mental model (Nersessian, 2008), initial model (Clement, 1989), or proto-model (Gilbert & Justi, 2016) and is performed on a mental level. First, an attempt is made to identify a known suitable or analogous (professional) model by means of an observation. If this cannot be achieved or is insufficient, new model elements and links are generated on this basis. In the development of the model, the internal consistency and fit to the phenomenon are examined. The process results in one or more externalizations, which can be referred to as the model object(s) (Mahr, 2015). The model object as a medium focuses on selected variables of the system. In addition, hypotheses can be derived from the conceptual model or model object about how the system will behave under certain conditions. Experimental investigations, comparisons, or systematic observations then lead to results that confirm or falsify the hypotheses that are being considered (Krell et al., 2016; Figs. 1.1 and 1.2).

As explained above, there are several approaches for describing modeling. So far, the process of modeling has been described rather generally. An attempt to develop a unified theory for this purpose was made by Ritchey (2012). He first defined a scientific model as consisting of at least two mental constructs (e.g. light as a physical variable and photosynthesis rate as a chemical variable) that can be interpreted as variables or dimensions and can be experimentally investigated. The modeler has to build relationships between these constructs or variables, e. g. a causal relationship. In addition, Ritchey (2012) characterized five features of modeling: The constructs can take on values or be nominal (no value), the contexts can be directed or not, their relationships can but do not have to be quantified, the relations can be cyclic or acyclic, and the type of relation can be mathematical/functional, probabilistic, quasi-causal, or non-causal (logical, normative). Ritchey (2012), however, allowed additional attributes to be assigned to a modeling process (e.g., continuous/discrete), but he left it with these five properties and identified 42 plausible modeling types with specific combinations of these properties.

In summary, models and modeling in science have two main functions. By externalizing a conceptual model in the form of a model object, scientists can communicate their ideas about a phenomenon and discuss it with others. Models are primarily used as media that transport and communicate the state of research. In addition, science gains new knowledge by applying and testing these models. In this sense, models are used as research tools for gaining knowledge and allow to reflect about the inquiry process.

1.3 The Framework for Modeling Competence

The FMC was developed for science education purposes and involves the use of models as research tools and modeling as a research practice. This notion of models and the reflection of the modeling process are interdisciplinary and considered part of a scientific understanding (Gobert et al., 2011; Reinisch & Krüger, 2018) that has been conceptualized as "a type of nature of science understanding" and encompasses "how models are used, why they are used, and what their strengths and limitations are in order to appreciate how science works and the dynamic nature of knowledge that science produces" (Schwarz et al., 2009, pp. 634–635). Therefore, modeling competence includes the ability to gain insightful knowledge with models, to be able to judge models with regard to their purpose, and to reflect on the process of gaining knowledge through models and modeling (Krüger et al., 2018; Upmeier zu Belzen & Krüger, 2010). Thus, the framework provides a theory-based overview of how students and pre- and in-service science teachers should understand models and modeling in science.

1.3.1 Competence as an Ability to Reflect on Models and Modeling

Building on different structural approaches in the natural sciences (e.g., Crawford & Cullin, 2005; Grosslight, Unger, Jay, & Smith, 1991; Justi & Gilbert, 2003), Upmeier zu Belzen and Krüger (2010) differentiated between five theoretical aspects of modeling competence (Fig. 1.3). These aspects were based on the results of international studies on students' (e.g., Grosslight et al., 1991) and teachers' (e.g., Crawford & Cullin, 2005) conceptions of models and modeling: *nature of models, multiple models, purpose of models, testing models,* and *changing models* (Krell et al., 2016; Krüger et al., 2018; Upmeier zu Belzen & Krüger, 2010). Further, for each aspect, they identified levels that are based on Mahr's (2015) conceptualization of model-being. The proposed structure (five aspects with three levels each)

Aspects	Level I	Level II	Level III
Nature of Models	Replication of the phenomenon	Idealized representation of the phenomenon	Theoretical reconstruction of the phenomenon
Multiple Models	Different model objects	Different foci on the phenomenon	Different hypotheses about the phenomenon
Purpose of Models	Describing the phenomenon	Explaining the phenomenon	Predicting something about the phenomenon
Testing Models	Testing the model object	Comparing the model and the phenomenon	Testing hypotheses about the phenomenon
Changing Models	Correcting defects in the model object	Revising due to new insights	Revising due to the falsification of hypotheses about the phenomenon

Fig. 1.3 Framework for modeling competence with five aspects and three levels

has been extensively investigated (cf. Krell et al., 2016); nevertheless, it should be interpreted as a nominal category system until it can be regarded as an empirically validated developmental model (cf. Kauertz, Fischer, Mayer, Sumfleth, & Walpuski, 2010). The levels (Fig. 1.3) are theoretically described as follows:

Level I: The ability to assess the appearance of the model object (cf. Mahr, 2015) from an aesthetic point of view or technical functionality without putting the phenomenon in relation to the model object, except in its capacity as a copy or for the purpose of illustrating; the model object is judged as such.

Level II: The ability to assess the process of model construction; primarily, there is a focus on the model as media use of the model object as a more or less accurate representation of a phenomenon; the model object is representative *of* something already known in the natural sciences.

Level III: The ability to use a model in an application as a tool for investigating a phenomenon and thereby assessing its productivity; the model object as a model *for* something leads to the processing of new, thus far unexplained, scientific questions.

The aspects and their gradations can be described as follows:

Nature of models: The ratio of the similarity between the model and the phenomenon is assessed as a model *of* something. Competence is expressed in the different meanings of the model object as a true-to-life replica (level I), as an idealized representation (level II), or as a theoretical reconstruction of a phenomenon (level III).

Multiple models: Reasons are assessed for the existence of several models that represent one phenomenon. The variety of models is characterized by differences between the model objects (level I), different areas of focus in the construction of the models (model *of* something, level II), and various assumptions about a phenomenon and the application of the models in further examinations (model *for* something, level III).

Purpose of models: The purpose of models is to guide the corresponding process of modeling. If the purpose of models is to illustrate (level I) and explain (level II) something with educational intentions, then models are used as media (models *of* something). However, if the purpose of models is to derive a prediction from them, they become a model *for* something with the perspective of application as a tool in the generation of knowledge (level III).

Testing models and *changing models*: The levels describe different ways and reasons to test and to change models. Level I is about tests and optimizations at the model object only. On level II, the model object is often parallelized with the phenomenon and is improved in the case of misfit. On level III, the model object as a model *for* something is tested through the verification of previously derived hypotheses and changed when the hypotheses are rejected.

The aspects and levels represent perspectives of reflection, which not only receive their meaning in an abstract, cognitive reflection on the term model but are relevant under the competence-based perspective in subject-related problem-solving situations at different stages of the cyclic process of modeling (Fig. 1.2).

1.3.2 Empirical Investigations

The theoretically based FMC (Fig. 1.3) has been empirically examined and the results have been incorporated into its further development (cf. Krell et al., 2016). The framework is based on qualitative interview studies on the perceptions of students and teachers with regard to models and modeling and the roles of models and modeling in an inquiry process (e.g., Crawford & Cullin, 2005; Grosslight et al., 1991; Krüger et al., 2018). Furthermore, when using open-ended tasks, initial levels have been identified for the aspects of *multiple models* (rejecting the existence of multiple models), *testing models* (rejecting the testing of models), and *changing models* (rejecting the changing of models; Grünkorn et al., 2014).

Using quantitative methods, the extent to which the structure of the FMC (aspects, levels) can be empirically supported (e.g. Terzer, Hartig, & Upmeier zu Belzen, 2013) has been examined. From an educational point of view, the organization into the aspects has great diagnostic potential (Fleige et al., 2012). Empirically, however, it has not yet been conclusively clarified whether modeling competence can be viewed as a five-dimensional (Krell, 2013) or one-dimensional construct (Terzer, 2013). By contrast, the assumption of three ordinal levels was substantiated except for the aspect of *testing models* (Krell, 2013; Terzer, 2013).

A longitudinal study for evaluating the FMC as a development model in Grades 7 to 10 has shown that students' (13–16 years) modeling competence results in a significant development, but the effect sizes were small (Patzke, Krüger, & Upmeier zu Belzen, 2015). Also, the modeling competence of pre-service biology, chemistry, and physics teachers has demonstrated development throughout several studies in the aspects *purpose of models, testing models*, and *changing models* (Hartmann et al., 2015; Mathesius, Upmeier zu Belzen, & Krüger, 2014).

Successfully training pre-service biology teachers with an explicit reflection of the FMC (Fig. 1.3) led to a significant increase in modeling competence in all five aspects with average effect sizes (Günther, Fleige, Upmeier zu Belzen, & Krüger, 2019). However, students who were taught by these trained teachers did not benefit from the increase in their teachers' modeling competence. The results showed that teachers with an elaborate modeling competence did not have adequate diagnostic competences to foster students' modeling competence (Günther et al., 2019).

Additionally, a tool with forced-choice tasks to receive immediate feedback was developed and validated in order to diagnose students in the aspects *nature of models* and *purpose of models* (Gogolin & Krüger, 2015, 2018). The tool makes it possible to offer individual support measures and to evaluate students' success directly.

1.4 Conclusion

In summary, with the FMC, we structure the different theoretical aspects and levels of modeling competence as a basis for teaching and learning. In order to use the FMC for evaluation purposes in certain domains, it has to be adapted with regard to content because the FMC is content free. Bearing in mind the presented perspectives (Fig. 1.3), it is possible to evaluate whether students or pre-service or in-service biology teachers exhibit more or less elaborated performances while solving tasks with certain contents. The FMC allows a person's potential to solve problems in varying situations with models and modeling to be assigned to different levels of the five aspects (Upmeier zu Belzen & Krüger, 2010). Whereas cognitive aspects are considered in the FMC, volitional and behavior-related components are not directly included although they are needed to show modeling competence.

The FMC is located between the theory of competence and competence-oriented teaching in special domains, (Upmeier zu Belzen et al., 2019). It is derived from teaching methodology, the psychology of learning, and the philosophy of science. Although the FMC has been conceptualized as a structural model, empirical evidence that the levels are hierarchically ordered still has to be provided before it can be considered a developmental model (cf. Schecker & Parchmann, 2006).

The FMC provides a strong foundation for empirically testing the structure of modeling competence, and it can support the understanding of the aspects and levels of modeling competence and student learning as well as the development of curricular materials (Fleige et al., 2012; Rahmenlehrplan Berlin/Brandenburg, 2015). Two main functions of models need to be highlighted in this context: By developing a model object as a representation of the model, scientists are able to communicate their conceptions about a phenomenon and discuss it with others. In this case, models are primarily used as media (level I and level II) that carry the state of scientific knowledge. In addition, science is gaining new knowledge by applying and testing models. In this sense, models are used to generate hypotheses about unknown phenomena (level III). Models in this sense are research tools that are used to gain new knowledge. The FMC provides an integration of ontological, procedural and epistemological functions of models and allows researchers to determine students' and pre-service and in-service teachers' modeling competence.

References

Agassi, J. (1995). Why there is no theory of models? In W. Herfel, W. Krajewski, I. Niiniluoto, & R. Wójcicki (Eds.), *Theories and models in scientific processes. Proceedings of AFOS '94 workshop, August 15–26, Madralin and IUHPS '94 conference, August 27–29, Warszawa* (pp. 17–26). Rodopi, France: Amsterdam & Atlanta.

Alonzo, A. C. (2011). Learning progressions that support formative assessment practices. *Measurement: Interdisciplinary Research and Perspectives, 9*(2–3), 124–129.

Bailer-Jones, D. (1999). Tracing the development of models in the philosophy of science. In L. Magnani, N. J. Nersessian, & P. Thagard (Eds.), *Model-based reasoning in scientific discovery. Proceedings of an international conference on model-based reasoning in scientific discovery, held December 17–19, 1998, in Pavia, Italy* (pp. 23–40). New York: Kluwer Academic.

Bailer-Jones, D. (2003). When scientific models represent. *International Studies in the Philosophy of Science, 17*(1), 59–74.

Campbell, T., & Oh, P. S. (2015). Engaging students in modeling as an epistemic practice of science: An introduction to the special issue of the journal of science education and technology. *Journal of Science Education and Technology, 24*, 125–131.

Clement, J. (1989). Learning via model construction and criticism. In J. Glover, C. Reynolds, & R. Royce (Eds.), *Handbook of creativity* (pp. 341–381). Berlin, Germany: Springer.

Clement, J. (2009). *Creative model construction in scientists and students*. Dordrecht, The Netherlands: Springer.

Crawford, B., & Cullin, M. (2005). Dynamic assessments of preservice teachers' knowledge of models and modelling. In K. Boersma, M. Goedhart, O. de Jong, & H. Eijkelhof (Eds.), *Research and the quality of science education* (pp. 309–323). Dordrecht, The Netherlands: Springer.

Fleige, J., Seegers, A., Upmeier zu Belzen, A., & Krüger, D. (2012). Förderung von Modellkompetenz im Biologieunterricht [Fostering modeling competence in biology classes]. *Der mathematische und naturwissenschaftliche Unterricht, 65*, 19–28.

Giere, R. N. (2004). How models are used to represent reality. *Philosophy of Science, 71*(5), 742–752.

Giere, R. N. (2010). An agent-based conception of models and scientific representation. *Synthese, 172*(2), 269–281.

Giere, R. N., Bickle, J., & Mauldin, R. F. (2006). *Understanding scientific reasoning* (5th ed.). Belmont, CA: Thomson.

Gilbert, J. K., & Justi, R. (2016). *Modelling-based teaching in science education* (Vol. 9). Cham, Switzerland: Springer.

Gobert, J., O'Dwyer, L., Horwitz, P., Buckley, B. C., Levy, S., & Wilensky, U. (2011). Examining the relationship between students' understanding of the nature of models and conceptual learning in biology, physics, and chemistry. *International Journal of Science Education, 33*(5), 653–684.

Gogolin, S., & Krüger, D. (2015). Nature of models: Entwicklung von Diagnoseaufgaben [nature of models: Development of diagnostic tasks]. In M. Hammann, J. Mayer, & N. Wellnitz (Eds.), *Lehr- und Lernforschung in der Biologiedidaktik 6* (pp. S.27–S.41). Innsbruck: Studienverlag.

Gogolin, S., & Krüger, D. (2018). Students' understanding of the nature and purpose of models. *Journal of Research in Science Teaching, 55*(9), 1313–1338.

Gotwals, A. W. (2012). Learning progressions for multiple purposes: Challenges in using learning progressions. In A. C. Alonzo & A. W. Gotwals (Eds.), *Learning progressions in science: Current challenges and future directions* (pp. 461–472). Rotterdam, The Netherlands: Sense Publishers.

Gouvea, J., & Passmore, C. (2017). Models of' versus 'models for. *Science & Education, 26*(1–2), 49–63.

Grosslight, L., Unger, C., Jay, E., & Smith, C. (1991). Understanding models and their use in science: Conceptions of middle and high school students and experts. *Journal of Research in Science Teaching, 28*(9), 799–822.

Grünkorn, J., Upmeier zu Belzen, A., & Krüger, D. (2014). Assessing students' understandings of biological models and their use in science to evaluate a theoretical framework. *International Journal of Science Education, 36*(10), 1651–1684.

Günther, S. L., Fleige, J., Upmeier zu Belzen, A., & Krüger, D. (2019). Using the case method to Foster preservice biology Teachers' content knowledge (CK) and pedagogical content knowledge (PCK) related to models and modeling. *Journal of Science Teacher Education, 30*(4), 321–343. https://doi.org/10.1080/1046560X.2018.1560208

Hartmann, S., Upmeier zu Belzen, A., Krüger, D., & Pant, H. A. (2015). Scientific reasoning in higher education: Constructing and evaluating the criterion-related validity of an assessment of pre-service science teachers' competencies. *Zeitschrift für Psychologie, 223*, 47–53.

Justi, R., & Gilbert, J. K. (2002). Philosophy of chemistry in university chemical education: The case of models and modelling. *Foundations of Chemistry, 4*(3), 213–240.

Justi, R. S., & Gilbert, J. K. (2003). Teachers' view on the nature of models. *International Journal of Science Education, 25*(11), 1369–1386.

Kauertz, A., Fischer, H., Mayer, J., Sumfleth, E., & Walpulski, M. (2010). Standardbezogene Kompetenzmodellierung in den Naturwissenschaften der Sekundarstufe I [Standards-based competence modeling in natural science of lower secondary levels]. *Zeitschrift für Didaktik der Naturwissenschaften, 16*, 135–153.

Kircher, E. (2015). Modellbegriff und Modellbildung in der Physikdidaktik [Concept of models and model construction in physics education]. In E. Kircher, R. Girwidz, & P. Häußler (Eds.), *Physikdidaktik: Theorie und Praxis* (3rd ed., pp. 783–807). Berlin/Heidelberg, Germany: Springer.

Klieme, E., Hartig, J., & Rauch, D. (2008). The concept of competence in educational contexts. In J. Hartig, E. Klieme, & D. Leutner (Eds.), *Assessment of competencies in educational contexts* (pp. 3–23). Göttingen, Germany: Hogrefe.

KMK [Sekretariat der Ständigen Konferenz der Kultusminister der Länder in der BRD]. (2005). *Bildungsstandards im Fach (a) Biologie, (b) Chemie, (c) Physik für den Mittleren Schulabschluss* [Educational standards in (a) biology, (b) chemistry, (c) physics for middle school graduation]. München/Neuwied, Germany : Wolters Kluwer.

Koeppen, K., Hartig, J., Klieme, E., & Leutner, D. (2008). Current issues in competence modeling and assessment. *Zeitschrift für Psychologie* [Journal of Psychology], *216*, 61–73.

Krell, M. (2013). *Wie Schülerinnen und Schüler biologische Modelle verstehen (dissertation)* [How students understand biological models (dissertation)]. Berlin, Germany: Logos.

Krell, M., & Krüger, D. (2016). Testing models: A key aspect to promote teaching-activities related to models and modelling in biology lessons? *Journal of Biological Education, 50*, 160–173.

Krell, M., Upmeier zu Belzen, A., & Krüger, D. (2016). Modellkompetenz im Biologieunterricht [Modeling competence in biology classes]. In A. Sandmann & P. Schmiemann (Eds.), *Biologiedidaktische Forschung: Band 1* (pp. 83–102). Berlin: Logos: Schwerpunkte und Forschungsstände.

Krüger, D., Kauertz, A., & Upmeier zu Belzen, A. (2018). Modelle und das Modellieren in den Naturwissenschaften [Models and Modeling in Natural Sciences]. In D. Krüger, I. Parchmann, & H. Schecker (Hrsg.), *Theorien in der naturwissenschaftsdidaktischen Forschung* (pp. 141–157). Berlin/Heidelberg, Germany: Springer.

Lehrer, R., & Schauble, L. (2015). The development of scientific thinking. In R. M. Lerner (Ed.), *Handbook of child psychology and developmental science* (pp. 671–714). Hoboken, NJ: Wiley.

Louca, L. T., Zacharia, Z. C., & Constantinou, C. P. (2011). In quest of productive modeling-based learning discourse in elementary school science. *Journal of Research in Science Teaching, 48*(8), 919–951.
McClelland, D.C. (1973). Testing for competence rather than for "intelligence." *American Psychologist*, 28, 1–14.
Mahr, B. (2012). On the epistemology of models. In G. Abel & J. Conant (Eds.), *Berlin studies in knowledge research* (*Rethinking epistemology*) (Vol. 1, pp. 301–352). Berlin, Germany/Boston: De Gruyter.
Mahr, B. (2015). Modelle und ihre Befragbarkeit: Grundlagen einer allgemeinen Modelltheorie [Questioning models: Basis of a general model theory]. *Erwägen Wissen Ethik, 26*(3), 329–342.
Mäki, U. (2005). Models are experiments, experiments are models. *Journal of Economic Methodology, 12*, 303–315.
Mathesius, S., Upmeier zu Belzen, A., & Krüger, D. (2014). Kompetenzen von Biologiestudierenden im Bereich der naturwissenschaftlichen Erkenntnisgewinnung: Entwicklung eines Testinstruments [Biology students' competences in scientific reasoning: Development of a test instrument]. *Erkenntnisweg Biologiedidaktik, 13*, 73–88.
Mittelstraß, J. (2005). Anmerkungen zum Modellbegriff [Annotations of the Model Term]. *Modelle des Denkens: Streitgespräch in der Wissenschaftlichen Sitzung der Versammlung der Berlin-Brandenburgischen Akademie der Wissenschaften am 12. Dezember 2003*, 65–67.
Morgan, M. (2005). Experiments versus models: New phenomena, inference and surprise. *Journal of Economic Methodology, 12*(2), 317–329.
Morrison, M., & Morgan, M. (1999). Introduction. In M. Morgan & M. Morrison (Eds.), *Models as mediators: Perspectives on natural and social science* (pp. 1–9). Cambridge, UK: Cambridge University Press.
Nersessian, N. J. (2008). *Creating scientific concepts*. Cambridge, MA: MIT Press.
Nersessian, N. J. (2013). Mental modeling in conceptual change. In S. Vosniadou (Ed.), *International handbook of research on conceptual change* (2nd ed., pp. 395–411). New York: Taylor & Francis.
NGSS Lead States (Ed.). (2013). *Next generation science standards: For states, by states*. Washington, DC: National Academies Press.
Nowak, K. H., Nehring, A., Tiemann, R., & Upmeier zu Belzen, A. (2013). Assessing students' abilities in processes of scientific inquiry in biology using a paper-and-pencil test. *Journal of Biological Education, 47*, 182–188.
NRC [National Research Council]. (2012). *A framework for K-12 science education*. Washington, DC: The National Academy Press.
Oh, P., & Oh, S. (2011). What teachers of science need to know about models: An overview. *International Journal of Science Education, 22*, 1109–1130.
Osborne, J. (2018). Styles of scientific reasoning: What can we learn from looking at the product, not the process, of scientific reasoning? In F. Fischer, C. A. Chinn, K. Engelmann, & J. Osborne (Eds.), *Scientific reasoning and argumentation* (pp. 162–186). New York: Taylor & Francis.
Passmore, C., Gouvea, J. S., & Giere, R. N. (2014). Models in science and in learning science: Focusing scientific practice on sense-making. In M. R. Matthews (Ed.), *International handbook of research in history, philosophy and science teaching* (pp. 1171–1202). Dordrecht: Springer.
Patzke, C., Krüger, D., & Upmeier zu Belzen, A. (2015). Entwicklung von Modellkompetenz im Längsschnitt [Longitudinal development of modeling competence]. In M. Hammann, J. Mayer, & N. Wellnitz (Eds.), *Lehr- und Lernforschung in der Biologiedidaktik* (pp. 43–58). Innsbruck: Studienverlag.
Popper, K. (2005). *The logic of scientific discovery*. London, New York: Taylor & Francis e-Library.
Rahmenlehrplan Berlin/Brandenburg. (2015). Senatsverwaltung für Bildung, Jugend und Wissenschaft Berlin, Ministerium für Bildung, Jugend und Sport des Landes Brandenburg (2015); Teil C: Biologie, Jahrgangsstufen 7–10 [Curriculum Berlin/Brandenburg, Part C: Biology, Grades 7–10].

Reinisch, B., & Krüger, D. (2018). Preservice biology teachers' conceptions about the tentative nature of theories and models in biology. *Research in Science Education, 48*, 71–103.

Reusser, K. (2014). Kompetenzorientierung als Leitbegriff der Didaktik [Competence orientation as guiding concept of didactics]. *Beiträge zur Lehrerinnen- und Lehrerbildung, 32*(3), 325–339.

Ritchey, T. (2012). Outline for a morphology of modelling methods: Contribution to a general theory of modelling. *Acta Morphologica Generalis, 1*, 1–20.

Ropohl, M., Nielsen, J. A., Olley, C., Rönnebeck, S., & Stables, K. (2018). The concept of competence and its relevance for science, technology and mathematics education. In J. Dolin & R. Evans (Eds.), *Transforming assessment* (pp. 3–25). Cham, Switzerland: Springer.

Russ, R. S. (2014). Epistemology of science vs. epistemology for science. *Science Education, 98*, 388–396.

Rychen, D. S., & Salganik, L. H. (2003). A holistic model of competence. In D. S. Rychen & L. H. Salganik (Eds.), *Key competencies* (pp. 41–62). Cambridge, MA/Toronto, Canada: Hogrefe & Huber.

Schecker, H., & Parchmann, I. (2006). Modellierung naturwissenschaftlicher Kompetenz [Modeling of scientific competence]. *Zeitschrift für Didaktik der Naturwissenschaften, 12*, 45–66.

Schwarz, C., Reiser, B., Davis, E., Kenyon, L., Achér, A., Fortus, D., et al. (2009). Developing a learning progression for scientific modeling. *Journal of Research in Science Teaching, 46*(6), 632–654.

Stachowiak, H. (1973). *Allgemeine Modelltheorie*. Wien, Austria: Springer.

Terzer, E. (2013). *Modellkompetenz im Kontext Biologieunterricht* (Dissertation) [Modeling competence in the context of biology classes]. Humboldt-Universität zu Berlin. Verfügbar unter. http://edoc.hu-berlin.de/dissertationen/terzer-eva-2012-12-19/PDF/terzer.pdf

Terzer, E., Hartig, J., & Upmeier zu Belzen, A. (2013). Systematisch Konstruktion eines Tests zur Modellkompetenz im Biologieunterricht unter Berücksichtigung von Gütekriterien [Systematic construction of a test to assess modeling competence in biology classes considering quality criteria]. *Zeitschrift für Didaktik der Naturwissenschaften, 19*, 51–76.

Upmeier Zu Belzen, A. (2013). Unterrichten mit Modellen [Teaching with models]. In H. Gropengießer, U. Harms, & U. Kattmann (Eds.), *Fachdidaktik Biologie* (pp. 325–334). Hallbergmoos, Germany: Aulis.

Upmeier zu Belzen, A., Alonzo, A. C., Krell, M., & Krüger, D. (2019). Learning progressions and competence models – a comparative illustration through models and modeling in science education. In: E. McLoughlin, O. Finlayson, S. Erduran, & P. Childs (Eds.), *Bridging research and practice in science education* (Selected Papers from the ESERA 2017 Conference). Chapter 16. Springer.

Upmeier zu Belzen, A., & Krüger, D. (2010). Modellkompetenz im Biologieunterricht [Modeling competence in biology classes]. *Zeitschrift für Didaktik der Naturwissenschaften, 16*, 41–57.

Chapter 2
Semantic Views on Models: An Appraisal for Science Education

Agustín Adúriz-Bravo

2.1 Introduction

The importance of models in scientific activity can hardly be overrated; indeed, current philosophy of science has recognized that

> [s]cientists spend a great deal of time building, testing, comparing and revising models, and much journal space is dedicated to introducing, applying and interpreting these valuable tools. In short, models are one of the principal instruments of modern science. (Frigg & Hartmann, 2012)

The contemporary depiction of science as a model-based enterprise provides theoretical foundation to understand the role that scientific models are assigned in science education; such foundation are also crucial for the notion of modeling competence (Chap. 1), defended in this book. At the same time, models used in science classes at different educational levels are considered "an integral part of the understanding of the nature of science [...], effective means for teaching scientific literacy [and] effective tools for teaching [science] content knowledge" (Krell, Upmeier zu Belzen, & Krüger, 2012, p. 2). In addition, the ability to effectively use models for specific purposes in specific contexts coupled with a robust understanding of such use is beginning to be considered one of the key aims of science education.

In tune with this perceived importance of modeling competence in science and in science education, meta-theoretical analyses of science have been devoting careful attention to the nature and use of models for six decades now:

> Given the ubiquity of models as well as their variety in form and content, major [philosophical] questions that arise from model-based scientific research concern the nature of

A. Adúriz-Bravo (✉)
Instituto CeFIEC, Facultad de Ciencias Exactas y Naturales, Universidad de Buenos Aires, Buenos Aires, Argentina
e-mail: aadurizbravo@cefiec.fcen.uba.ar

A. Upmeier zu Belzen et al. (eds.), *Towards a Competence-Based View on Models and Modeling in Science Education*, Models and Modeling in Science Education 12, https://doi.org/10.1007/978-3-030-30255-9_2

models, their relationship with real-world phenomena, and their reliability as a source of knowledge. (Sanches de Oliveira, 2013)

In science education, there is also "intense research [...] using models and modeling" (Chamizo, 2013, p. 1616) along different lines, but if we aim to understand, investigate, and foster modeling competence in science teaching, there is an unavoidable prerequisite of a theoretical nature: We need to ascertain which views of all the former issues about models – views coming from the many historically or currently available conceptualizations of models in the philosophy of science – can be valuable for investigation, innovation, curriculum, teaching, evaluation, and teacher education (cf. Grandy, 2003).

Some authors in our field (cf. Adúriz-Bravo, 2013; Chamizo, 2013; Develaki, 2007; Izquierdo-Aymerich, 2013; Koponen, 2007) have tackled the previous question by arguing in favor of a 'semantic view' on scientific models, coming from the so-called 'semanticist family' – a philosophical tendency from the last quarter of the twentieth century. But there are a wide variety of semantic views: How can we characterize them in order to generate *educational* criteria to compare and choose?

With all these considerations in mind, the aims of this chapter are: (a). to discuss what counts as a semantic conception of a model in science and in science education, (b). to distinguish between different semantic conceptions available in recent and contemporary philosophy of science, (c). to locate these conceptions within the various epistemological characterizations of scientific models that have been produced in the history of the philosophy of science, (d). to identify 'transpositions' of the semantic views on models circulating in our research community of science education, and (e). to draw some inferences for the study of modeling competence.

2.2 Archaeology of the Concept of Model

Arising as an idea in the late medieval period, 'model' is a relatively recent construct (cf. R. Müller, 1983, 2009). Its origins and history are connected to the areas of architecture, design, and engineering (Ammon & Capdevila-Werning, 2017; Mahr, 2009), areas in which it has conveyed some sense of canonical measure that should be copied (cf. Mahr, 2009, 2011; Müller & Müller, 2003).

The initial conception of models, stabilized in the Renaissance, stems more or less directly from the technical developments of the Roman Empire. Thus, in ancient Greece, there was no full equivalent of this notion, which was covered by a wide range of terms: prototype, archetype, icon, image, paradigm, epitome, canon, metaphor, analogy, representation, allegory, and simulacrum, among a host of others (cf. Müller, 2000, 2004).

The term 'model' was derived from the classical Latin term 'modus.' 'Modus' and its diminutive 'modulus' were employed in the first centuries A.D. in a diversity of fields, such as music, rhetoric, and architecture (Müller, 2004, 2009). These words were used to describe the way something is or is done and, much more frequently, to refer to something that can be measured (Müller & Müller, 2003, p. 31).[1] The eventual convergence of these two complementary meanings of the Latin stem lies at the center of the current use of the idea of model in scientific activity. For instance, one of the early applications of the term designated a dressmaker's dummy, which was at the same time a *simulacrum* of a client's body and a *prototype* for the client's garments (Müller & Müller, 2003, pp. 32–35). Analogously, a scientific model –as we understand it in contemporary science – is a stylized capture of a phenomenon that serves as a mold or cast that is used to understand other phenomena.

During the two millennia of their prehistory and history, the concepts of module and model progressed toward emphasizing representational power, i.e. the capacity to stand for something else –the entity being modeled (cf. Mahr, 2011; Müller, 2004). A model came to be seen as an exemplar entity that serves as a measure to shape and bind the existence of other entities. In its current sense, it *re*presents (i.e. presents again: *acts in the place of*) a whole abstract class even when no member of such class (no 'specimen') is physically present.

According to the previous analysis, the concept of model was readily available during the institutionalization of the philosophy of science as an academic discipline. In spite of this, such a concept is notably under-represented in the literature of classical positivism, the 'orthodox' epistemology of logical positivism, and the ensuing 'received view' of the 1950s and 1960s (cf. Mahr, 2009; Müller, 2004).[2] This shows that models – unlike theories – were not regarded until very recently as a key element in philosophers' understanding of the functioning of science (Chap. 1).

During the whole nineteenth century and the first half of the twentieth century, the canonical conceptualization of models was based on a 'derivative' definition of the concept, influenced by developments of logic and meta-mathematics (cf. Suppes, 1961). The usual practice was to define models as derived from theories: a model (*of a theory*) was considered to be "a structure constructed by means of the theory's concepts" (Moulines, 2010, p. 20). Thus, a pre-existing theory (a completely formal object) could be later 'interpreted' in a particular domain of experience, and such a domain subsequently became a model of the theory's axioms.

In the second half of the twentieth century, more sophisticated definitions of 'model' ensued. Philosophers' attentions were driven toward scientific models mainly as a result of major changes in scientific activity, which led to a 'discrepancy' between the actual practices of science and the reconstructions of such practices circulating in the philosophy of science, which were still *theory*-based (cf. Sal, 2013, pp. 29–56). Thus, models became the most appropriate form of representation of natural or artificial systems, and this brought the theoretical category of model to the forefront of philosophical meta-analyses.

---

[1] In English, we find that 'modality' (a particular mode in which something exists) and 'modularity' (the quality of being composed of standardized units) are derivatives of these two separate meanings.

[2] e.g. in Carl Hempel's famous textbook, *Philosophy of Natural Science* (Hempel, 1966), scientific models are not mentioned as an important meta-theoretical concept.

2.3 Meta-theoretical Approaches to Models in the Twentieth Century

The historical evolution of philosophical 'models on models' during the twentieth century was intricate and eventually led to a proliferation of conceptualizations in the last few decades (cf. Frigg & Hartmann, 2012; Moulines, 2010; Müller, 2009; Sal, 2013). Three of these conceptualizations of scientific models will be compared here through a non-technical reconstruction of their definitions of model using the much more intuitive notion of 'example.'

'Example' has two distinguishable meanings: It can be understood as an instance (constituting a mere specimen) or as an epitome (setting a general pattern). An example understood as an instance, case, illustration, occurrence, specimen, and so forth, is an element from a class or a kind, merely *conforming to* the 'rules' that determine and delimitate such a kind. Thus, we could say that "an apple is an example of fruit." In turn, an example understood as an epitome, exemplar, paragon, embodiment, pattern, and so forth, is an element that stands out in its kind and is selected in terms of its fitness *to be imitated*. Thus, we could say that "Mother Teresa of Kolkata is an example of compassion." In this second sense, examples are seen as more abstract, idealized, and prototypical than in the first sense (Adúriz-Bravo, 2013).

In the conceptualizations pertaining to logical positivism –the first 'professional' school of philosophy of science– and to the so-called 'received view' that subsequently settled after the dissolution of the Vienna Circle (i.e. during the lengthy period spanning from the 1920s to the 1970s), a scientific model was identified with *any* example of a theory (i.e., a mere instance that satisfies the mandates of that theory –typically, its laws). This reduction of models to more or less irrelevant parts of theories gave way, for instance, to the proposal "to collapse the distinctions between models, theories, analogies, and to take all of these, and more besides, as species of the genus representation; and to take representation in the most direct sense of image or copy" (Wartofsky, 1966, p. 1). In extreme cases of this reductive tendency that was consubstantial to the *syntactic* approach to theories, models were "superfluous additions [to those theories] that are at best of pedagogical, aesthetical or psychological value" (Frigg & Hartmann, 2012).

In the 1950s and 1960s, in the context of an emerging 'new' philosophy of science and especially through the works of Thomas S. Kuhn and some of his contemporaries, a first crevasse to this analytic and formalist conception of models opened. A scientific model began to be portrayed as a *paradigmatic*[3] example of a theory, serving as a theoretical epitome worthy of imitation for problem solving during 'normal science.'

[3]The adjective 'paradigmatic' comes from the Greek term for 'example': a paradigmatic example is thus an 'exemplary example', i.e. example in the second sense. Kuhn advocated for the use of the category 'exemplars' (as a noun) to denote models.

A model à la Kuhn can be seen as a particular socio-historical achievement of a scientific community, outside scientists' heads, contained –in a very stylized version– in disciplinary textbooks, and embodying operative rules to be followed (Nickles, 2003). This idea of model, which stresses its analogical nature, was extremely influential until the 1980s.

Finally, in the last quarter of the twentieth century, a semantic conception of scientific theories (which had noteworthy antecedents from the early 1950s, see Suppe, 1977) gained momentum, rapidly shifting the interest from form to content, from structure to meaning. Within this new theoretical framework, a scientific model began to be identified with an *intended* example of a theory (i.e., a phenomenon that the theory itself was purposefully conceived to account for). This idea that *all* models are "models-for" is fundamental to the conception of modeling competence presented in this book.

Such a semantic characterization of models purported to offer a 'third way' between the received view and the new philosophy of science, explicitly welding together the Kuhnian reconstruction of models as exemplar cases with the conservative analytical requirement that they can all be represented in (semi)formal ways, formulating them as generally and as abstractly as possible.[4] This 'hybrid' semantic view of models was soon shown to be "the only serious contender to emerge as a replacement for the received view analysis of theories" (Suppe, 1977, p. 709) and eventually became the most widely held view among philosophers of science, at least in the communities of strong Anglo-Saxon influence (Frigg, 2006; Suppe, 1989). It is the contention of the author of this chapter that a semantic approach to models is the most useful for the idea of modeling competence.

2.4 Semantic Views of Models in the Late Twentieth Century

From this point on, the umbrella title of 'semantic views of models' is used to encompass a large number of relatively recent characterizations of the concept of 'scientific model' proposed by a range of philosophers that can be situated in what has come to be called the *semantic conception of scientific theories*, by opposition to the hegemonic syntactic conception (cf. Portides, 2005). Semantic views in a broad sense have existed since the 1950s (with the early structuralism of Patrick Suppes and even previous meta-models influenced by the Polish logician Alfred Tarski); in a strict sense, the term refers to the well-known 'model-based views' on science that hail from the 1980s and 1990s.

The following three objectives will be pursued: a. to ascertain some common traits shared by the diverse semantic views of models, especially the most recent ones, b. to make a few distinctions between the theoretical frameworks of the best-known semanticist philosophers of science of the last quarter of the twentieth

[4] This strategy of recovering the best of each of the two preceding periods constitutes a key feature of the semanticist approach (Lorenzano, 2001).

century: Ronald Giere, Frederick Suppe, and Bas van Fraassen, and c. to briefly point toward the existence of conceptualizations of models that can be considered semantic (or post-semantic) but are cited much less often in the science education literature (e.g. proposals by Roman Frigg, Margaret Morrison, Michael Weisberg).

The lists of commonalities and differences between the semantic views that are presented here have emerged from previous work of elucidation and argumentation and from literature reviews (cf. Adúriz-Bravo, 2013; Ariza, Lorenzano, & Adúriz-Bravo, 2016); such work has been based on different sources: textbooks of philosophy of science written by authors with a 'bias' toward semanticism (e.g. Díez & Moulines, 1997; Rosenberg, 2000), reviews of the emergence of the semantic view (e.g. Díez & Lorenzano, 2002; Sal, 2013), general overviews of the field of model studies in academic books by semanticists (e.g. Giere, 1988; Weisberg, 2013), and 'transpositions' of the semantic approach made by researchers in science education (e.g. Izquierdo-Aymerich, 2000; Passmore, Gouvea, & Giere, 2014).

Working on all these sources, at least five 'common pillars' of all semantic views on models can be recognized[5]:

1. *The focus of theory meta-analysis is displaced from syntax to semantics.* The philosophical interest of the semanticist family has been placed on how scientific theories give meaning to the world and make sense to their users. Attention moves from the structure of theories to the functioning of models. The concept of model itself and all its related constructs that this new approach considers essential for meta-analyses (e.g., truth, predication, correspondence, homology, meaning, use, context) are markedly semantic (Guerrero Pino, 2000). Most of the first post-classical (1945–1975) analyses on scientific models are directly shaped by Tarski's semantic theory of truth (Glennan, 2000). Additionally, the more contemporary representational, cognitive, or mediation-based approaches to the concept of model (1975-today), which are overtly model-theoretical, fully embody the 'semantic turn' in the philosophy of science, and thus move much closer to the theses in the 'second Wittgenstein' of the *Philosophische Untersuchungen*.
2. *Empirical theories are, at their very fundamentals, families of models.* From the point of view of philosophical analyses, a scientific theory, even though it is a complex entity with various components, can be fruitfully characterized as a family of models (cf. Suppe, 2000). The very identity of a theory could be in principle determined by that family (or e.g. class, set, population, collection, cluster). A theory defines, through a diversity of mechanisms, the family of its models; accordingly, *presenting* a theory (for philosophical and also most probably for educational purposes) mostly means specifying its models, which are understood as structures (van Fraassen, 1980, p. 64).
3. *An empirical theory admits 'equivalent' presentations through different symbolic resources.* Semanticists do not assume the primacy or superiority of some of

[5] Readers can compare this presentation with other lists of 'common elements' shared by the members of the semanticist family: Díez, 1997; Echeverría, 1999; Estany, 1993.

these forms of theory (re)presentation (e.g., the axiomatic, which was the preferred in classical philosophy of science) over the others. In this sense, non-rigidly formalized knowledge can be considered theoretical and can be expressed ('defined') with very different languages –scale models, drawings, paradigmatic facts, cases, metaphors, gestures, etc.– conserving their explanatory power (Izquierdo-Aymerich, 2007).

4. *Empirical theories explicitly intend to relate models to the real world.* A theory unequivocally states that there is a substantive relationship between the models that belong to it and the phenomena it intends to 'cover.' The theory 'empirically asserts' that some phenomena are adequately accounted for by its models, and such an assertion, which has a linguistic nature, can be deemed (approximately) true or false. In turn, models are seen as non-linguistic items that are

> true by definition. An ideal gas is by definition just what behaves in accordance with the ideal-gas law. [Thus, the] empirical or factual question about a model is whether it 'applies' to anything closely enough to be scientifically useful –to explain and predict its behavior. […] Once we specify [what is meant by] 'well enough' […], this is a hypothesis […]. A theory is a set of hypotheses claiming that particular sets of things in the world are satisfied to varying degrees by a set of models which reflect some similarity or unity. (Rosenberg, 2000, p. 98)

5. *Empirical theories contain the phenomena explained by the models.* The semanticist characterization of theories leaves behind the neo-positivistic metaphorical portrayal of a theory as a 'safety net' connected by poles to the floor and projecting its shadow onto it (i.e., a network of formal, axiomatic elements and relations that only afterwards are 'projected' onto reality through interpretation rules; cf. Sijuwade, 2007). In opposition to such a metaphor, semantic views include the class of theoretical models *and* the 'intended applications' of such models (i.e., the set of real systems that these models pretend to account for) within the theory. In this conception, models can be seen as idealized, reconstructed, or interpreted facts:

> Models show in which phenomenological context theoretical entities make sense and how they are used to intervene in it and to explain what happens. The set of theoretical models can be described through axioms and entities (this is what textbooks usually do), but neither the former nor the latter have meaning without the phenomena from which they emerged; thus, [theories] are action, not only mental representation or language (Izquierdo-Aymerich, 2013, p. 1636).

Of all the previous commonalities in the semantic portrayal of models, the first and fourth ones are the most in tune with the idea of modeling competence as it is approached in this book. On the one hand, Bernd Mahr's analysis of 'model-being' (Chap. 1, providing the foundation of the framework for modeling competence: FMC) emphasizes *pragmatic* aspects that are typical of the semantic turn: it is the *users* who identify an entity as a model through a process of constructive operations ("relationships of creation") of clear semantic nature:

> An object M is not a model in itself, but only if it is conceived of as a model by a judging subject. Through the judgment by which the object M is conceived of as a model, M is placed in a context in which, according to the judging subject, M presents itself as a model (Mahr, 2011, p. 371).

On the other hand, the seminal conception of modeling competence (cf. Krell, Upmeier zu Belzen, & Krüger, 2016) that is being fully developed in this book consistently highlights the process through which models become models-for-something (the process that Mahr calls the 'application'). The FMC locates such a process at the highest level of competence that students (and teachers) should ideally achieve. The semantic pretension that models are created to *account for* systems, an idea that is theoretically captured in the notion of models as intended examples, is in accordance with these ideas.

As stated at the initial paragraph, it is also possible to identify several very notable differences between the various semantic 'versions' inscribed in the semanticist family. It might be useful to organize these differences into the following categories (Ariza et al., 2016): a. the ways in which the notion of model is formally captured; b. the ways in which models and model classes are identified; c. the ways in which the 'pieces/portions of reality' (we can call them, for the sake of simplicity, 'real systems,' see Ruttkamp, 2002, pp. 90–140) that theories intend to account for are characterized; d. the ways in which these real systems are related to models; and e. the constituents of a scientific theory beyond its family of models.

For the sake of space, only category d. will be developed here as an illustrative example of the disagreements that exist among authors within the semanticist family. Afterwards, we present the three quite distinct theoretical conceptions of the relationships between models and systems held by van Fraassen, Suppe, and Giere –which are shaped by their commitment (or lack thereof) to a realist stance. Van van Fraassen (1980) talked about *embeddability*: the different actual and observable aspects of a phenomenon are 'saved' by a single model allowed by the theory. Suppe (1989) resorted to the idea of *homomorphism*, a 'mapping relationship' between a real system and model that can be established within the scope of the theory (i.e., disregarding the influence of variables that are not contemplated in such a theory). Giere (1988) introduced a relationship of *similarity* of type and degree between model and system, which was indebted to Wittgenstein's notion of 'family resemblance.'

Additionally, in order to demarcate between various semantic conceptions of models, the "precise nature of [the] entities called models" (Lorenzano, 2010, p. 46), a most noteworthy point of divergence between semanticist philosophers of science, is also very useful. The definition of theoretical model (in empirical sciences) used by the different authors in the semanticist family could be arranged from the earliest, most formal approaches, resorting to model theory, through conceptions analogically drawing from the natural sciences (considering models as 'phase-' or 'state-spaces,' as van Fraassen or Suppe did: cf. Thompson, 1989, Chap. 5), to much more informal characterizations (e.g., the one by Ronald Giere; see Ariza et al., 2016; Lorenzano, 2010). In all the aforementioned cases, nevertheless, more or less close relationships to classical conceptions of models in mathematics, meta-mathematics, and logic are conserved (cf. Downes, 1992). Of all these 'models of models,' the ones that are more flexible in setting conditions for an entity to be a model seem the most suitable for a model-based science education and in order to go deeper into the notion of modeling competence.

The state of affairs described above is something that could be rapidly changing in the twenty-first century, when even more sophisticated semantic reconstructions of models are emerging. Indeed, more flexible and theory-independent depictions of what models are and how they work are available (e.g. Frigg, 2006; Herfel, Krajewski, Niiniluoto, & Wójcicki, 1995; Morgan & Morrison, 1999; Suárez, 2003; Weisberg, 2013). New meta-models assign to models the function of connecting the theoretical and empirical realms. Models would then 'mediate' between these realms, and it would not be possible to completely reduce them to concrete items or to linguistic enunciations, conserving a high degree of epistemic autonomy.

In order to explain these emerging conceptions of models as theory-independent mediators, it is useful to resort to the 'clementine analogy' (clementines being a hybrid of oranges and mandarins). When a clementine lies next to an orange, it looks like a small, dried version of the latter; when lying next to a mandarin, the clementine appears to be a particularly big, turgid specimen thereof. Analogically, models can be imagined as ontological hybrids participating in the 'fabric' of theoretical frameworks and of real systems. According to this view, a model would act at the same time as a well-formed applicative restriction of theoretical principles and as an idealized, concept-laden portion of the world.

The general notion of model incorporated into the FMC for this book, drawing on ideas by Stachowiak (1973) and described in Chap. 1, finely adjusts the 'mediating' conception, while being less radical concerning the *ontological* nature of models.

2.5 Semantic Characterizations of Models in Science Education

The starting point here is the recognition of two consensuses within our community of science education. First, even though the notions of model and modeling have been implicitly present for some time in the science curricula of all educational levels, it is only recently that curriculum designers, science education researchers, and science teachers have begun to advocate for an explicit treatment of the meta-theoretical concept of model in science teaching (cf. Gilbert & Boulter, 2000; Harrison & Treagust, 2000; Justi & Gilbert, 2002; Khine & Saleh, 2011). Second, academic production on models and modeling in science education has reached significant levels of depth and sophistication, but, in spite of this, our community still needs further discussions of fundamental issues about the epistemology of models. We may have adopted a standard definition of the construct of model –of neo-positivistic filiation– that has barred more careful elucidation around some basic issues (cf. Johsua & Dupin, 1993; Koponen, 2007).

In addition to this, in the academic field of science education, there seems to be a very timid materializing of a new portrayal of models for research and practice that –with more or less awareness from us science education researchers– can be located in the arch of 'model-based' or 'model-theoretical' conceptualizations (e.g. Gouvea & Passmore, 2017; Grandy, 2003; Justi, 2006, 2009; Koponen, 2007; Oh & Oh, 2011). We could thus talk about the emergence of a 'model-based science education' (Adúriz-Bravo, 2010). Within this emergent approach to research and innovation that focuses on models, modeling competence could be considered a new and promising line.

Semanticism still remains a philosophical school that is far from being understood within our discipline; hence, carefully reviewing what counts as a semantic view on models and drawing implications of such a view for science education continues to be a necessary task. In addition, the existence of a variety of semantic understandings of models in the community of the philosophy of science makes it complex to straightforwardly pick out a 'definition' that is ready for educational use. This also holds for the discussions in this book around the new idea of modeling competence.

Authors in science education have undertaken the aforementioned review by looking into some fundamental epistemological aspects of models and modeling (e.g. Erduran & Duschl, 2004; Izquierdo-Aymerich, 2004; Johsua & Dupin, 1993; Lombardi, 1998, among many others). Of all these antecedents, it may be interesting to focus on three texts—by Chamizo (2006), Oh and Oh (2011), and Krell et al. (2016). In these texts, the authors explain what they regard as the most important issues around models for the purpose of educational discussion, and they do this from theoretical positions that can be considered more or less semantic.

In his article, Chamizo (2006) identified what he considers the eight "least controverted" (p. 476) characteristics of scientific models: (1) models are representations (of, e.g., objects, systems, phenomena, processes); (2) models are instruments that can provide an answer to scientific problems; (3) models constitute analogies of the phenomena they represent; (4) models differ from reality because their construction follows a particular aim; (5) models are constructed by compromising between the similarities and differences that they have with their represented reality; (6) models are developed and changed along history; (7) models undergo a process of acceptance in the scientific community; and (8) models can be classified into types.

In turn, Oh and Oh (2011) presented "an overview of the nature of models and their uses in the science classroom for science teacher educators and subsequently for science teachers" (p. 1111). Through an analysis of specialized literature and empirical research on different groups of experts, they identified "five subtopics concerning the nature of models and modelling" (p. 1111), and, similar to Chamizo, they found some consensus among philosophers of science and science education researchers surrounding such subtopics: (1) models are usually meant to refer to representations; (2) the usual purposes of models are to describe, explain, predict, and communicate; (3) scientists use a multiplicity of models when engaged in scientific problem solving; (4) models are developed and changed in history; (5) models are usually used in science teaching with the justification that "external presentations of visual representations provide support for constructing and reasoning with internal representations" (Oh & Oh, 2011, p. 1120).

Finally, Krell et al. (2016) identified five important aspects that should be taken into account when reflecting on models and modeling: (1) models are *of* and *for* something; (2) scientists use a multiplicity of models for the same phenomenon; (3) models serve different purposes (to describe, explain, and hypothesize: Krell et al., 2012, 2016; Krüger, Krell, & Belzen, 2017); (4) in scientific practice, models undergo rigorous testing; and (5) models are developed and changed along history.

Practically all the characteristics, topics, or aspects (collectively, 'facets' of model meta-analysis) that were proposed in the previous three texts have been incorporated into the theoretical FMC. Such facets cover issues such as the nature, use, and evolution of scientific models, the processes of formulation and evaluation of such models, and their purposes and value in science (see Grünkorn, Hänsch, Upmeier zu Belzen, & Krüger, 2012).

2.6 Teaching Modeling Competence from a Semantic Perspective

As stated in the Introduction, many contemporary philosophers of science asseverate that modeling ('acting-with-models') is arguably the most important intellectual activity in contemporary science (cf. Herfel et al., 1995; Magnani, Nersessian, & Thagard, 1999). The idea of organizing science teaching around modeling has also gained momentum in research in science education (cf. Justi & Gilbert, 2016). But it can be contended that such an idea crucially depends on our conception of the nature of modeling competence in scientists' science and in school science. The following paragraphs briefly tackle the issue of the implications of infusing a semantic view of models in a competence-based approach to school scientific modeling.

What counts as 'modeling' when it is *understood as a scientific competence*? Just as with the construct of 'model', there are important theoretical disagreements around this issue. We can consider at least four main senses with which the idea of 'modeling' is used in science education (Adúriz-Bravo, 2012):

1. The creation of an original theoretical model to face the study of a phenomenon. In extreme cases, a model may be completely new with respect to the body of established knowledge in a particular historical moment; more commonly, it is new only from the point of view of the learners' knowledge base.
2. The process of subsuming a puzzling fact that is being investigated in the science classroom under an already available model that can account for it, in a process of *inference to the best explanation* (i.e., reasoning backwards).
3. The interactive adjustment of an established model after the emergence of new, unexpected, or anomalous elements during investigation.
4. The intellectual exercise of reconstructing well-known 'couplings' between models and facts in the context of learning the scope and use of a theory.

In the first comprehension of modeling competence proposed here, students, through scientific activities in school, can develop more or less innovative theoretical models in order to tackle 'scientific problem solving'; such models can be generated from previous models through analogy, combination, or refinement or they can also appear through rather intricate cognitive mechanisms (including: dreaming, illumination, and intuition; accident, coincidence, or serendipity). For this sense of modeling, undoubtedly very ambitious for science education, the semanticist analogy of theoretical models as maps to navigate a territory may be useful.

In the second sense of modeling competence, established models, available culturally, can be deliberately applied to the explanation of puzzling facts through very elaborate ampliative (e.g. abductive and analogical) reasoning. The aim of such modeling processes would be to show that, in some way, the facts to be explained are 'similar' to those models that are prospective candidates to explain them through the establishment of a case-rule relationship (Adúriz-Bravo, 2005). In this second sense of modeling, the semanticist insistence that models must be understood as "models-for" could be illuminating.

As for the third meaning of modeling competence, in the process of explaining families of phenomena in scientific research, new phenomena, observations, and results, more and better empirical data, additional theoretical knowledge, or new modes of representation and communication may force the need for adjustments in the accepted models; in this way, details, expansions, and corrections would be added, allowing models to be refined and improved. These iterative sequences may be captured by the semanticist idea that scientists continually evaluate whether their models satisfactorily account for phenomena.

Finally, a more modest –and yet educationally powerful– conception of modeling competence in the science classrooms of all educational levels is available in our discipline. It consists of understanding modeling as the process of reconstructing the established ('normative') linkage between facts and models. Although such a linkage is transparent in scientists' science, it certainly appears as new to students. Students, aided by the class group and the teacher, would put into action robust school scientific models in order to shed light on problems that are of interest to them and, at the same time, constitute the intended applications of those models.

According to this conception, the ultimate aim of modeling competence in science education would be that students use the models that they are learning in order to explain to themselves and to others some issues of interest in the natural world, aware that such an explanation already exists in science. In this last scenario, a fully semantic (as opposed to syntactic) approach to the process seems to be necessary. Additionally, a conception of models as epitomes that 'guide' new applications of knowledge may turn out to be appropriate.

In this last, albeit conservative, conception of modeling competence, theoretical models could be introduced with an explicit emphasis on their analogical nature, thus leading to learning about the *nature of models* (understood as what, in Chap. 1, is designated as a set of abilities to reflect on models and modeling). A learning goal –which complements 'pure' science content– would be to recognize that the extremely abstract way in which a scientific model of a phenomenon used in school can be described ensures the possibility to project it onto other phenomena under study, between which similarity is perceived. As Hernán Miguel stated:

> [An] abstract model can have two interpretations: one in which the abstract entities of the model correspond to [the model-for, taken as analogans] (...) and another in which the abstract entities are assigned [the meanings of the new model, taken as analogandum]. Evidencing this double interpretation of a same abstract model permits teachers to generate (in students) the idea that they can have structural knowledge of [a phenomenon] and that perhaps, within the limitations of the analogy, other [phenomena] could be well-represented using the same abstract model (Miguel, 1999, p. 95, translated).

In the semantic approach to modeling competence introduced in this chapter, and compatible with the more general characterization of such a competence in the rest of the book, school models are construed and taught as models-of and models-for at the same time (cf. Adúriz-Bravo, 2012, 2013; Giere, 1988; Gouvea & Passmore, 2017; Krell et al., 2016; Mahr, 2009, 2011). On the one hand, they are introduced as the abstract counterparts of the systems modeled in 'interventions' (observations, experiences, experiments, simulations); on the other hand, they are tested as exemplars in order to create new models that are more specific or more general and that can be meaningfully linked to the initial ones in 'families.' Together, these two epistemic processes, when enacted *and reflected upon*, constitute modeling competence as a whole.

Thus, modeling competence would imply the conscious use of scientific models as paradigmatic and intended examples: students would be acquainted with a theoretical model as a stylized case standing for a larger and more abstract reality and as a robust example of a type, thus setting a norm. For example, the 'school model' of a cell would serve in science teaching as a highly schematic version of something that can be 'identified' under a microscope and also as a blueprint (in the architectural sense) that guides our description, understanding, and manipulation of different cell types (e.g. neurons, liver cells, white cells, skin cells).

The notion of modeling competence proposed here can be understood as the testing of explicit hypotheses on the degree of adjustment between our ideas and our interventions. A 'new,' less dogmatic, scientific method could thus be introduced in the science classroom (cf. Adúriz-Bravo, 2008; Giere, Bickle, & Mauldin, 2006; Izquierdo-Aymerich, 2013); such a method would consist of making critical decisions about the 'convergence' between consequences derived from our theoretical ideas (after 'putting models to work') and data obtained from carefully planned observations and experiments. The aim would be to compare the results of these two coordinated sets of activities and assess the extent to which our ideas 'talk about the world' (see level III in Chap. 1).

According to this approach to modeling competence, school science would be analogous to scientists' science in an 'irreducible' epistemic aspect: Science students would work in a way that is similar to that of scientists, who

> use abstract thinking in a way that gives rise to a set of 'idealized facts' about which they speak using the entities that they define as 'theory,' [and such] facts (constructed with actions, representations and language) [become] the 'models' of the theories. (Izquierdo-Aymerich, 2013, p. 1636)

References

Adúriz-Bravo, A. (2005). Methodology and politics: A proposal to teach the structuring ideas of the philosophy of science through the pendulum. In M. Matthews, C. Gauld, & A. Stinner (Eds.), *The pendulum: Scientific, historical, philosophical and educational perspectives* (pp. 277–291). Dordrecht, The Netherlands: Springer.

Adúriz-Bravo, A. (2008). ¿Existirá el "Método Científico"? [Does the scientific method exist?]. In L. Galagovsky (Ed.), *¿Qué Tienen de 'Naturales' las Ciencias Naturales?* (pp. 47–59). Buenos Aires, Argentina: Biblos.

Adúriz-Bravo, A. (2010). Hacia una Didáctica de las Ciencias Experimentales Basada en Modelos [Towards a Model-Based Didactics of Science]. In L. Álvarez, R. Rickenmann, & J. Vallès (Eds.), *II Congrés Internacional de Didàctiques: Llibre d'Actes* (CD-ROM, n° 248, n/pp). Girona, Spain: Universitat de Girona.

Adúriz-Bravo, A. (2012). Algunas Características Clave de los Modelos Científicos Relevantes para la Educación Química [Some key characteristics of scientific models that are relevant for chemistry education]. *Educación Química, XXIII*(E2), 248–256.

Adúriz-Bravo, A. (2013). A semantic view of scientific models for science education. *Science & Education, 22*(7), 1593–1612.

Ammon, S., & Capdevila-Werning, R. (Eds.). (2017). *The active image: Architecture and engineering in the age of modeling*. Dordrecht, The Netherlands: Springer.

Ariza, Y., Lorenzano, P., & Adúriz-Bravo, A. (2016). Meta-theoretical contributions to the constitution of a model-based didactics of science. *Science & Education, 25*(7), 747–773.

Chamizo, J. A. (2006). Los Modelos de la Química [Models in Chemistry]. *Educación Química, 17*(4), 476–482.

Chamizo, J. A. (2013). A new definition of models and modeling in Chemistry's teaching. *Science & Education, 22*(7), 1613–1632.

Develaki, M. (2007). The model-based view of scientific theories and the structuring of school science Programmes. *Science & Education, 16*(7), 725–749.

Díez, J. A. (1997). La Concepción Semántica de las Teorías Científicas [The semantic view of scientific theories]. *Éndoxa: Series Filosóficas, 8*, 41–91.

Díez, J. A., & Lorenzano, P. (2002). La Concepción Estructuralista en el Contexto de la Filosofía de la Ciencia del Siglo XX [The structuralist view in the context of 20th century philosophy of science]. In J. A. Díez & P. Lorenzano (Eds.), *Desarrollos Actuales de la Metateoría Estructuralista: Problemas y Discusiones* (pp. 13–78). Quilmes, Argentina: UNQui/Universidad de Zacatecas/Universitat Rovira i Virgili.

Díez, J. A., & Moulines, C. U. (1997). *Fundamentos de filosofía de la ciencia* [Foundations of the philosophy of science]. Barcelona, Spain: Ariel.

Downes, S. M. (1992). *The importance of models in theorizing: A deflationary semantic view* (PSA 1992) (Vol. 1, pp. 142–153). East Lansing, MI: Philosophy of Science Association.

Echeverría, J. (1999). *Introducción a la Metodología de la Ciencia: La Filosofía de la Ciencia en el Siglo XX* [Introduction to the methodology of science: The philosophy of science in the 20th century]. Madrid, Spain: Cátedra.

Erduran, S., & Duschl, R. (2004). Interdisciplinary characterizations of models and the nature of chemical knowledge in the classroom. *Studies in Science Education, 40*, 111–144.

Estany, A. (1993). *Introducción a la Filosofía de la Ciencia* [Introduction to the philosophy of science]. Barcelona, Spain: Crítica.

Frigg, R. (2006). Scientific representation and the semantic view of theories. *Theoria, 55*, 49–65.

Frigg, R. & Hartmann, S. (2012). Models in science. In: Zalta, E.N. (ed.). *The Stanford encyclopedia of philosophy* (Fall 2012 Edition, n/pp). On-line: http://plato.stanford.edu/archives/fall2012/entries/models-science/

Giere, R. N. (1988). *Explaining science: A cognitive approach*. Chicago: University of Chicago Press.

Giere, R. N., Bickle, J., & Mauldin, R. F. (2006). *Understanding scientific reasoning* (5th ed.). Belmont, CA: Thomson/Wadsworth.
Gilbert, J., & Boulter, C. (Eds.). (2000). *Developing models in science education*. Dordrecht, The Netherlands: Kluwer.
Glennan, S. (2000). A model of models. Unpublished paper [On-line: http://philsci-archive.pitt.edu/1134/1/Models1.4.pdf].
Gouvea, J. S., & Passmore, C. (2017). 'Models of' versus 'models for'. *Science & Education, 26*(1–2), 49–63.
Grandy, R. E. (2003). What are models and why do we need them? *Science & Education, 12*(8), 773–777.
Grünkorn, J., Hänsch, J., Upmeier zu Belzen, A., & Krüger, D. (2012). Determination of students' model competence using open-ended and hands-on tasks. In C. Bruguière, A. Tiberghien, & P. Clément (Eds.), *E-Book proceedings of the ESERA 2011 conference: Science learning and citizenship, part 5* (pp. 39–45). Lyon, France: European Science Education Research Association.
Guerrero Pino, G. (2000). Determinismo, Modelos y Modalidades [Determinism, models, and modalities]. *Revista de Filosofía, tercera época, XIII*(24), 191–216.
Harrison, A. G., & Treagust, D. F. (2000). A typology of school science models. *International Journal of Science Education, 22*(9), 1011–1026.
Hempel, C. G. (1966). *Philosophy of natural science*. Englewood Cliffs, NJ: Prentice-Hall.
Herfel, W., Krajewski, W., Niiniluoto, I., & Wójcicki, R. (Eds.). (1995). *Theories and models in scientific processes*. Amsterdam, The Netherlands: Rodopi.
Izquierdo-Aymerich, M. (2000). Fundamentos Epistemológicos [Epistemological foundations]. In F. J. Perales & P. Cañal (Eds.), *Didáctica de las Ciencias Experimentales: Teoría y Práctica de la Enseñanza de las Ciencias* (pp. 35–64). Alcoy, Spain: Marfil.
Izquierdo-Aymerich, M. (2007). Enseñar ciencias, una nueva ciencia [Teaching science: A new science]. *Enseñanza de las Ciencias Sociales, 6*, 125–138.
Izquierdo-Aymerich, M. (2013). School chemistry: An historical and philosophical approach. *Science & Education, 22*(7), 1633–1653.
Johsua, S., & Dupin, J.-J. (1993). *Introduction à la Didactique des Sciences et des Mathématiques* [Introduction to didactics of science and mathematics]. Paris: PUF.
Justi, R. (2006). La Enseñanza de Ciencias Basada en la Elaboración de Modelos [Science teaching base on model elaboration]. *Enseñanza de las Ciencias, 24*(2), 173–184.
Justi, R. (2009). Learning how to model in science classroom: Key teacher's role in supporting the development of students' modelling skills. *Educación Química, 20*(1), 32–40.
Justi, R. S., & Gilbert, J. K. (2002). Modelling, teachers' views on the nature of modelling, and implications for the education of Modellers. *International Journal of Science Education, 24*(4), 369–387.
Justi, R. S., & Gilbert, J. K. (2016). *Modelling-based teaching in science education*. Dordrecht, The Netherlands: Springer.
Khine, M. S., & Saleh, I. M. (Eds.). (2011). *Models and modeling: Cognitive tools for scientific enquiry*. Dordrecht, The Netherlands: Springer.
Koponen, I. T. (2007). Models and modelling in physics education: A critical re-analysis of philosophical underpinnings and suggestions for revisions. *Science & Education, 16*(7–8), 751–773.
Krell, M., Upmeier zu Belzen, A., & Krüger, D. (2012). Students' understanding of the purpose of models in different biological contexts. *International Journal of Biology Education, 2*(2), 1–34.
Krell, M., Upmeier zu Belzen, A., & Krüger, D. (2016). Modellkompetenz im Biologieunterricht [Modeling competence in biology classes]. In A. Sandmann & P. Schmiemann (Eds.), *Biologiedidaktische Forschung: Schwerpunkte und Forschungsgegenstände* (pp. 83–102). Berlin, Germany: Logos.

Krüger, D., Krell, M., & Upmeier zu Belzen, A. (2017). *Competence-based view on models and modelling*. Summary of the paper presented at the ESERA 2017 international conference. On-line: https://keynote.conference-services.net/resources/444/5233/pdf/ESERA2017_1045_paper.pdf

Lombardi, O. (1998). La Noción de Modelo en Ciencias [The notion of model in science]. *Educación en Ciencias, 2*(4), 5–13.

Lorenzano, P. (2001). La Teorización Filosófica sobre la Ciencia en el Siglo XX [Philosophical theorizing on science in the 20th century]. *Boletín de la Biblioteca del Congreso de la Nación, 121*, 29–43.

Lorenzano, P. (2010). Logic and methodology of science: An introduction to the philosophy of science. In P. Lorenzano, H.-J. Rheinberger, E. Ortiz, & C. Galles (Eds.), *History and philosophy of science and technology* (Vol. 1, pp. 1–50). Oxford, UK: EOLSS Publishers.

Magnani, L., Nersessian, N. J., & Thagard, P. (Eds.). (1999). *Model-based reasoning in scientific discovery*. New York: Kluwer Academic/Plenum Press.

Mahr, B. (2009). Information science and the logic of models. *Software & Systems Modeling, 8*, 365–383.

Mahr, B. (2011). On the epistemology of models. In A. Günter & J. Conant (Eds.), *Rethinking epistemology* (pp. 301–352). Berlin, Germany: De Gruyter.

Miguel, H. (1999). La Analogía como Herramienta en la Generación de Ideas Previas [Analogy as a tool in the generation of conceptions]. *El Caldero de la Escuela, 73*, 85–97.

Morgan, M. S., & Morrison, M. (Eds.). (1999). *Models as mediators: Perspectives on natural and social science*. Cambridge, UK: Cambridge University Press.

Moulines, C. U. (2010). The nature and structure of scientific theories. *Meta, 1*(1), 15–29.

Müller, R. (1983). Zur geschichte des Modelldenkens und des Modellbegriffs [On the history of modeling and of the concept of model]. In H. Stachowiak (Ed.), *Modelle: Konstruktion der Wirklichkeit* (pp. 17–86). Munich, Germany: Wilhelm Fink Verlag.

Müller, R. (2000). The concept of model and its triple history. In E. Neuenschwander (Ed.), *Scientific models: Their historical and philosophical relevance* (pp. 7–19). Zurich, Switzerland: Universität Zürich.

Müller, R. (2004). Model: The history of the concept and of its use in science. In M. Negrotti (Ed.), *Models in contemporary science: Yearbook of the artificial* (Vol. 2, pp. 239–262). Bern, Switzerland: Peter Lang.

Müller, R. (2009). The notion of a model: A historical overview. In A. Meijers (Ed.), *Philosophy of technology and engineering sciences* (pp. 637–664). Amsterdam, The Netherlands: Elsevier.

Müller, T., & Müller, H. (2003). *Modelling in natural sciences: Design, validation and case studies*. Berlin, Germany: Springer.

Nickles, T. (2003). Normal science: From logic to case-based and model-based reasoning. In T. Nickles (Ed.), *Thomas Kuhn* (pp. 142–177). Cambridge, UK: Cambridge University Press.

Oh, P. S., & Oh, S. J. (2011). What teachers of science need to know about models: An overview. *International Journal of Science Education, 33*(8), 1109–1130.

Passmore, C., Gouvea, J. S., & Giere, R. N. (2014). Models in science and in learning science: Focusing scientific practice on sense-making. In M. Matthews (Ed.), *International handbook of research in history, philosophy and science teaching* (pp. 1171–1202). Dordrecht, The Netherlands: Springer.

Portides, D. P. (2005). Scientific models and the semantic view of scientific theories. *Philosophy of Science, 72*(5), 1287–1298.

Rosenberg, A. (2000). *Philosophy of science: A contemporary introduction*. London: Routledge.

Ruttkamp, E. B. (2002). *A model-theoretic realist interpretation of science* (Synthese library series, Vol. 311). Dordrecht, The Netherlands: Kluwer.

Sal, A. S. (2013). *Scientific progress on the semantic view: An account of scientific progress as objective logical and empirical strength increments*. Master thesis. Oslo, Norway: University of Oslo [On-line: https://www.duo.uio.no/bitstream/handle/10852/37757/MasterThesis_Sal.pdf?sequence=1].

Sanches de Oliveira, G. (2013). Review – Simulation and similarity using models to understand the World by Michael Weisberg. *Metapsychology Online Reviews, 17* (40). On-line: http://metapsychology.mentalhelp.net/poc/view_doc.php?type=book&id=6985
Sijuwade, P. O. (2007). Recent trends in the philosophy of science: Lessons for sociology. *Journal of Social Sciences, 14*(1), 53–64.
Stachowiak, H. (1973). *Allgemeine Modelltheorie* [General model theory]. Vienna, Austria: Springer.
Suárez, M. (2003). Scientific representation: Against similarity and isomorphism. *International Studies in the Philosophy of Science, 17*, 225–244.
Suppe, F. (1977). Afterword. In F. Suppe (Ed.), *The structure of scientific theories* (2nd ed., pp. 617–730). Urbana, IL: University of Illinois Press.
Suppe, F. (1989). *The semantic conception of theories and scientific realism*. Urbana, IL: University of Illinois Press.
Suppe, F. (2000). Understanding Scientific theories: An assessment of developments, 1969–1998. *Philosophy of Science*, *67*, Supplement of Proceedings, S102–S115.
Suppes, P. (1961). A comparison of the meaning and uses of models in mathematics and the empirical sciences. In H. von Freudenthal (Ed.), *The concept and role of the model in mathematics and natural and social sciences* (pp. 163–177). Dordrecht, The Netherlands: Reidel.
Thompson, P. (1989). *The structure of biological theories*. Albany, NY: SUNY Press.
van Fraassen, B. (1980). *The scientific image*. Oxford, UK: Clarendon Press.
Wartofsky, M.W. (1966). The model muddle: Proposals for an immodest realism. In: Cohen, R.S., & Wartofsky, M.W. (eds.) (1979). *Models: Representation and the scientific understanding* (pp. 1–11). Dordrecht, The Netherlands: Reidel.
Weisberg, M. (2013). *Simulation and similarity: Using models to understand the world*. Oxford, UK: Oxford University Press.

Chapter 3
A Framework for Modeling-Based Learning, Teaching, and Assessment

Constantinos P. Constantinou, Christiana Th. Nicolaou, and Marios Papaevripidou

3.1 Introduction

Interest in encouraging learners to engage in modeling is grounded in the premise that models help learners learn more robustly. First, models can facilitate an improved understanding of the behavior of systems (Crawford & Cullin, 2004; Hokayem & Schwarz, 2014; Maia & Justi, 2009; Nicolaou & Constantinou, 2014) by acting as intermediates between learners' efforts to describe and represent a phenomenon and their endeavors to interpret it. Second, by engaging in modeling-based learning, we can improve our ability to construct, revise, compare, evaluate, and validate models, all of which are important science practices (National Research Council, 2012; Nicolaou & Constantinou, 2014; Schwarz et al., 2009).

Many scientists, science educators, and philosophers consider modeling to be the backbone of knowledge construction with regard to systems and natural phenomena (Bunge, 1983; Chapman, 2000; Gilbert, Boulter, & Rutherford, 1998; Gilbert, 1991). Educational reform documents identifying the value of engaging learners in constructing and using models (National Research Council, 2012, 2013), also highlight the need to promote teaching interventions that are aimed at developing modeling competence. To accomplish this, it is important to support teachers with robust frameworks for teaching and assessment methods that are related to active modeling. Additionally, in this chapter, we will argue that the development of such frameworks, as well as processes for supporting and guiding teachers in their efforts to help learners engage with modeling, could contribute to overcoming the problem of the relative scarcity of modeling-based learning in schools (Duschl, Schweingruber, & Shouse, 2007).

C. P. Constantinou (✉) · C. T. Nicolaou · M. Papaevripidou
University of Cyprus, Nicosia, Cyprus
e-mail: c.p.constantinou@ucy.ac.cy

A. Upmeier zu Belzen et al. (eds.), *Towards a Competence-Based View on Models and Modeling in Science Education*, Models and Modeling in Science Education 12, https://doi.org/10.1007/978-3-030-30255-9_3

In this chapter, we present an approach for teaching science learning through model construction, refinement, and validation. We begin by describing the epistemological underpinnings and the rationale for a modeling-based teaching and learning approach that is designed to develop knowledge of natural phenomena. We proceed by describing the modeling-based learning framework (MLF) in terms of the modeling practices (model construction, model use, model revision, model comparison, and model validation) and the modeling meta-knowledge (knowledge about models and metacognitive knowledge of the modeling process) that emerge alongside the development of expertise in scientific modeling. We then present a process for identifying the attainment levels of each component of the modeling-based learning framework as well as examples of these attainment levels. Our core argument refers to the interconnectedness of the practical and epistemological aspects of modeling-based learning. On the one hand, the MLF seems to deviate from other frameworks for models and modeling. On the other hand, the MLF emphasizes implementation into meaningful learning and teaching practice in ways that have a positive influence on science teaching and learning. First, the MLF has the potential to help researcher-teacher teams bring out the nature of scientific modeling in the classroom. Second, it will facilitate more empirical research in this area by better informing both curriculum design and teaching practices with respect to how each aspect of the MLF can potentially be interwoven into teaching-learning sequences and also be assessed (Constantinou & Papadouris, 2004).

Science can be thought of as a complex and dynamic network of models that are interrelated through a system of theoretical principles (Constantinou, 1999). A model is the outcome of an application of a theory to a phenomenon. As noted by Upmeier zu Belzen, Krüger, and van Driel in the introductory chapter of this book (Chap. 1), there is no uniform definition of the concept of a model or the modeling process in the natural sciences. However, for communication purposes and as a basis for the development of the MLF, we assume that a *scientific model*[1] is an epistemological entity that represents a phenomenon (Giere, 1999, 2004; Hughes, 1997; Passmore & Stewart, 2002), provides the mechanism behind how this phenomenon operates (Berland & Reiser, 2009; Braaten & Windschitl, 2011), and can be used to make predictions about the future behavior of the phenomenon (Bunge, 1983; Hughes, 1997; Raftopoulos, Kalyfommatou, & Constantinou, 2005). This definition emphasizes the idea of a model as an external representation, it provides an interpretation of a phenomenon or system, and it can be used to predict the future behavior of the system. The *representational aspect* of the model is shaped by its constituent components (i.e., objects, variables, and processes). The *interpretive aspect* of the model is related to the provision of an interpretation that posits one or more mechanisms that underlie the observable behavior of a phenomenon. This can take the form of a story that elaborates on how a phenomenon operates and highlights

[1] It is important to clarify the idea that scientific models are different from mental models (Gentner & Stevens, 1983), which are Cognitive Psychology constructs that refer to "transient representations that are activated usually when one is exposed to a new situation and act as structural analogies to situations or processes" (Greca & Moreira, 2002, p. 108).

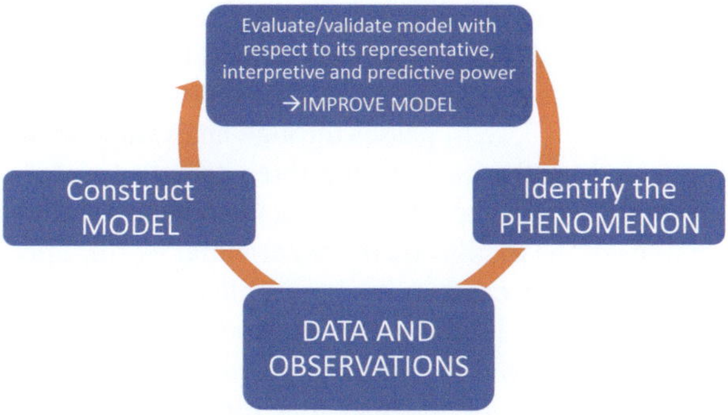

Fig. 3.1 The modeling-based learning cycle

the relationships between the objects, the variables, and the processes. Specifically, an interpretation provides one or more mechanisms and elaborates on how the phenomenon emerges from the mechanism(s). Finally, a model needs to have predictive strength by facilitating the formulation and *testing of predictions* for new manifestations of the phenomenon it represents.

Learning by modeling refers to the modeling practice and more specifically to the idea of learning through the construction, revision, and validation of models (Crawford & Cullin, 2004; Nicolaou, Nicolaidou, & Constantinou, 2009; Schwarz et al., 2009). Learning by modeling should be differentiated from learning with models (De Jong, Van Driel, & Verloop, 2005; van der Meij & de Jong, 2006). The latter denotes learners who are using previously constructed models with the aim of gaining insights into the phenomenon represented by the model. By contrast, learning by modeling is about constructing interpretive representations with predictive power (often in symbolic form).

Central to learning by modeling is the modeling-based learning cycle (Constantinou, 1999; Nicolaou et al., 2009), an iterative process that engages the learner in a continuous comparison of the model with the represented phenomenon (Fig. 3.1). The purpose of this comparison is to obtain feedback for improving the model so that it accurately represents as many desired details of the original system/phenomenon as required. It is also a cyclical procedure (Mendonça & Justi, 2014) that could involve the generation of models of various forms until the model generator finds one that successfully emulates the observable behavior of the system. Hence, the outcome of engaging in a modeling-based learning cycle is a series of successive models, usually ranging from more superficial versions to more scientifically coherent entities.

The cyclical nature of the modeling process is also stressed by the framework for modeling competence (FMC; Chap. 1), according to which the phenomenon (i.e., the experiential world) is distinct from the model (i.e., the model world). However, the two worlds are interconnected. In the framework, the identification of the phenomenon under study is followed by the data collection process, which leads to the construction of the first model. Likewise, the authors of the FMC propose that, at the experiential level, the modeler observes and investigates the phenomenon with the aim of constructing multiple models as part of the process of model development.

Simultaneously, the identified purpose of the model, which is influenced by the modeler's experiences, sets the basis for the construction of multiple models, also as part of the process of model development. In the MLF, each model prototype (Version 1, 2, 3, …n) passes through an evaluation and validation process, which is conducted on the basis of the three criteria (model representation, interpretation, and prediction). The FMC also acknowledges the connection between the various versions of the model and the experiential world. This is done through model testing aiming to verify conformity. It is implied through the MLF that by implementing the modeling-based learning cycle, a student-constructed model is continuously evaluated and improved through the practice of model revision. At the same time, students are aware of the fact that different model prototypes are constructed by different students in their class (or scientists in the field), which could lead to different models of the same phenomenon, often based on different ideas/hypotheses. This helps the teacher overcome the problem identified by Grünkorn, Upmeier zu Belzen, and Krüger (2014), who proposed that students encounter difficulty in accepting the existence of multiple models. These researchers suggested that users should avoid the reflective use of historical models in school because it might lead students to fail to accept the existence of multiple models.

A detailed description of the MLF is the focus of the next section. We propose that the MLF is a tool that can be used by science educators to structure instructional designs and assessment methods to support the development of the modeling competence.

3.2 A Framework for Modeling-Based Learning

We view modeling as a competence because this term is comprehensive enough to indicate that modeling entails more than just knowledge or skills, a claim that has been emphasized in the literature (e.g., Klieme, Hartig, & Rauch, 2008). For instance, in prior research, modeling has been conceptualized as (a) an ability or a skill, which consists of modeling sub-skills (Dori & Kaberman, 2012; Papaevripidou, Constantinou, & Zacharia, 2007), (b) a practice (National Research Council, 2012; Pluta, Chinn, & Duncan, 2011; Schwarz et al., 2009), (c) a scientific process (Crawford & Cullin, 2004; Sins, Savelsbergh, & van Joolingen, 2005), or (d) an instructional approach (Louca, Zacharia, & Constantinou, 2011). Our definition of modeling is related to Weinert's (2001) definition of competence, which refers to the successful mastery, through an appropriate understanding and practices, of a concise range of demands, tasks, problems, and goals that are related to some performance that is of interest to both the community and society. Additionally, as Weinert purports, competence draws on combinations of the cognitive, motivational, moral, and social underpinnings that are available to (or can potentially be learned by) a person or a community and that underlie the concise mastery in question. Consequently, the concept of competence designates a complex action system encompassing all types of knowledge, including cognitive skills, attitudes, and other non-cognitive components. Additionally, it involves the ability to meet complex demands by drawing on and mobilizing epistemological resources in a particular context (OECD, 2003; Rychen & Salganik, 2003). The MLF is primarily a structural construct that describes the constituent components that should be in place for a modeler to be competent. However, different teaching interventions suggest that a developmental layout for each of the components presented in Fig. 3.2 could be compared with the learning progression presented by Schwarz and her colleagues (Schwarz et al., 2009; Schwarz, Reiser, Acher, Kenyon, & Fortus, 2012).

Before describing the proposed MLF, it is important to point out that it differs from "model-based learning," which has been used extensively in prior research to denote learning with mental models (Clement, 2000; S. Gilbert, 1991; Gobert & Buckley, 2000). By contrast, the MLF concerns the active participation of learners in modeling-based learning instruction, which engages them in the construction, revision, refinement, and validation of external representations or artifacts that purport to meet the criteria necessary to ensure their classification as scientific models.

Efforts to design modeling-based learning instruction have relied on an understanding of modeling competence with constituent components in two broad categories, namely *modeling practices and modeling meta-knowledge* (Nicolaou, 2010; Papaevripidou, 2012; Papaevripidou, Nicolaou, & Constantinou, 2014; Fig. 3.2). Attempts to validate such designs have led to the claim that learners' modeling competence can emerge and evolve as a result of their active participation in modeling practices with the concurrent development of meta-knowledge about modeling. Model construction (Stratford, Krajcik, & Soloway, 1998), model use (NRC, 2012), model comparison (Penner, Giles, Lehrer, & Schauble, 1997), model revision (Wu, Krajcik, & Soloway, 2001), and model validation (Halloun, 1996) have been identified as the main *practices* that learners engage in, during modeling. Metacognitive knowledge about the modeling process, which refers to a learner's ability to

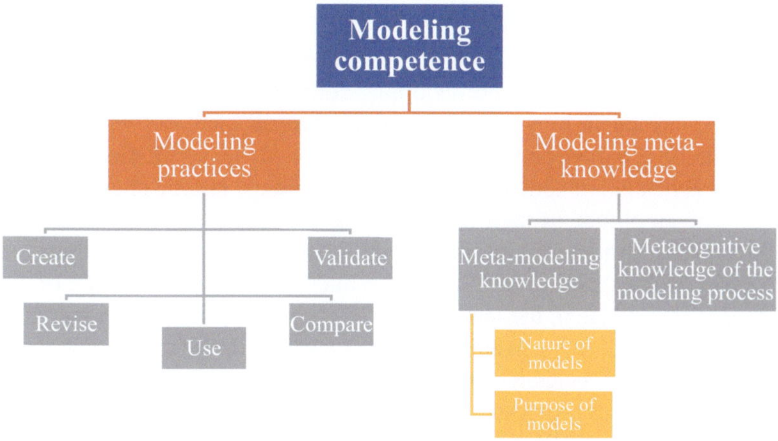

Fig. 3.2 Constituent components of the modeling-based learning framework (Nicolaou & Constantinou, 2014)

explicitly describe and reflect on the actual process of modeling, and meta-modeling knowledge (Schwarz & White, 2005) (i.e., an epistemological awareness of the nature and the purpose of models) together shape the meta-knowledge about modeling.

3.2.1 Modeling Practices

Model construction pertains to a learner's ability to develop an external representation of a physical phenomenon, a system, or an object (Constantinou, 1999; Namdar & Shen, 2015) after he or she has collected data by directly observing the phenomenon or indirectly by using secondary sources. In doing this, the learner needs to consider the modeling medium, his or her familiarity with it, the type of information that is available, and data on the phenomenon. A competent modeler is one who can construct a model that entails clear representational, interpretive, and predictive power.

First, a model with representational power includes objects, variables, and processes:

(a) Objects or entities constitute the core components of a model because they form the basis on which the rest of the components will be based (e.g., animals, plants, air, and water in a forest ecosystem, or the earth, sun, and moon in the solar system).
(b) Variables are the changing aspects characterizing the objects or the phenomenon as a whole (e.g., size, population, velocity).
(c) Processes are usually series of occurrences that produce change. In a model of how thermal equilibrium is attained, "heat flow" is a process, driven by the difference in temperature between two objects and causing change in the internal energy of the interacting objects.

Second, interpreting a phenomenon is about providing a story of how a phenomenon comes to manifest itself. For example, consider the following "story," which explains the mechanism by which the volume of a ball increases when it is heated: "The volume of a ball depends on the amplitude of the oscillation of its atoms. When a ball is heated, its temperature increases, and the kinetic energy of its atoms increases. The amplitude of their oscillations gets bigger. Therefore, the volume of the ball increases." An interpretation will typically include the relationships between the objects, variables, and processes. These interrelationships could be of a causal nature (e.g., the increase in the temperature of an iron cube causes an increase in its volume) or non-causal (e.g., an animal interacts with plants to eat them or with other animals of the same species to reproduce). The interpretive power of the model is related to its efficacy in providing one or more mechanisms that underlie the behavior of the phenomenon (Berland & Reiser, 2009; Braaten & Windschitl, 2011). Mechanisms are organized so that they produce regular changes when comparing the initial with the final conditions of the phenomenon under consideration

(Machamer, Darden, & Carver, 2000). Mechanisms tell us how the various processes and the interrelationships work together to manifest the observable aspects of the phenomenon.

Third, a model has predictive power when it allows the formulation and testing of predictions for new or future aspects of the phenomenon it represents (Bunge, 1983; Hughes, 1997). A model is an epistemological object that allows the user to change the input variables and record different outputs. Hence, a constructed model allows at least some of its elements to be changed and the resulting changes in the behavior of the phenomenon to be observed. This aspect of a model is of particular importance as predictions form a significant aspect of the usability of scientific models (Bunge, 1983) and also provide a clear means for testing and validating models.

Model use is a practice that is often closely connected with model construction. Like scientists and engineers, in the MLF, learners use the models they construct to express their current understanding of the system (or parts of the system). The purpose is to use the models to develop questions and interpretations and to communicate ideas to others (National Research Council, 2012; Nersessian, 2008). Learners who become competent in using models gain a purposeful, meaningful, and fruitful understanding of scientific knowledge (Xiang & Passmore, 2015) with respect to its content as well as its procedural and epistemological aspects.

Learners, like scientists and engineers, typically formulate various models of different forms. In order to make a decision about how and which of these prototype models to reject in favor of the most appropriate model that satisfies a set of specific criteria, learners need to be engaged in another practice: *model comparison* (Stratford et al., 1998). The importance of this modeling practice was delineated by Penner et al. (1997), who declared that "understanding the possibility of different models, and thinking about the advantages and disadvantages of various alternatives, might in turn support children's progress from a primarily descriptive use of models to a beginning recognition that models can serve as instantiations of rival hypotheses" (p.126). To practice model comparison, the learner should be capable of selecting the most appropriate model from among a series of models for the same hypothesis, and this model should be related to a phenomenon under study that satisfies certain criteria with respect to its representational, interpretive, and predictive power.

During modeling, the learner revisits the phenomenon under study and identifies the discrepancies that appear when comparing the phenomenon with the model that is being constructed. In doing so, the learner is engaged in the practice of *model revision* (Stratford et al., 1998; Wu et al., 2001). Model revision pertains to the learner's ability to (i) contrast a model with its corresponding phenomenon, (ii) evaluate it on the basis of the absence or presence of a model's basic components, and (iii) find ways to integrate missing parts or remove redundant parts in order to produce a revised model. Fretz et al. (2002) stated that the scientific practice of evaluating a model involves several actions, such as predicting what would happen, identifying anomalies, interpreting and critiquing the results, and proposing solutions. Similarly, Stratford et al. (1998) asserted that *testing and debugging* are examples of modeling activities that learners can productively engage in when testing the model, trying different possibilities, identifying problems with its behavior, and searching for solutions.

The constructed model is complete when it is also validated (often with other phenomena from the same class). *Model validation* refers to the learner's ability to abstract the model from the phenomenon and apply it in a new situation, possibly in phenomena of the same class. If the model fails to account for a new phenomenon, the learner needs to formulate a new model that will successfully describe, represent, and predict the observable patterns of both phenomena. Halloun (1996) asserted that "validation includes different forms of assessment that provide learners with opportunities to fulfill a major objective of science education: *critical thinking*" (p. 1028). To better describe this practice and differentiate it from model revision, consider an elastic collision of two balls (A, B) of equal mass (Ma = Mb), one moving at a constant horizontal velocity (Va) and one at rest (Vb = 0) (Fig. 3.3a). A modeler builds a "transfer model" (Ball A transfers its velocity to Ball B), which includes a mechanism for an exchange of velocity between the two balls. Then, a new phenomenon of the same class is presented (Fig. 3.3b). The "transfer" model is not consistent. Ball A does not transfer its velocity to Ball B. The model's validity is therefore tested. To account for the new phenomenon of the same class, the modeler constructs a "swap" model with a different mechanism that now pertains to the swapping of velocity during the collision. The validation process can continue with new phenomena (Fig. 3.3c) until the model validation process leads to the construction of the momentum model.

The validation of a model is an important part of the modeling process. It serves as a confirmation of viability or as an indication of room for improvement in the learner's model. However, it is often ignored during instruction. There is a need for improved scaffolds and more elaborate designs for teaching-learning sequences to facilitate the practice of model validation.

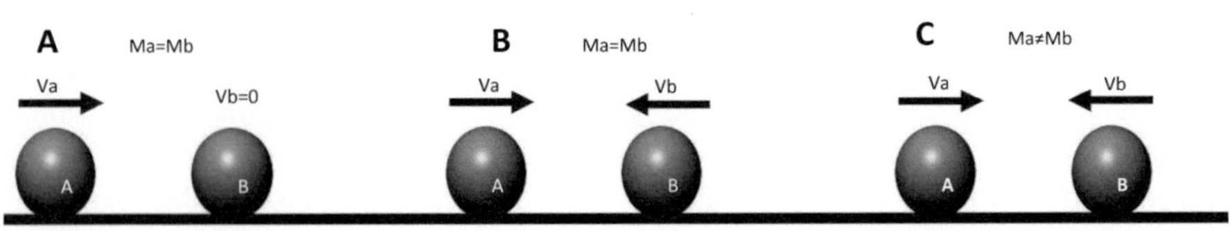

Fig. 3.3 Elastic collisions: Three phenomena from the same class. M stands for the masses of the two balls, V for their velocities. The three collisions could be modeled with three models of increasing sophistication and validity: velocity transfer, velocity swapping, momentum exchange. (**a**) moving ball colliding with stationary ball of equal mass; (**b**) balls of equal masses colliding with equal speeds in opposite directions; (**c**) collision between balls of different mass moving with different speeds in opposite directions

3.2.2 Modeling Meta-Knowledge

Learners engaging systematically with the five modeling practices will have mastered modeling competence to a great extent. However, there are other aspects of knowledge that are equally important to the successful modeler, namely, the modeling meta-knowledge (Schwarz & White, 2005, p. 167). With respect to modeling, meta-knowledge is shaped by metacognitive knowledge about the modeling process as well as meta-modeling knowledge.

Metacognitive knowledge about the modeling process refers to the ability of a learner to explicitly describe and reflect on the major actions undertaken to model a phenomenon under study (the modeling-based learning cycle; Fig. 3.1). For instance, one starts by observing the phenomenon that is of interest, collects information from the phenomenon, formulates a model by implementing the collected information, contrasts the formulated model with the phenomenon as a means of evaluating the model, revises the model in the light of new information that was not implemented in the original formulation, and then repeats the process in an iterative and cyclical manner with the purpose of refining the model to make it consistent, rigorous, and usable for testing hypotheses and making predictions.

The second aspect of meta-knowledge is "meta-modeling knowledge," which refers to developing an understanding of the nature of models and an appreciation of the purpose of scientific modeling. Engaging learners to simply develop models is not enough for developing an epistemological awareness of models and modeling. Deviating from Schwarz and White (2005),[2] the MLF distinguishes between metacognitive knowledge about the modeling process (which is metacognitive knowledge about how to construct and validate scientific models) and meta-modeling knowledge (which is epistemic knowledge about the *nature* and *purpose* of models in science). Consequently, *meta-modeling knowledge about the nature of models* entails a definition of models in terms of their *representational*, *interpretive*, and *predictive powers*. Likewise, *meta-modeling knowledge about the purpose and use of models* entails an epistemic understanding of the purposes of models in science. For instance, models (i) serve as sense-making tools for constructing knowledge, (ii) are used as communication platforms for conveying understanding or knowledge, (iii) can be used to develop new understandings by predicting new aspects of phenomena or showcasing mechanisms, and (iv) are used to illustrate, interpret, and predict phenomena (Schwarz et al., 2009).

---

[2]These researchers propose that meta-modeling knowledge consists of (i) the *nature of models*, (ii) the *nature or process of modeing*, (iii) the *evaluation of models*, and (iv) the *purpose or utility of models*.

3.2.3 The Role of Reflection and Metacognition in Modeling-Based Learning

The relationship between the epistemological underpinnings of modeling and actual modeling practices has been investigated in only a few studies (Cheng & Lin, 2015; Gobert et al., 2011), even though it has been emphasized as an important goal of science education (National Research Council, 2012, 2013; Nicolaou & Constantinou, 2014).

The MLF presents a conceptualization of what scientists do with modeling as well as what science educators expect learners to do with modeling. Therefore, for someone to be competent in modeling, he or she needs to be able to practise modeling as well as to exhibit meta-modeling knowledge and metacognitive knowledge about the modeling process. Stated differently, both the practical aspect (modeling practices) and the epistemological aspect (meta-knowledge) of modeling are important to a competent modeler.

For scientists who engage in authentic inquiry as established members of a scientific community, it is not necessarily sufficient in and of itself to ensure informed epistemological conceptions or conceptions that are identical to the conceptions of other members of the scientific community (Constantinou & Papadouris, 2012). Those who engage in authentic scientific inquiry might or might not develop epistemological views that are aligned with philosophically informed perspectives on scientific practice (Papadouris & Constantinou, 2014). Those views may be bound to the context of the individual scientist, and individual contexts may vary considerably across and within scientific disciplines (Schwartz & Lederman, 2008).

Educational research findings have demonstrated that for learners to develop the epistemological bases of scientific knowledge, implicit instruction is not sufficient. There is a need for explicit epistemological discourse that places features of the epistemology of science at the center of instruction and is both taught and assessed (Schwartz, Lederman, & Abd-El-Khalick, 2012). In the same line of reasoning are findings from studies by Gobert et al. (2011) who investigated the effects of modeling activities performed by students in three subject areas (Physics, Biology, and Chemistry) and found no significant relationship between students' understanding of models and their modeling practices in biology and physics. This was not the case for Chemistry, where a weak but statistically significant relationship was identified. In Chemistry lessons, students were explicitly taught about the nature and purpose of models, whereas in Physics and Biology, no support for the teaching of the nature of models and modeling was implemented.

In the same vein, Cheng and Lin (2015) conducted a study to explore the relationship between students' views of scientific models and their ability to generate their own models. Their study shed light on the relationship between students' model construction practices and their epistemological views on models and modeling. More specifically, they found that a few students who had shown above-average science learning performance and interest in science were able to develop coherent microscopic models. By contrast, students with lower science learning performance and interest were only able to develop observational or fragmented models. With regard to the relationship between students' views of the nature of models and their self-developed models, these researchers found that students who could develop coherent microscopic models had a better understanding of some aspects of the epistemology of models (i.e., representations of models, models as explanatory tools, and the use of scientific models) than students who had developed models at the observational level. Nevertheless, this study did not find any statistical evidence that the sub-factors of "models as exact replicas" and the "changing nature of models" were associated with the development of students' modeling competence.

Finally, the FMC (Chap. 1) suggests that this competence is the ability to reflect on models and modeling but leaves the role of practicing modeling somewhat unclear, perhaps implying that a competent modeler is the one reflecting correctly and successfully on the process of modeling and the nature and purpose of models regardless of his or her ability to really construct, use, compare, revise, and validate models (Krell et al., 2012).

On the basis of the conflicting discourse presented in this chapter on the connection of modeling practices, the epistemology of models, and the theoretical underpinnings of modeling competence, we propose that:

(a) A modeler who is competent in modeling practices is not necessarily an epistemologically competent modeler (Cheng & Lin, 2015; Gobert et al., 2011; Guisasola, Almudí, & Zubimendi, 2004).
(b) An epistemologically competent modeler is more likely to be competent in modeling practices (Sandoval, 2015; Schwartz & Lederman, 2008).
(c) An epistemologically incompetent modeler, however, is not necessarily incompetent in modeling practices.

Taken together, these three claims support the existence of the dual nature of the MLF and the need to develop both instruction and assessment that will support both modeling practices and the modeling of meta-knowledge. Additional research is needed to further clarify the interconnectedness between the reflection aspects of the MLF and using it in practice.

3.3 Monitoring the Development of the Modeling-Based Learning Framework

Several assessment tasks have been designed to be consistent with the MLF, and they were used to assess learners' modeling competence in various domains and in different instructional situations. These formative and summative techniques were employed in the framework of a series of teaching interventions (Papaevripidou et al., 2014) and through the use of a variety of data collection tools (Nicolaou, 2010; Nicolaou & Constantinou, 2014; Papaevripidou, 2012). This perspective enabled us to examine learners' modeling competence in a comprehensive manner and to arrive at a holistic view of how it emerges. Specifically, the designed interventions, which followed modeling-centered scientific inquiry principles, differed with respect to (a) the content of the curriculum, (b) the age of the participants, and (c) the modeling tool used by learners. However, they maintained the same format and duration, with each intervention lasting for about 8–10 90-min sessions.

At the beginning and end of each intervention, each component of the MLF was evaluated through a set of two assessment tasks. The two assessment tasks had the same structure (e.g., they consisted of a scenario and open-ended questions). Both the scenarios and the questions were comprised of short and simple statements. The assessment tasks were grouped in such a way that each modeling competence component was evaluated by two tests in two different subject areas (e.g., Test 1: free fall and Test 2: evaporation). The purpose of designing and administering two tasks for each of the components of the MLF was to explore whether modeling competence is content-dependent or not (Papaevripidou et al., 2014).

Students' responses to each diagnostic test were subjected to phenomenographic analysis, which led to the construction of different attainment levels for each component of modeling competence. The results of phenomenography (Marton, 1981) are a set of logically interrelated category conceptions (in this case, comprising the modeling competence), which are usually created on the basis of their content and their correctness (scientific level) and are differentiated from each other on the basis of qualitative criteria (Nicolaou, 2010). Here, the categories qualitatively describe the different ways in which the participants responded to each component of modeling competence prior to and after the teaching intervention.

The analysis of the collected data (from the whole series of interventions and assessment tools) revealed different levels of increased sophistication that exist for each component of the MLF among learners. Figure 3.4 provides a summary of the most superior level that emerged for each component of the MLF. As such, it also serves as an illustration of how the content of the MLF was reframed on the basis of the most superior level that emerged from the analysis of the data collected during the interventions. It is notable that the most superior levels were most commonly found to emerge after learners participated in modeling-based instruction.

For each component of the MLF presented in Fig. 3.4, specific hierarchical levels with increased sophistication that illuminate the degree of development of learners' modeling practices and their modeling of meta-knowledge emerged. Figure 3.5 presents the hierarchical levels that emerged from the data analysis with regard to the practice of revising the model. The tests for evaluating model revision asked students to first observe a specific model (e.g., a diagram presenting the photosynthetic growth of a plant) and state whether the model was complete. If they considered the model to be complete, they were asked to describe the ways in which the model appeared to be complete. Otherwise, they were asked to state how they would improve the model.

Component	Highest attainment level
I. Modeling practices	
• *Model construction*	Construction of a model that (i) provides a *comprehensive representation* of the phenomenon (e.g., all types of the components of the phenomenon are represented), (ii) encompasses both a *mechanistic interpretation* of how the phenomenon functions and a *causal interpretation* that explains why the phenomenon functions the way it does, and (iii) has *strong predictive power*.
• *Model use*	Efficient use of a model to (a) describe a phenomenon, and/or (b) interpret how the phenomenon functions, and/or (c) predict its future behavior or state.
• *Model comparison*	Detection of the best or worst model based on the model's (i) *representational comprehensiveness,* (ii) *interpretive potential*, and (iii) *predictive power.*
• *Model revision*	Proposal of specific model revision measures after identifying the limitations of (i) the representational completeness of the model (e.g., absence of objects, variables, or processes among the components of the model), (ii) the interpretive potential of the model (e.g., the model is missing a mechanism that explains how the phenomenon functions), or (iii) the predictive power of the model.
• *Model validation*	Validation of the model on the basis of the comparison of the two phenomena with respect to the model's components (e.g., The two phenomena do not share the same variables, so the new data cannot be used with this model).
II. Meta-knowledge	
• *Metacognitive knowledge about the modeling process*	The process of modeling involves (i) collecting information about the phenomenon (e.g., performing observations and collecting data, identifying objects, variables, processes, and interactions), (ii) selecting the most appropriate means for building the model, (iii) building a model on the basis of the data that were collected, (iv) comparing the model and the phenomenon or the model with other models, (v) evaluating the model according to its representational completeness, interpretive potential, and predictive power, (vi) improving the model, (vii) testing the validity of the model, (viii) repeating steps (iv) through (vii).
• *Meta-modeling knowledge*	
- *Nature of models*	A model describes, represents, and explains a phenomenon under study (e.g., provides a possible mechanism for how the phenomenon functions) and can be used to test predictions about specific aspects of the phenomenon.
- *Purpose of models*	The models serve as (i) instructional aids, (ii) simulations, (iii) facilitators of the conceptual understanding of the phenomenon under study, (iv) communication tools, (v) external representations of a phenomenon under study, and (vi) vehicles for formulating and testing predictions.

Fig. 3.4 Summary of the highest levels of attainment in school for each of the MLF components

3.4 Conclusions and Discussion

The implementation of a framework that is grounded in contemporary perspectives of learning science through modeling has great potential for classroom use because it can promote significant aspects of modeling-centered inquiry teaching and learning. The MLF presents a conceptualization of what scientists do while modeling as well as what science educators expect learners to do when developing and using models. Based on the MLF, learners are expected to practise modeling (to construct, use, compare, revise, and validate models) but also to develop modeling meta-knowledge, that is, to explicitly describe and reflect on the actual process of modeling as well as to become epistemologically aware of the nature and the purpose of models.

Level and Description
Level 6*. The learner identifies the need for in-depth improvements with regard to the representational, interpretive, and predictive power of the model The model is not complete because (1) it does not provide a strong representation of the phenomenon. The learners identify deficiencies with respect to: (a) its objects (carbon dioxide, oxygen, starch, etc.), (b) its variables (intensity of light, humidity, air composition, etc.), (c) its processes (photosynthesis, transformation of carbon dioxide and water into glucose and oxygen, transport of water in the plant, etc.), and (d) its relations (relation between the quantities of carbon dioxide and oxygen, relations between light, chlorophyll, and the process of photosynthesis, etc.), (2) it does not fully interpret the phenomenon (it does not reveal how photosynthesis happens, how the plant takes in water and other resources, what factors are important, or what the processes are), (3) it does not have predictive power (What will happen if the humidity increases or if the sun is not present for some period?).
Level 5. The learner identifies the need for in-depth improvements with regard to two of the model's utilities Level 5.3. ….the interpretive and predictive power of the model. Level 5.2. ….the representational and predictive power of the model. Level 5.1. ….the representational and interpretive aspects of the model.
Level 4. The learner identifies the need for in-depth improvements with regard to one of the model's utilities Level 4.3. ….the model's representational power. Level 4.2. ….the model's interpretive power. Level 4.1. ….the model's predictive power.
Level 3. The learner identifies that the model needs superficial representational and interpretive improvements
Level 2. The learner identifies that the model needs superficial representational improvements
Level 1. The learner identifies that the model needs unspecified improvements (which may result from their personal experience or from focusing on the superficial features of the model)
Level 0. Irrelevant or no answer (The model is incomplete. No improvements are needed)

* Level six includes learners' responses that are closer to the scientifically correct perspective.

Fig. 3.5 Levels of attainment for the practice of model revision (using photosynthesis as an example)

When comparing the MLF to the FMC, specific differences arise. The latter describes the aspects of modeling competence (i.e., the nature of models, multiple models, purpose of models, testing models, changing models), and it provides a description of three different levels for each of these five aspects, which are based on whether the modeler considers a model to be a "model for something" or a "model of something" (Mahr, 2009; in Krell, Reinisch, & Krüger, 2015). At the first level of each aspect, the FMC describes modelers' beliefs with respect to models of something, whereas at the third level, the FMC states modelers' reflection with respect to the essence of models for something.

The MLF distinguishes between modeling practices and meta-knowledge (learners' reflection on models and modeling) and considers them both equally important for teaching and learning, whereas the FMC refers only to learners' cognitive reflections about models and modeling. Additionally, each of the five modeling practices is important and has its own levels of competence according to the MLF. For example, model construction is one of the five constituent components of the practices that need to be developed by a competent modeler and is at the same level of importance as the remaining four practices (use, compare, revise, validate). The levels of attainment for each modeling practice (Fig. 3.5) are based on (a) what students actually do while practicing modeling, that is, when developing models, and (b) how students reflect on their competence to do so. By contrast, the model construction practice is not referenced in the FMC. Moreover, the validation practice seems to be absent from this framework. These researchers acknowledge that the testing and modifying of models to resolve inconsistencies emerge when comparing the model to the phenomenon, but no reference is made to phenomena of the same class. However, this practice is also important to the development of modeling competence in accordance with research claims that challenging students to defend the validity of their models results in significant improvements in their scientific discourse (White & Frederiksen, 1990).

With respect to the modeling of meta-knowledge, the FMC entails the notion of metacognitive knowledge of the modeling process through the aspects of "testing models" by considering that at the second level, modelers are able to show the correspondence between the model and the initial object (i.e., test a model of something). Additionally, at Level III, modelers test the model of something by verifying hypotheses during the application of the model. Therefore, the notable difference between the two frameworks with respect to meta-knowledge is that the MLF considers the idea that a metacognitively competent modeler is the one who can reflect on the process that *the modeler him- or herself followed* when engaging with the five modeling practices (i.e., creating, using, comparing, revising, and validating) a model. It is therefore considered to be an externalization of the steps already followed by the modeler when constructing and revising a model.

The MLF considers practices and meta-knowledge to be equally important and necessary for the development and assessment of modeling competence. On the other hand, the FMC maintains a rather different view. It emphasizes a theoretical understanding of models and the reflection on the process of modeling and acknowledges modeling competence as an ability to reflect on models and modeling. This underlines the importance of gaining insightful knowledge with models, judging models with regard to their purpose, and reflecting on the process of gaining knowledge through models and modeling (Upmeier zu Belzen & Krüger, 2010). This emphasis on meta-modeling competence will affect the design of teaching interventions as well as assessment efforts to evaluate modeling competence. Following the principles of the FMC, when developing modeling-based learning and teaching interventions, the focus should be (only) on developing learners' meta-modeling competence, and hence, the actual extent of "hands-on" modeling could be underestimated. Additionally, it is implied that an assessment of a learner's meta-knowledge is sufficient to help the assessor understand whether the learner is a competent modeler or not.

The presented MLF can be used to track the development of learners' modeling competence. The levels of increased sophistication that emerged for every component of the MLF (Fig. 3.5) provide a useful guide that instructors can use to better understand students' progress and even predict many of the difficulties that might emerge when implementing modeling-based learning. It can also assist in the design and organization of learning experiences and assessment tools that recognize and take advantage of the most likely trajectories that are typically followed by students as they move toward expertise (learning progressions; Schwarz et al., 2009, 2012). Because textbooks rarely include modeling assignments that invite students to actively practice modeling (van der Valk, van Driel, & de Vos, 2007; VanLehn, 2013), the clarification of each constituent component of the MLF can inform science educators and curriculum designers how to design and teach learning sequences for modeling and also additional assessment tasks to evaluate students' modeling competence in unison.

The need for a coherent framework that can help define what is being assessed as well as the subdimensions of modeling has been suggested before (Nicolaou & Constantinou, 2014). In previous work, modeling competence was not defined or assessed in a unified manner. Each study presented by this review paper has defined and assessed only one part of what can be conceptualized as modeling competence on the basis of available theoretical frameworks (National Research Council, 2012; Penner et al., 1997; Schwarz et al., 2009; Stratford et al., 1998). Even in cases where one aspect of modeling was under investigation, researchers have often used different definitions and consequently different assessment approaches. The MLF serves as a means for overcoming the fragmented diversity identified by Nicolaou and Constantinou (2014). The unifying nature of the MLF with respect to the complementary and interconnected relationship between modeling practices and the modeling of meta-knowledge as a combination of the nature/role of models and the modeling process is the most powerful characteristic of this framework. As such, the framework can serve as a basis for conceptualizing the teaching and assessment of modeling competence in a holistic manner, in contrast to focusing, as most published research has done, on one part of the modeling meta-knowledge (i.e., meta-modeling knowledge). Metacognitive knowledge about the modeling process is equally important for a robust development of the modeling competence. Additionally, this meta-knowledge should not be examined in isolation from modeling practices because being a competent modeler is not based exclusively on learners' modeling meta-knowledge or on learners' modeling practices. On the contrary, it is based on a learner's ability to both practise modeling and to demonstrate an understanding of the modeling process and the nature of models as epistemological entities.

Further research is needed to investigate and clarify the role of reflection and metacognition in modeling practices, both for scientists who work with models in their everyday activities and for learners who use and develop models in the framework of structured teaching interventions aimed at enhancing modeling competence.

References

Berland, L. K., & Reiser, B. J. (2009). Making sense of argumentation and explanation. *Science Education, 93*(1), 26–55. https://doi.org/10.1002/sce.20286

Braaten, M., & Windschitl, M. (2011). Working toward a stronger conceptualization of scientific explanation for science education. *Science Education, 95*(4), 639–669. https://doi.org/10.1002/sce.20449

Bunge, M. (1983). *Treatise on basic philosophy: Volume 5: Epistemology & methodology I: Exploring the world*. Dordrecht, The Netherlands: Springer/Reidel.

Chapman, O. L. (2000). Learning science involves language, experience and modeling. *Journal of Applied Developmental Psychology, 21*(1), 97–108.

Cheng, M.-F., & Lin, J.-L. (2015). Investigating the relationship between students' views of scientific models and their development of models. *International Journal of Science Education, 0693*(September), 1–23. https://doi.org/10.1080/09500693.2015.1082671

Clement, J. (2000). Model based learning as a key research area for science education. *International Journal of Science Education, 22*(9), 1041–1053.

Constantinou, C. P. (1999). The cocoa microworld as an environment for developing modeling skills in physical science. *International Journal of Continuing Education and Life-Long Learning, 9*(2), 201–213.

Constantinou, C. P., & Papadouris, N. (2004). Potential contribution of digital video to the analysis of the learning process in physics: A case study in the context of electric circuits. *Educational Research and Evaluation, 10*(1), 21–39. https://doi.org/10.1076/edre.10.1.21.26300

Constantinou, C. P., & Papadouris, N. (2012). Teaching and learning about energy in middle school: An argument for an epistemic approach. *Studies in Science Education, 48*(2), 161–186. https://doi.org/10.1080/03057267.2012.726528

Crawford, B., & Cullin, M. (2004). Supporting prospective teachers' conceptions of modelling in science. *International Journal of Science Education, 26*(11), 1379–1401.

De Jong, O., Van Driel, J. H., & Verloop, N. (2005). Preservice teachers' pedagogical content knowledge of using particle models in teaching chemistry. *Journal of Research in Science Teaching, 42*(8), 947–964. https://doi.org/10.1002/tea.20078

Dori, Y., & Kaberman, Z. (2012). Assessing high school chemistry students' modeling sub-skills in a computerized molecular modeling learning environment. *Instructional Science, 40*(1), 69–91. https://doi.org/10.1007/s11251-011-9172-7

Duschl, R. A., Schweingruber, H. A., & Shouse, A. W. (2007). *Taking science to school: Learning and teaching science in grades K-8*. Washington, DC: National Academy Press.

Fretz, E. B., Wu, K., Zhang, B., Davis, E., Krajcik, J., & Soloway, E. (2002). An invstigation of software scaffolds supporting modelling practices. *Reseach in Science Education, 32*, 567–589.

Giere, R. N. (1999). Using models to represent reality. In L. Magnani, N. J. Nersessian, & P. Thagard (Eds.), *Model-based reasoning in scientific discovery* (pp. 41–57). New York: Kluwer/Plenum.

Giere, R. N. (2004). How models are used to represent physical reality. *Philosophy of Science, 71*(5), 742–752.

Gilbert, J., Boulter, C., & Rutherford, M. (1998). Models in explanations, part 2: Whose voice, whose ears? *International Journal of Science Education, 20*(2), 187–203.

Gilbert, S. (1991). Model building and a definition of science. *Journal of Research in Science Teaching, 28*(1), 73–79.

Gobert, J. D., & Buckley, C. B. (2000). Introduction to model-based teaching and learning in science education. *International Journal of Science Education, 22*(9), 891–894.

Gobert, J. D., O'Dwyer, L., Horwitz, P., Buckley, B. C., Levy, S. T., & Wilensky, U. (2011). Examining the relationship between students' understanding of the nature of models and conceptual learning in biology, physics, and chemistry. *International Journal of Science Education, 33*(5), 653–684. https://doi.org/10.1080/09500691003720671

Grünkorn, J., zu Belzen, A. U., & Krüger, D. (2014). Assessing students' understandings of biological models and their use in science to evaluate a theoretical framework. *International Journal of Science Education, 36*(10), 1651–1684. https://doi.org/10.1080/09500693.2013.873155

Guisasola, J., Almudí, J. M., & Zubimendi, J. L. (2004). Difficulties in learning the introductory magnetic field theory in the first years of university. *Science Education, 88*(3), 443–464. https://doi.org/10.1002/sce.10119

Halloun, I. (1996). Schematic modeling for meaningful learning of physics. *Journal of Research in Science Teaching, 33*, 1019–1041.

Hokayem, H., & Schwarz, C. (2014). Engaging fifth graders in scientific modeling to learn about evaporation and condensation. *International Journal of Science and Mathematics Education, 12*(1), 49–72. https://doi.org/10.1007/s10763-012-9395-3.

Hughes, R. (1997). Models and representation. *Philosophy of Science, 64*, S325–S336.

Krell, M., Upmeier zu Belzen, A., & Krüger, D. (2012). Students' understanding of the purpose of models in different biological contexts. *International Journal of Biology Education, 2*(2), 1–34.

Krell, M., Reinisch, B., & Krüger, D. (2015). Analyzing students' understanding of models and modeling referring to the disciplines biology, chemistry, and physics. *Research in Science Education, 45*(3), 367–393. https://doi.org/10.1007/s11165-014-9427-9

Louca, L., Zacharia, Z., & Constantinou, C. P. (2011). In quest of productive modeling-based learning discourse in elementary school science. *Journal of Research in Science Teaching, 48*(8), 919–951. https://doi.org/10.1002/tea.20435

Machamer, P., Darden, L., & Carver, C. F. (2000). Thinking about mechanisms. *Philosophy of Science, 67*(67), 1–25. https://doi.org/10.1086/392759

Mahr, B. (2009) Information science and the logic of models. *Software and Systems Modeling, 8*:365–383.

Maia, P. F., & Justi, R. (2009). Learning of chemical equilibrium through modelling-based teaching. *International Journal of Science Education, 31*(5), 603–630.

Marton, F. (1981). Phenomenography-describing conceptions of the world around us. *Instructional Science, 10*, 177–200.

Mendonça, P. C. C., & Justi, R. (2014). An instrument for analyzing arguments produced in modeling-based chemistry lessons. *Journal of Research in Science Teaching, 51*(2), 192–218. https://doi.org/10.1002/tea.21133

Namdar, B., & Shen, J. (2015). Modeling-oriented assessment in K-12 science education: A synthesis of research from 1980 to 2013 and new directions. *International Journal of Science Education, 37*(7), 993–1023. https://doi.org/10.1080/09500693.2015.1012185

National Research Council. (2012). *A framework for K-12 science education: Practices, crosscutting concepts, and core ideas*. Washington, DC: The National Academies Press. http://www.nap.edu/catalog.php?record_id=13165

National Research Council. (2013). Next generation science standards: For states, by states. Washington, DC: The National Academies Press. doi.org/https://doi.org/https://doi.org/10.17226/18290.

Nersessian, N. J. (2008). Model-based reasoning in scientific practice. In R. A. Duschl & R. E. Grandy (Eds.), *Teaching scientific inquiry: Recommendations for research and implementation* (pp. 57–79). Sense Publishers.

Nicolaou, C. T. (2010). *Modelling and collaboration in learning environments (in Greek). Department of Education*. Doctoral dissertation. University of Cyprus Nicosia, Cyprus (ISBN: 978-9963-689-84-2).

Nicolaou, C. T., & Constantinou, C. P. (2014). Assessment of the modeling competence: A systematic review and synthesis of empirical research. *Educational Research Review, 13*(3), 52–73.

Nicolaou, C. T., Nicolaidou, I. A., & Constantinou, C. P. (2009). Scientific model construction by pre-service teachers using Stagecast creator. In P. Blumstein, W. Hung, D. Jonassen, & J. Strobel (Eds.), *Model-based approaches to learning: Using systems models and simulations to improve understanding and problem solving in complex domains* (pp. 215–236). Rotterdam, The Netherlands: Sense Publishers.

OECD. (2003). The definition and selection of key competencies – Executive summary. *DeSeCo*, 1–20. https://doi.org/10.1080/2159676X.2012.712997

Papadouris, N., & Constantinou, C. P. (2014). An exploratory investigation of 12-year-old students' ability to appreciate certain aspects of the nature of science through a specially designed approach in the context of energy. *International Journal of Science Education, 36*(5), 755–782. https://doi.org/10.1080/09500693.2013.827816

Papaevripidou, M. (2012). *Teachers as learners and curriculum designers in the context of modeling-centered scientific inquiry*. Doctoral dissertation, Learning in Science Group, Department of Educational Sciences, University of Cyprus, Nicosia, Cyprus (ISBN: 978-9963-700-56-1).

Papaevripidou, M., Constantinou, C. P., & Zacharia, Z. C. (2007). Modeling complex marine ecosystems: An investigation of two teaching approaches with fifth graders. *Journal of Computer Assisted Learning, 23*(2), 145–157. https://doi.org/10.1111/j.1365-2729.2006.00217.x

Papaevripidou, M., Nicolaou, C. T., & Constantinou, C. P. (2014). *On defining and assessing learners' modeling competence in science teaching and learning*. In Annual Meeting of American Educational Research Association (AERA), Philadelphia, Pennsylvania, USA.

Passmore, C., & Stewart, J. (2002). A modeling approach to teaching evolutionary biology in high schools. *Journal of Research in Science Teaching, 39*(3), 185–204.

Penner, D. E., Giles, N. D., Lehrer, R., & Schauble, L. (1997). Building functional models: Designing an elbow. *Journal of Research in Science Teaching, 34*(2), 125–143.

Pluta, W. J., Chinn, C. A., & Duncan, R. G. (2011). Learners' epistemic criteria for good scientific models. *Journal of Research in Science Teaching, 48*(5), 486–511. https://doi.org/10.1002/tea.20415

Raftopoulos, A., Kalyfommatou, N., & Constantinou, C. P. (2005). The properties and the nature of light: The study of Newton's work and the teaching of optics. *Science & Education, 14*(7–8), 649–673. https://doi.org/10.1007/s11191-004-5609-6

Rychen, D. S., & Salganik, L. H. (2003). Key competencies for a successful life and a well-functioning society.

Sandoval, W. A. (2015). Epistemic goals (and scientific reasoning). *Encyclopedia of Science Education*, (January 2015), 694698. https://doi.org/10.1007/978-94-007-2150-0

Schwartz, R. S., & Lederman, N. G. (2008). What scientists say: Scientists' views of nature of science and relation to science context. *International Journal of Science Education, 30*(6), 727–771. https://doi.org/10.1080/09500690701225801

Schwartz, R. S., Lederman, N. G., & Abd-El-Khalick, F. (2012). A series of misrepresentations: A response to Allchin's whole approach to assessing nature of science understandings. *Science Education, 96*(4), 685–692. https://doi.org/10.1002/sce.21013

Schwarz, C., Reiser, B. J., Acher, A., Kenyon, L., & Fortus, D. (2012). MoDeLS: Challenges in defining a learning progression for scientific modeling. In A. C. Alonso & A. W. Gotwals (Eds.), *Learning progressions in science: Current challenges and future directions* (pp. 101–137). Rotterdam, The Netherlands: Sense Publishers.

Schwarz, C., Reiser, B. J., Davis, E. A., Kenyon, L., Achér, A., Fortus, D., et al. (2009). Developing a learning progression for scientific modeling: Making scientific modeling accessible and meaningful for learners. *Journal of Research in Science Teaching, 46*(6), 632–654. https://doi.org/10.1002/tea.20311

Schwarz, C., & White, B. (2005). Metamodeling knowledge: Developing students' understanding of scientific modeling. *Cognition and Instruction, 23*(2), 165–205.

Sins, P. H. M., Savelsbergh, E. R., & van Joolingen, W. R. (2005). The difficult process of scientific modelling: An analysis of novices' reasoning during computer-based modelling. *International Journal of Science Education, 27*(14), 1695–1721.

Stratford, S. J., Krajcik, J., & Soloway, E. (1998). Secondary students' dynamic modeling processes: Analyzing, reasoning about, synthesizing, and testing models of stream ecosystems. *Journal of Science Education and Technology, 7*(3), 215–234.

Upmeier zu Belzen, A., & Krüger, D. (2010). Modellkompetenz im Biologieunterricht [Model competence in biology classes]. *Zeitschrift Für Didaktik Der Naturwissenschaften, 16*, 41–57. http://archiv.ipn.uni-kiel.de/zfdn/pdf/16_Upmeier.pdf

van der Meij, J., & de Jong, T. (2006). Supporting students' learning with multiple representations in a dynamic simulation-based learning environment. *Learning and Instruction, 16*(3), 199–212.

van der Valk, T., van Driel, J. H., & de Vos, W. (2007). Common characteristics of models in present-day scientific practice. Research in Science Education, 37(4), 469–488. https://doi.org/10.1007/s11165-006-9036-3.

Van Lehn, K. (2013). Model construction as a learning activity: A design space and review. *Interactive Learning Environments, 21*(4), 371–413. https://doi.org/10.1080/10494820.2013.803125

Weinert, F. E. (2001). Concept of competence: A conceptual clarification. In *Definition and selection of competencies: Theoretical and conceptual foundation (DeSeCo)* (pp. 44–65). https://doi.org/10.1073/pnas.0703993104

White, B. Y. and Frederiksen. J. R. (1990) Causal model progressions as a foundation for intelligent learning environments. *Artificial Intelligence, 42*(1), 99–157.

Wu, H. K., Krajcik, J., & Soloway, E. (2001). Promoting understanding of chemical representations: Students' use of a visualization tool in the classroom. *Journal of Research in Science Teaching, 38*(7), 821–842.

Xiang, L., & Passmore, C. (2015). A framework for model-based inquiry through agent-based programming. *Journal of Science Education and Technology, 24*(2–3), 311–329. https://doi.org/10.1007/s10956-014-9534-4

Chapter 4
Modeling Competence in the Light of Nature of Science

Renee S. Schwartz

4.1 Introduction

All scientists use models, and if they say they do not, then they are failing to understand what they are doing. (Aquatic ecologist)

Competence is described as "domain-specific cognitive dispositions that are required to successfully cope with certain situations or tasks, and that are acquired by learning processes" (Koeppen, Hartig, Kleime, & Leutner, 2008, p. 62). With respect to a biological context specifically, Upmeier zu Belzen and Krüger (2010) state that modeling competence includes (1) the ability to get purposeful new insights into biological topics with models, (2) the ability to judge on models and the process of modeling in relation to the purpose, (3) the ability to reflect upon the process of getting insights with models, (4) the motivational and social willingness to use these abilities in problem based situations.

By these descriptions, then, a competence-based view of models in science can be considered a contextually-based cognitive function needed to understand a scientific concept or scientific process. In other words, modeling competence in science involves the epistemic practices of knowledge generation through the understanding and use of scientific models. Scientists rely on modeling competence to gain an understanding of nature and natural phenomena. As expressed in the introductory quote above, if a scientist says they do not use models, they do not understand what they are doing (or, perhaps, they are not doing science). The epistemic functions of scientific models have been described as "a bridge between scientific theory and the world-as-experienced ('reality')" (Gilbert, 2004, p. 116). Gilbert drew from the

R. S. Schwartz (✉)
Department of Middle and Secondary Education, College of Education and Human Development, Georgia State University, Atlanta, GA, USA
e-mail: rschwartz@gsu.edu

A. Upmeier zu Belzen et al. (eds.), *Towards a Competence-Based View on Models and Modeling in Science Education*, Models and Modeling in Science Education 12, https://doi.org/10.1007/978-3-030-30255-9_4

literature to describe several depictions of models and model use within science, including abstractions of theory, idealized reality, visible or simplified phenomena/abstractions, and explanations that enable predictions. These descriptions have been further elaborated and extended in more recent writings (e.g. Gilbert & Justi, 2016). Moreover, various frameworks for how learners' understand models and modeling have been discussed and refined (i.e. Grünkorn, Upmeier zu Belzen, & Krüger, 2014; Krell, Upmeier zu Belzen, & Krüger, 2014; Mahr, 2011). A synthesis of this work yielded the competence-based view on models and modeling.

Understanding what science is and what scientists do requires epistemic knowledge, which includes the nature of science [NOS]. Scientists generate scientific knowledge through practices that are uniquely scientific, grounded in empirical observations of the natural world. Because models and modeling are both practices and products of science, modeling competence necessitates an understanding of NOS and the practices of scientific inquiry. This chapter explores these connections and their place within science education. To gain additional perspective on a competence-based view of models and representations of NOS in authentic contexts, this chapter also examines how practicing scientists describe the role of models and modeling in their research. This perspective provides insights into the meaning of the opening quote, as well as how modeling competence reflects the epistemic nature of science and scientific practices.

4.2 Theoretical Background

4.2.1 Nature of Science: A Cognitive Construct

The phrase "nature of scientific knowledge" [NOS] refers to characteristics of scientific knowledge inherently derived from the manner in which that knowledge is produced through scientific practices (Lederman & Lederman, 2014). These qualities, or characteristics, are what make scientific knowledge *science*, as opposed to other forms of knowledge. The specifics of NOS have been described in various ways; all delineating scientific knowledge as foundational to how we can understand the natural world through empirical observation. With respect to what is relevant and appropriate for science teaching and learning, there is broad consensus within the literature that strongly supports the view that NOS is a *cognitive* construct (Lederman & Lederman, 2014; Osborne, Collins, Ratcliffe, Millar, & Duschl, 2003; Schwartz, Lederman, & Abd-El-Khalick, 2012) rather than a skill, attitude, or activity. Others have provided more general descriptions of science through emphasis on broader epistemic and philosophical commonalities across knowledge domains, such as "features of science" (Matthews, 2012) and "family resemblance" (Dagher & Erduran, 2016; Irzik & Nola, 2014).

Despite the debates, there is a developmentally appropriate level of generality regarding NOS that is accessible to pre-university students and relevant to their daily lives. These general, cross-cutting characteristics representing the nature of scientific knowledge as a cognitive outcome are described here (modified from Schwartz, Lederman, & Crawford, 2004; Schwartz, Lederman, & Abd-El-Khalick, 2015). Figure 4.1 provides a description of each of the aspects that hold consistent agreement amongst science educators.

These aspects are not a definitive or privileged listing of NOS, but rather a compilation of aspects commonly advocated within empirical research dating back to the 1960s. These aspects are considered cross-cutting because regardless of the science domain, one can find representative examples (McComas, 2008; Schwartz & Lederman, 2008). Furthermore, these aspects should be considered *a collection*, as opposed to isolated features. Understanding NOS includes understanding how these aspects are intricately connected and derived from the scientific enterprise. For example, due to the inherent subjective and socio-cultural features of scientific knowledge, that knowledge is inherently tentative, yet robust due to the empirical foundation upon which the knowledge is generated.

NOS has been advocated for scientific literacy for decades (Lederman & Lederman, 2014). Driver, Leach, Millar, and Scott (1996) argued that scientific literacy comprises understanding scientific inquiry; understanding the social nature of science; understanding that people produce, validate, consume, and benefit from scientific knowledge; and understanding some aspects of science content. Each of these can be connected to why understanding NOS is relevant, as scientific knowledge is developed through inquiry practices, in a social context, by real people, and leads to further understanding of the natural world. Because many natural concepts are complex and not directly observable; and because relationships among components of natural phenomena are not always directly accessible, models and modeling are essential to inquiry practices and the generation of scientific knowledge. Thus, as detailed in the theoretical Section A of this book, understanding and being able to utilize models for learning and decision-making; or, in other words, engaging in and understanding modeling competence is essential to scientific literacy.

4.2.2 Scientific Models and Modeling in Science Education

To promote epistemological views of science, learners should experience science through engaging in scientific practices (AAAS, 1993; Lead States, 2013; NRC, 1996, 2000, 2012).

> Inquiry is a critical component of a science program at all grade levels and in every domain of science, and designers of curricula and programs must be sure that the approach to content, as well as the teaching and assessment strategies, reflect the acquisition of scientific understanding through inquiry. Students then will learn science in a way that reflects how science actually works. (NRC, 1996, p. 214)

NOS aspect	Description
Empirically-based	Scientific knowledge, including scientific models, is based on and/or derived from observations of the natural world. These observations are made directly or indirectly through use of senses, tools, measuring devices, and other technological instruments that offer detection of natural phenomena.
Distinction between observation and inference	Observations are descriptive statements about natural phenomena that are directly or indirectly (through instruments) accessible to the senses. Inferences serve to explain or extend observations but are not directly accessed. The notion of gravity is inferential in the sense that it can only be accessed and/or measured through its manifestations or effects. Models, as theoretical constructs, are based on observations and provide inferential explanations of relationships and functionality.
Creativity	Science, as a human endeavor, involves the invention of explanations, negotiation of meaning from data, and the generation of ideas. This aspect of science, coupled with its inferential nature, entails that scientific concepts are functional theoretical models rather than faithful copies of reality. For example, the observation of birds in flight inspired studies of aerodynamics and the eventual invention of flight mechanisms.
Distinction between scientific theories and laws	There is an epistemic distinction between scientific theories and laws, stemming from the type of evidence and functional purpose of the knowledge. Laws are descriptions of relationships among features of observable phenomena. Theories, by contrast, are inferred explanations for observable phenomena. Put simply, laws are statements of what is observed; theories are statements of why something occurs. For the example of gravitational force, Newton's law of gravity states that there is an attraction between two masses. Gravitational theory attempts to explain why this occurs. Models can depict relationships (such as in mathematical terms) and present theoretical constructs that explain interactions and functionality.
Subjectivity/theory-driven	Scientists' beliefs, previous knowledge, training, experiences, and expectations, in addition to theoretical commitments affect what problems scientists investigate, how they conduct their investigations, what they observe (and do not observe), and how they make sense of, or interpret observations. An evolutionary developmental biologist and an ecologist will interpret components of an ecosystem differently based on the lenses through which they work. Multiple models arise due to varying perspectives and purposes. The models scientists use to explain and test the system are necessarily informed by different theoretical lenses.
Socially and culturally situated	Science as a human enterprise is practiced in the context of a larger culture. Scientists are the product of that culture. Scientific knowledge affects and is affected by the various elements and intellectual spheres of the culture in which it is embedded. These elements include, but are not limited to, social fabric, power structures, politics, socioeconomic factors, philosophy, and religion. Related to the subjective and creative NOS, social and cultural influences are unavoidable. They reflect what questions are asked, how science is practiced, and what knowledge is generated and accepted. Through the ages, scientific models have reflected cultural and social positions (i.e. geocentric model), with society being resistant to changes that require profound paradigm shifts that run counter to societal and religious convictions (i.e. heliocentric model).
Inherently tentative	Scientific knowledge is necessarily subject to change, yet due to the empirical nature, scientific knowledge is also robust (not likely to change on a whim or without substantial evidence). Attaining absolute truth is outside the realm of science (Chalmers, 1982; Kuhn, 1962). Scientific claims change as new evidence is brought to bear on existing theories or laws, or as old evidence is reinterpreted in the light of new theoretical advances. A look through the history of science provides myriad examples of change in how we understand the natural world. Model revision can result from falsifying hypotheses through predicting and testing.

Fig. 4.1 Nature of science aspects and descriptions

4.2.3 Nature and Purpose of Scientific Models

Scientific models are integral to the development and exploration of scientific knowledge (e.g. Gilbert, 1991; Khine & Saleh, 2011; Mahr, 2011). Models have been described in a variety of ways, and in several chapters of the present book (Sec. A). Consistent among them is that models are representations that serve to describe, explain, or predict (Grünkorn et al., 2014; Schwarz et al., 2009; Van Der Valk, Van Driel, & De Vos, 2007; van Driel & Verloop, 2002). Gilbert (2004) describes models as "simplified depictions of a reality-as-observed, produced for specific purposes, to which the abstractions of theory are then applied" (p. 116), "idealizations of a possible reality" (p. 116), visualizations of abstract phenomena or of something too small or too big to see otherwise, simplifications of something complex, and "the basis for both scientific explanations of and predictions about phenomena" (p. 116). Models can represent myriad of phenomena including: objects, abstractions, systems, parts of systems, entities, relationships among entities, an event, a behavior, and a process (Gilbert, 2004; Mahr, 2011). Further, models are products of investigations, frameworks for investigations, and tools for predictions and testing (Gouvea & Passmore, 2017; Krell et al., 2014; Upmeier zu Belzen & Krüger, 2010; Van Der Valk et al., 2007).

Ontologically, models can be mathematical, physical, analogical, or mental constructs (representations) of the natural world (*what are models of?) (*Gouveu & Passmore, 2017; Mahr, 2011, among others). Models also have epistemic purposes (*what are models for?)* including that models: (1) explain or organize observations that then enable prediction and testing through further observation; (2) simplify a complex phenomenon or render an abstract concept visible; and (3) provide a framework for guiding further investigation. A model-object *of* an object (Mahr, 2011) is not an exact replica of the actual phenomenon or process; but serves as a representation of the model of the phenomenon (target, or object) and features deemed important and applicable to the structure and function of the target (object). Models are epistemic tools for explaining, predicting, visualizing, simplifying, testing, and showing relationships in the development of scientific knowledge.

4.2.4 Scientific Models in Science Standards

In the United States, the prominence of models and modeling competence in science education has increased in recent years with the inclusion of models within the essential scientific practices that learners should understand and be able to perform (Lead States, 2013). The scientific and engineering practices in the Next Generation Science Standards [NGSS] explicitly list "*Developing and using models*" among them. However, one can stress that developing and using models are integral to other scientific practices as well. For example, models and the modeling process are useful tools for conducting investigations (model a stream system to study impacts of erosion), analyzing and interpreting data (using a model of inheritance patterns to interpret generational data related to a genetically-based disease), and communicating information (demonstrating impacts of climate change).

To represent this idea further, the Framework for K12 Science Education (NRC, 2012), upon which the NGSS is based, discusses the importance of developing scientific proficiencies through engaging in the practices and epistemic thinking that generates scientific knowledge, including the role of models as products and modeling as practices within science. The Framework states, "Students' opportunities to immerse themselves in these practices and to explore why they are central to science and engineering are critical to appreciating the skill of the expert and the nature of his or her enterprise" (NRC, 2012, p. 47). Clearly, in order to achieve this goal, learners must have a meaningful understanding of models and modeling as well as how these concepts relate to targeted features of NOS. The dynamic nature of scientific practices and the role of theories and models in the actions are parallel with the framework for the modeling process (Fig. 1.2). The competences associated with models and modeling are evident within the figure. Scientific practices embrace the role of models as epistemic tools used to formulate hypotheses, test and propose solutions, as well as generate arguments through analysis and evaluation of data.

The nature of scientific knowledge, including models, is also emphasized on an international level. The 2015 PISA Assessment and Analytical Framework (OECD, 2017) includes the nature and purpose of scientific models as part of their recommendations. The three competencies for scientific literacy are that learners should be able to (1) explain phenomena scientifically, (2) evaluate and design scientific enquiry, and (3) interpret data and evidence scientifically. The PISA Framework states that these "… competencies, however, require more than a knowledge of what is known; they depend on an understanding of how scientific knowledge is established and the degree of confidence with which it is held" (OECD, 2017, p. 21). They also state that the competencies require epistemic knowledge: "Epistemic knowledge includes an understanding of the function that questions, observations, theories, hypotheses, models, and arguments play in science" (p. 21).

Despite being targeted as an essential learning outcome for decades, students (Grosslight, Unger, Jay & Smith, 1991; Krell, Upmeier zu Belzen, & Krüger 2012), preservice teachers (Schwartz & Skjold, 2012, Hartmann, Upmeier zu Belzen, Krüger, & Pant 2015) and practicing teachers (Crawford & Cullen, 2004; Justi & Gilbert, 2003; van Driel & Verloop, 2002; Krell & Krüger 2015), typically hold narrow or varying conceptions of models. Yet, with scaffolding and experience, learners can develop understandings of scientific models and modeling (e.g. Akerson, White, Colak, & Pongsanon, 2011; Gilbert & Justi, 2016; Schwarz et al., 2009; Windschitl & Thompson, 2006). For example, Schwarz and colleagues used a model-centered, meta-modeling approach to engage learners in modeling activities as well as develop learners' epistemologies of science (Schwarz & White, 2005). They attest to the effectiveness of the "meta" component when engaged with models and modeling as essential for fostering epistemic knowledge.

If children are to learn science in a way that reflects how science really works, it is important for teachers to have an understanding of these real workings of science and instructional strategies that are effective in developing modeling competence and epistemological views of science. Other chapters in Section A dive more deeply into the meaning and nature of scientific models and modeling, as well as the research on teachers' and learners' conceptions (Sec. C; Sec. D).

4.2.5 *What Scientists Say About Models and Modeling in the Scientific Community*

To get a sense of how models are defined and used in the scientific community, as products and practices of science, and then how models and modeling reflect NOS features, we can explore how scientists think about and use models in their own research (Chap. 5). Van Der Valk et al. (2007) conducted a study to test and revise a comprehensive description of "features of scientific models." Their study asked practicing scientists who had recently published research involving scientific models, to comment on the extent of their agreement or disagreement with the features. The study provides an empirically supported description of features of scientific models that represented views of contemporary scientists. In similar form, Schwartz (2004) conducted a study of scientists' views of NOS and scientific models. The following section presents partial results from this study, as they relate to how scientists' conceive the purpose of models. These descriptions reveal scientists' thinking about a competence-based view of models in their work. For the purpose of this chapter, the discussion draws clear connections between model descriptions and the aforementioned NOS aspects.

The current study reports on scientists' views of the purpose of scientific models and their use in authentic science practice. This study provides descriptions and examples of models and connections to NOS aspects. Results of a larger study on scientists' views of NOS have been reported elsewhere (Schwartz & Lederman, 2008; Schwartz, 2011). Participants were experienced scientists from four science disciplines (life science, earth science, physics, and chemistry) and who employed various approaches to research (e.g. experimental; descriptive; theoretical). The research question focused on here is "What are practicing scientists' views on the purpose of scientific models?" and "Do views vary based on science discipline and/or investigative approach?"

4.3 Method

Participants were 24 practicing scientists (6 female, 18 male) from across the United States and representing four primary science disciplines and a variety of sub-disciplines and investigative approaches (ten life scientists; five earth and space scientists, five physicists (four theoretical), four chemists). All of the participants were currently engaged in research and publishing. With an average of 25 years research experience since earning their doctorate, the participants were clearly experienced within their respective communities. With the exception of one participant (an aquatic ecologist with 22 years post PhD research experience, currently in a non-academic institution), all held tenured academic positions at universities. All were educated and currently employed within the United States. Most had extended international experiences through post-docs, sabbaticals, or collaborative programs.

4.3.1 Data Collection and Analysis

For the larger study, participants responded to two open-ended surveys [VNOS-Sci and VOSI-Sci] (Lederman et al., 2002; Schwartz, Lederman, & Lederman, 2008). These were modified to prompt the scientists to consider the NOS and inquiry aspects within the context of their research. Two questions were added to the VNOS-Sci survey that directly addressed ideas about the purpose of scientific models:

(a) What is the purpose of a scientific model?
(b) Describe a scientific model from your own area of research, if appropriate. If you do not use scientific models, describe a scientific model from another area of research. Describe why your example is a scientific model.

Semi-structured interviews served to elicit additional information as well as validate scientists' responses to questionnaire items (Lederman et al., 2002).

Through a process of analytic induction, participants' questionnaires and interviews were analyzed separately to generate individual profiles of scientists' views. Analysis specifically sought reference to models and model use. All instances of the words "models" or "use models" or similar phrases were coded. The sub-codes that emerged are descriptors of how the scientists talked about models (their own words), model construction, and model use. Each participant could have provided statements consistent with multiple sub-codes. Thus, results are presented as number of participants and % of participants who made reference to each sub-code. Results are reported based on emergent descriptions, trends, and patterns.

4.4 Results

The following results represent how the scientists describe models within their field (Fig. 4.2). The top descriptors are listed, with representative quotes to provide context and elaboration of meaning. It is important to note that the descriptors are what emerged from the voices of the scientists. They were not asked if they agree/disagree with a particular descriptor. Thus, the results are considered

Sub-Code	Total #	Total %
explain or organize observations/predict/test	17	70.8
complex made simple/abstract made visual	9	37.5
mathematics	9	37.5
directing framework	3	12.5
more specific than a theory	2	8.3
analogy	1	4.2
mental construct	1	4.2
representation of reality	1	4.2

Fig. 4.2 Scientists' descriptions of the purpose of models ($N = 24$)

"first ideas" of models that occurred to scientists as they provided their responses. Whether they agree or not with other descriptors is beyond the scope of this study. The sub-codes are not mutually exclusive. Many of the representative quotes included here fall within in multiple sub-codes. These results are discussed in terms of suggested patterns within this sample of scientists and should not be generalized beyond this sample.

4.4.1 Model Descriptions and Nature of Science Connections

The most common emergent themes are consistent with published descriptions of scientific models (Gilbert, 2004; Gilbert & Justi, 2016; Khine & Saleh, 2011; Van Der Valk et al., 2007). Moreover, the emergent themes are also consistent with many of the features of the modeling cycle (Fig. 1.2). Here, the scientists recognized the role of observation, testing, and influences. These results also reflect the aspects related to modeling competence. The findings described here are based on the emergent themes. Representative quotes are provided in nearly full form in order to provide voice to the scientists, which enriches the meaning through context and stories. Following the quotes, NOS connections are presented in *italics*. Where evident, connections to aspects and levels of the framework for modeling competence (Chap. 1) are made.

Explain or Organize Observations/Predict/Test Seventeen of the 24 scientists indicated models were explanations or ways to organize observations that also involved testing predictions (*purpose of models*: level III; Chap. 1). Most responses specifically related to the participant's research.

In my research I use the model of a trophic cascade that indicates how predator-prey interactions from the top of the food web propagate down the food web to affect lower trophic levels. This model explains some of the variability observed in food web dynamics and the relative abundance of predator and prey groups in ecosystems. [aquatic ecology] [*subjectivity/theory-driven*]

A scientific model is a description of a physical system that provides an understanding of what the system is and how it works. A scientific model allows us to organize our information about a system and to predict how the system might evolve or react… We use mathematical models of stellar atmospheres to compute what the spectrum of a star ought to look like. We compare the predicted stellar spectrum with the observed stellar spectrum to determine the composition of the star. [astronomy] [*subjectivity/theory-driven*]

An atmospheric scientist described the purpose of a model to provide understanding and predictability. In this way, he recognized the model function to be a tool to gain knowledge.

As models become more complex, such as general circulation models of the atmosphere and ocean, the models are used as predictive tools. They're used to predict how climate will change as we change the composition of the atmosphere. [atmospheric science2]

The other atmospheric scientist expanded on this perspective by discussing modeling of a system. In this remark, the subjective/theory-driven nature of scientific models is clearly connected to *multiple models* (Chap. 1).

You are probably aware that the treatment of clouds in climate models is one of the weakest links in the chain of things that we need to put together to say something sensible about global warming, and we don't do it very well. The models are all over the map, depending on how they parameterize the cloud process. [Atmospheric science1] [*subjectivity/theory-driven*]

Several responses within this sub-code demonstrated a connection between the scientists' views of models and their views of certainty of scientific knowledge (tentative NOS). These descriptions also related to the empirical NOS.

It [a model] is a mental or physical construct. [...] The model is a way to test whether we got our ideas right [...] Then you can test it and try a different set of conditions. If they do, then it means the model is working, at least for these conditions, and it has some predictive function. One is to test the input to see if I have my ideas straight and the other is to make predictions. [Models are] useful to guide experimentation and serve as a provisional understanding of a phenomenon. [environmental analytical chemistry] [*empirical & tentative NOS*]

One of the biologists elaborated on her view of models and modeling within her field. With respect to modeling competence, her response demonstrates level III for a set of aspects: *purpose of models*, *multiple models* and *changing models* (Chap. 1).

The theory of natural selection is also a model that explains much about the origin and behavior of biological systems. It provides a basis for making predictions about species responses to environmental changes … A lot of these conclusions are drawn from tests with models that show that if you create this kind of structure it accounts for the behavior that you measure. Again, just because you can come up with a model that explains it doesn't necessarily mean that is the only model. Just maybe we haven't thought of the model that works better … Models work at all these levels [hypothesis, theory, law]. A hypothesis is a model. The model becomes more robust as it becomes elevated to theory and then law. But a model initially is a hypothesis. [entomology] [*empirical & tentative NOS; theory & law* misconception]

This last statement shows a connection to the NOS aspect of "theory and law." However, the scientist held a hierarchical view of hypothesis, theory, and law; yet also saw a connection with scientific models at each "level" of scientific knowledge. According to this scientist, the more robust the model, the higher its status within the perceived hierarchy. In contrast to other scientists who described models as having predictive capabilities based on assigned parameters (and these parameters could change according to what the intent is), the position described above may suggest a view that models can approach certainty. Even though different scientists held differing views of certainty of models, they held the common view of models having predictive ability (*purpose of models*: level III, Chap. 1). This feature of models exemplifies the subjective and theory-driven NOS because they describe models as providing an explanation or system upon which to base further exploration. Moreover, these descriptions also provide links to the empirical and tentative NOS, as the models must be based on natural phenomena, yet they can be adjusted with further testing.

In response to a prompt to discuss the development of the atomic model, one chemist explained the historical development of the atomic model, along with the explanatory and predictive power of this model across disciplines. In this response, we see a connection of atomic models to the *empirical, tentative, subjective, and creative NOS.*

> Once the planetary model became acceptable, things that could be predicted from this model were consistent with what physicists were observing then it was quickly discovered that it was also consistent with the chemists, this whole body of knowledge that chemists were building. All of a sudden, the world was falling in place. Chemists could see very neatly how their atoms stuck together and begin to explain things. Linus Pauling came along and used the model, extended the model, to explain the chemical bond and all of modern chemistry … Of course over the years the model continues to be used and refined in ways we hadn't even imagined. We are comfortable with that until some day we bump up against something we can't explain with the model. At that time, we go back and try to adjust the model or come up with other explanations. It's progressive. [mass spectrometry]

Complex Made Simple/Abstract Made Visual Nine participants describe models more specifically as a means to simplify a complex process or system or a means to visualize an abstract concept. Most representatives from within this sub-code were distinct from the previous in that rather than considering models as explanations of observations that serve a predictive function; models here are considered limited, but useful, explanations because they serve to simplify natural phenomena that would otherwise be too complicated to investigate further. These views aligned with levels II and III of *nature of models* (Chap. 1). The descriptions exemplify the subjective/theory-driven NOS because they indicate choices made by the scientists in determining what features of the real phenomena to include in the models. These choices are based on what the scientists consider to be important for answering their questions of interest.

> A scientific model helps to explain a natural situation. Often it is a small scale general version of a more complex phenomenon. Scientific models help us to grasp a complex situation as a more watered-down version. In the field of landscape ecology, scientists often cut fields into different patch sizes and patterns and study animal movements in them to model (simulate) how larger animals move about in larger more complex landscapes. Models can be increased in scope and complexity to further explain the variability we often encounter in nature. [wildlife ecology] [*subjective/theory-driven & tentative NOS*]

> The models are okay as long as you understand the limitations of them. That isn't really how it is but it's the way we think about it… . We are showing pictures here that relate to certain aspects of an atom. That is what you do when you see an elephant. It depends where you are looking on the elephant and what scale. [high energy theoretical physics] [*subjective/theory-driven, creative, tentative NOS*]

Mathematics Nine participants referred to models as mathematical representations. Within this sub-code were statements to demonstrate the role of mathematics in dealing with complexity. Interestingly, the theoretical physicists had a higher tendency to explain models as mathematical entities. They described situations where as the complexity of the phenomenon increases, capabilities of mathematics become more important.

> So for particle physics there is a theory now known as quantum chromo dynamics, QCD. It is a field theory […] To solve that problem requires exchange of 16 different particles simultaneously. So it requires hundreds of equations to be solved simultaneously, and they are integral equations. That has taken years of computer time for most elementary, even models there, how to solve that. But in theory one has a complete mathematical description. In practice you say let's model it by limiting the number of particles. That makes it a model. [computational physics] [*theory-driven, tentative, socio-cultural NOS*]

Directing Framework Even though the majority of descriptions and examples provided by the scientists eluded to connections to the subjective/theory-driven NOS, a few scientists explicitly made reference to models as a theoretical framework that guides their work.

> Without models, observation would amount to cataloging data … There is a lot of data, and it doesn't mean anything until you have a model. If you have all these data and lots of satellites taking all these data ... it doesn't tell you what to look for. It just tells you whether a model you have is plausible or not. It is all indirect. [astrophysics].

> A gene network is a scientific model, postulating patterns of interacting among gene products following an analogy with a computer wiring diagram. It illustrates a mechanism, and helps develop hypotheses about other genes that must be involved to produce the observed phenotype. [evolutionary development].

4.4.2 Model Development, Model Use, and Nature of Science

The scientists in this sample discussed models in terms of development and use. The epistemic nature and functions of models are exemplified in their responses. Model development is described as the process of collecting information (empirical and/or theoretical), identifying relationships, and composing an explanation of the relationships. All but the theoretical physicists suggested the proposed relationships should lead to predictions that are empirically testable. Because of their reliance on mathematical models and complex computations, the theorists' ideas push the boundaries of how we might define "*empirical* observations." This is also a societal or cultural issue in that technological advances have enabled scientists to enter the realm of virtual reality to develop, test, and use their models. Van der Valk et al. (2007) also described the role of technological advances related to model development and use. The notion of "empirical" is changing as our perspectives of what is possible changes. In this way, NOS is like any other scientific concept – subject to change.

Model use, and thus modeling competence, involves testing predictions and identifying problems or cases where existing models do not work. The tentative yet empirical nature of model development and testing, among other aspects, are articulated within this description from one of the atmospheric scientists who works with cloud climate models:

> Most of my work is testing models. Model development is a whole other field. That might be the theoretical side. So I put myself in the observational side as opposed to the theoretical side. The models themselves are so complex. How do you build them in the first place? So what do they do to build these models? ... Real clouds don't behave this way. … it is easy to suspect these models. Doing the realistic calculations is very difficult. It takes a lot of number crunching and time. But we can test these ideas. … if we know what we are doing there should be no difference between the model and our observation of the clouds…. They don't [work]. Even the bumps on the tops of clouds are enough to throw it off…. *When we build these models and test them, we play games like this.* We try to develop a test where we know what we should expect. We predict the results and see whether we get them or not. We see the failure of the prediction and start probing and say" how come?" [*empirical, tentative, subjective/theory-laden, creative*].

This example also depicts competence within all five aspects at level III (Chap. 1).

4.4.3 Models and Anomalies

This quote raises the practices of predicting and testing, much like many of the other quotes have. Yet, here we go a step further to see what happens when predictions do not play out as expected. The idea of "playing games" to get the best fit with real data is rarely examined in science education. However, understanding how anomalous data are recognized and dealt with is a feature of the scientific enterprise (Chalmers, 1982; Kuhn, 1996). In this study, scientists often connected use of models with identification of anomalies. The question of "how come?" offered by the atmospheric scientist above marks the curiosity and exploration into why a model doesn't hold. From an epistemological perspective, this level of recognition would be essential for understanding the relationship between the empirical and tentative nature of science. For many of these scientists, it is in the testing of the models that anomalies are identified. Through exploration of anomalies, models are refined and/or new models are constructed.

The scientists indicated they would examine and attempt to explain an anomaly from the perspective of their existing framework (the subjective/theory-driven NOS). The cloud climate modeler quoted above fell within this category. In discussing competing models for the same anomaly, he described the need for better analysis and refinement of his model to explain the data. His statements also indicate a critical role of creativity and collaboration in how models can change:

> We are going to get better at our analysis of our data and when we do that it gets harder for people to say, "Ah ..." or how do you say, it motivates people to start looking at the model and ask what is really going on here. How do we understand this? Obviously, there is something strange going on here. By pursuing this and keeping the pressure up, I am hoping that people like John [colleague] will come along and start thinking again, "Well maybe if I did something else in *my* model … maybe we could pull this off." [atmospheric scientist].

4.5 Discussion and Implications of What Scientists Say

Creating and using scientific models is central to scientific inquiry, and included as one of the eight scientific practices learners should be able to do and understand (Lead States, 2013). In the "What scientists say"-study, there was overwhelming sentiment that models are used to explain or organize observations, then predict and test through further observations. The emphasis here is on empirical observation in the development and in the testing of models. In comparison, half as many scientists described models as a means to visualize something abstract or simplify a complex process. This latter view seems to place less emphasis on direct observation and incorporates theoretical entities, although these are not necessarily mutually exclusive. These results show that these scientists' perceptions and use of models fit broadly with published descriptions of functional roles of models in science, including descriptive, explanatory, and predictive characterizations (Justi & Gilbert, 2003, 2016; Van Der Valk et al., 2007; Van Driel & Verloop, 1999). The results demonstrate distinctions between *models of* and *models for* (Gouvea & Passmore, 2017). The multiple descriptors that the scientists used for models, such as mathematical, physical, and analogical, are also consistent with prior characterizations. In comparison to the range and multiple categories of meaning for the seven aspects of models identified in the Justi and Gilbert (2003) study of teachers' views of models, the present study suggests these scientists may hold more consistent or similar views of scientific models, with prioritizing the epistemic function of predictive ability.

These results suggest that a competence-based view of models and modeling relies on understanding the explanatory and predictive nature of models. Definitions of "model" used by scientists have been suggested (Justi & Gilbert, 2003; Van Der Valk et al., 2007). The study by Van Der Valk et al. (2007) produced a set of "features of scientific models" and solicited input from scientists as to the relevance of the features to their work. The present study provides additional information from constructed response data which are useful, in conjunction with scientists' examples, to understand the modeling cycle as well as the modeling competence (Chap. 1). To further emphasize the relevance and importance of input from the scientific community, Chap. 5 of this volume presents additional research on scientists' descriptions of models and modeling. These examples, and the discussions provided in this book, partially fulfil the request from Bernd Mahr when he wrote, "Because models are the most important epistemic tool of our knowledge and production, it is necessary to produce a methodological surplus when answering the question, 'What is a model?'" (2011, p. 296).

4.6 Modeling Competence and Science Instruction: Engaging Authentically by Connecting Scientific Models and Nature of Science

What can we learn from scientists about a competence-based view of models and nature of science? We can learn how scientists develop and use models through authentic scientific practices. This chapter details relationships between NOS principles and modeling competence. For example, the utility of models within scientific research relates to their representation of phenomena or systems. System features are selected based on subjective decisions stemming from scientific questions under study. A requisite for understanding NOS and exhibiting level III of modeling competence involves acknowledging that a scientific model is not an exact replica of reality, but a representation that serves to explain features and relationships that the scientists find curious and significant to their questions. Models also enable prediction and testing; thus, progressing scientific understanding. Acts of model construction and utility are inherently inferential, creative, and tentative; yet robust due to the empirical basis, and explanatory and predictive power. For meaningful understanding of models and modeling, and to achieve a competence-based view of models, epistemic knowledge is also essential (Mahr, 2011). If one can "do" modeling, but does not understand the epistemic nature of what they are doing, have they really achieved meaningful understanding? Have they reached a competence-based view of models?

The different descriptions noted here suggest models are not a "one size fits all" concept. Not all models explain direct observations and not all models take an abstract concept and make it more concrete. What constitutes a model is determined by the scientist and scientific community. As Mahr (2011) stated, "It turns out that the phenomenon of model-being can be understood if one stops looking for an answer to the question *of the nature of a model* and starts asking instead *what justifies conceiving of something as a model*" (p. 253). Further, Mahr explains that epistemically, "the model-being of an object will become the result of a judgment which is situated in contexts of invention and justification, and whose acceptance and reasoning may thus be questioned" (p. 253). The examples provided by the scientists in the "What scientists say"-study reinforce the notion of development and use of an object as a model-being to be contextualized and somewhat idiosyncratically judged.

Similar to the findings of Van Der Valk et al. (2007), the scientists shed light on the empirical NOS and the changing landscape of how scientists work with changing technology. There is a need to reconsider and, perhaps, reconceptualise how we define "empirical" within the realm of scientific practices. Furthermore, model development is described as a practice distinct from model use. In order to help students "learn science in a way that reflects how science actually works" (NRC, 1996, pg. 214), teachers should incorporate a variety of experiences that demonstrate models and model use in an authentic light (Chap. 3). That is, for a competence-based view of models, both model development and model use need to be addressed in multiple contexts, with clear objectives that align students to distinctions and similarities among models with respect to the contexts, functions, and NOS connections. Teachers need to consider how the many model examples and modeling activities they provide for their students are opportunities to explicitly address NOS aspects (Crawford & Cullin, 2004). This type of model-based instruction with explicit/reflective attention to epistemic connections has been shown to be effective (e.g. Akerson et al., 2011; Gilbert & Justi, 2016; Schwarz et al., 2009; Windschitl & Thompson, 2006). The use of historical examples of models and modeling has been recommended for some time (Grünkorn et al., 2014; Justi & Gilbert, 2002; Chap. 3). The scientists' narratives here and in other chapters provide contemporary examples of types of models and modeling functions within the scientific community and how they represent important NOS features. These can be adjusted for use in science instruction.

Another intriguing connection of model use and NOS has to do with the identification and role of anomalies in science. Are they mistakes? Are they opportunities? A competence-based view of models must recognize the subjective and theory-driven NOS and include their functional influence on predicting and testing. When expectations are not met, what happens next? There is potential to *model* the practice of model testing, anomaly identification, and scientific progress. How are anomalies typically identified and dealt with in the classroom? Are models used to make predictions and test them? Are students given opportunity to experience the excitement of finding a contradiction between prediction and observation? Are students given opportunity to refine models or develop a new model in light of contradictions? These are questions that should be considered in instructional design so that science learning might more closely reflect "*how science actually works*."

References

Akerson, V., White, O., Colak, H., & Pongsanon, K. (2011). Relationships between elementary teachers' conceptions of scientific modeling and the nature of science. In M. Khine & I. Saleh (Eds.), *Models and modeling: Cognitive tools for scientific enquiry*. Cham, Switzerland: Springer.

American Association for the Advancement of Science. (1993). *Benchmarks for science literacy: A project 2061 report*. New York: Oxford University Press.

Chalmers, A. (1982). *What is this thing called science?* (2nd ed.). University of Queensland Press.

Crawford, B., & Cullin, M. (2004). Supporting prospective teachers' conceptions of modeling in science. *International Journal of Science Education, 26*(11), 1379–1401.

Dagher, Z., & Erduran, S. (2016). Reconceptualizing the nature of science for science education: Why does it matter? *Science & Education, 25*(1–2), 147–164.

Driver, R., Leach, J., Millar, R., & Scott, P. (1996). *Young people's images of science*. Buckingham, UK: Open University.

Gilbert, J. (1991). Model building and a definition of science. *Journal of Research in Science Teaching, 28*, 3–17.

Gilbert, J. (2004). Models and modeling: Routes to more authentic science education. *International Journal of Science and Mathematics Education, 2*, 115–130.

Gilbert, J., & Justi, R. (2016). *Modelling-based teaching in science education*. Cham, Switzerland: Springer.

Gouvea, J., & Passmore, C. (2017). Models of versus models for. *Science & Education, 26*(1–2), 49–63.

Grosslight, L., Unger, C., Jay, E., & Smith, C. (1991). Understanding models and their use in science: Conceptions of middle and high school students and experts. *Journal of Research in Science Teaching, 28*, 799–822.

Grünkorn, J., Upmeier zu Belzen, A., & Krüger, D. (2014). Assessing students' understandings of biological models and their use in science to evaluate a theoretical framework. *International Journal of Science Education, 36*, 1651–1684.33.

Hartmann, S., Upmeier zu Belzen, A., Krüger, D., & Pant, H. A. (2015). Scientific reasoning in higher education: Constructing and evaluating the criterion-related validity of an assessment of preservice science teachers' competencies. *Zeitschrift für Psychologie, 223*(1), 47–53.

Irzik, G., & Nola, R. (2014). New directions for nature of science research. In M. Matthews (Ed.), *International handbook of research in history, philosophy and science teaching* (pp. 999–1021). Dordrecht, the Netherlands: Springer.

Justi, R., & Gilbert, J. (2002). Modelling, teachers' views on the nature of modelling, and implications for the education of modellers. *International Journal of Science Education, 24*(4), 369–387.

Justi, R., & Gilbert, J. (2003). Teachers' views on the nature of models. *International Journal of Science Education, 25*(11), 1369–1386.

Khine, M., & Saleh, I. (Eds.). (2011). *Models and modeling: Cognitive tools for scientific enquiry*. Cham, Switzerland: Springer.

Koeppen, K., Hartig, J., Klieme, E., & Leutner, D. (2008). Current issues in competence modeling and assessment. *Zeitschrift für Psychologie/Journal of Psychology, 216*(2), 61–73.

Krell, M., & Krüger, D. (2015). Testing models: A key aspect to promote teaching activities related to models and modelling in biology lessons? *Journal of Biological Education*. https://doi.org/10.1080/00219266.2015.1028570T

Krell, M., Upmeier zu Belzen, A., & Krüger, D. (2012). Students' understanding of the purpose of models in different biological contexts. *International Journal of Biology Education, 2*, 1–34.

Krell, M., Upmeier Zu Belzen, A., & Krüger, D. (2014). Students' levels of understanding models and modelling in biology: Global or aspect-dependent? *Research in Science Education, 44*, 109–132.

Kuhn, T. (1996). *The structure of scientific revolutions* (3rd ed.). Chicago: University of Chicago Press.

Lead States. (2013). *The next generation science standards*. Washington, DC: The National Academies Press.

Lederman, N., & Lederman, J. (2014). Research on teaching and learning of nature of science. In N. Lederman & S. Abell (Eds.), *Handbook of research on science education* (Vol. II). Routledge: New York.

Lederman, N. G., Abd-El-Khalick, F., Bell, R. L., & Schwartz, R. S. (2002). Views of nature of science questionnaire (VNOS): Toward valid and meaningful assessment of learners' conceptions of nature of science. *Journal of Research in Science Teaching, 39*(6), 497–521.

Mahr, B. (2011). On the epistemology of models. *Rethinking Epistemology, 1*, 301–352.

Matthews, M. (2012). Changing the focus: From nature of science (NOS) to features of science (FOS). In M. S. Khine (Ed.), *Advances in nature of science research* (pp. 3–26). Dordrecht, the Netherlands: Springer.

McComas, W. (2008). Seeking historical examples to illustrate key aspects of the nature of science. *Science & Education, 17*(2–3), 249–263.

National Research Council. (1996). *National science education standards*. Washington, DC: National Academic Press.

National Research Council. (2000). *Inquiry and the National science education standards*. Washington, DC: National Academic Press.

National Research Council. (2012). *Framework for K12 science teaching*. Washington, DC: National Academic Press.

OECD. (2017). *PISA 2015 assessment and analytical framework: Science, reading, mathematic, financial literacy and collaborative problem solving*. OECD Publishing, Paris. https://doi.org/10.1787/9789264281820-en

Osborne, J., Ratcliffe, M., Collins, S., Millar, R., & Duschl, R. (2003). What "ideas about science" should be taught in school science? A Delphi study of the "expert" community. *Journal of Research in Science Teaching, 40*(7), 692–720.

Schwartz, R. S. (2004). *Epistemological views in authentic science practice: A cross-discipline comparison of scientists' views of nature of science and scientific inquiry.* Unpublished doctoral dissertation. Oregon State University, Corvallis, OR.

Schwartz, R. S. (2011). The nature of scientists' nature of science views. In M. S. Khine (Ed.), *Advances in the nature of science research: Concepts and methodologies* (pp. 153–188). Springer.

Schwartz, R., & Lederman, N. (2008). What scientists say: Scientists' views of nature of science and relation to science context. *International Journal of Science Education., 30*(6), 727–771.

Schwartz, R., Lederman, N., & Abd-El-Khalick, F. (2012). A series of misrepresentations: A response to Allchin's whole approach to assessing nature of science understandings. *Science Education, 96*(4), 685–692.

Schwartz, R., Lederman, N., & Abd-El-Khalick, F. (2015). Measurement of nature of science. In R. Gunstone (Ed.), *Encyclopedia of science education*. Dordrecht, the Netherlands/Heidelberg, Germany/New York/London: Springer.

Schwartz, R., & Skjold, B. (2012). Teaching about scientific models in a science content course. *Educacion Quimica, 23*(4), 451–457.

Schwartz, R. S., Lederman, N., & Lederman, J. (2008, April). *An instrument to assess views of scientific inquiry: The VOSI questionnaire.* Paper presented at the annual meeting of the National Association for Research in Science Teaching, Baltimore, MD.

Schwartz, R. S., Lederman, N. G., & Crawford, B. A. (2004). Developing views of nature of science in an authentic context: An explicit approach to bridging the gap between nature of science and scientific inquiry. *Science Education, 88*(4), 610–645.

Schwarz, C., Reiser, B., Davis, E., Kenyon, L., Acher, A., Fortus, D., et al. (2009). Developing a learning progression for scientific modeling: Making scientific modeling accessible and meaningful for learners. *Journal of Research in Science Teaching, 46*(6), 632–654.

Schwarz, C., & White, B. (2005). Metamodeling knowledge: Developing students' understanding of scientific modeling. *Cognition and Instruction, 23*(2), 165–205.

Upmeier zu Belzen, A., & Krüger, D. (2010). Modellkompetenz im Biologieunterricht [Modeling competence in biology education]. *Zeitschrift für Didaktik der Naturwissenschaften [Journal of Science Education], 16*, 41–57.

Van Der Valk, T., Van Driel, J., & De Vos, W. (2007). Common characteristics of models in present-day scientific practice. *Research in Science Education, 37*, 469–488.

Van Driel, J., & Verloop, N. (2002). Experienced teachers' knowledge of teaching and learning of models. *International Journal of Science Education, 24*(12), 1255–1272.

Van Driel, J. H., & Verloop, N. (1999). Teachers' knowledge of models and modeling in science. *International Journal of Science Education, 21*(11), 1141–1153.

Windschitl, M., & Thompson, J. (2006). Transcending simple forms of school science investigation: The impact of preservice instruction on teachers' understandings of model-based inquiry. *American Educational Research Journal, 43*(4), 783–835.

Part II
Assessing and Diagnosing Modeling Competence

Chapter 5
Illuminating Scientists' Modeling Competence

Bev France

5.1 Introduction

There is the assumption that models and modeling are central to a scientist's work. In an attempt to find out how scientific knowledge is generated there have been efforts to monitor how they develop knowledge – for example with Latour and Woolgar's (1986) seminal work *Laboratory Life*. This ground-breaking research enabled non-scientists to glimpse into the organised chaos of the laboratory. Since then research on how science knowledge is developed has become central to science education at all levels.

Furthermore, this educational focus on understanding the epistemic role of models is justified when Gilbert, Boulter and Elmer (2000) wrote and edited the seminal book that argued for the central role of modeling in education about science. Their justification was based on the premise that models were one of the main products of science (Rosenblueth & Weiner, 1945). Consequently, this chapter's focus on describing and analyzing how scientists perceive the nature of models that they construct, test and adapt has the potential to frame some indication of competence.

5.2 Justification for Interpretation Rather Than Assessment

Adding to the complexity of analyzing scientist's use of models is that they employ diverse research approaches. Schwartz and Lederman (2008) identify four research approaches used by scientists – that is experimental, descriptive, experimental/descriptive and theoretical. Then there are the six styles of scientific reasoning

B. France (✉)
University of Auckland, Auckland, New Zealand
e-mail: b.france@auckland.ac.nz

A. Upmeier zu Belzen et al. (eds.), *Towards a Competence-Based View on Models and Modeling in Science Education*, Models and Modeling in Science Education 12, https://doi.org/10.1007/978-3-030-30255-9_5

proposed by Kind and Osborne (2017) underpinned by three forms of knowledge – i.e. ontic, procedural and epistemic constructs that scientists weave together as they pursue better explanations of phenomena. Coupled with the diversity of contexts that can affect the research focus within a domain, there are instances of applied research where the wider perspective encompasses practical usefulness as well as the epistemological scientific goal of truth (Olsson, 2015). Furthermore there is diversity in how the model is used. For example a model can provide a theoretical expression of a paradigm through which data is interpreted. Another more practical view is when a scientific model is thought of as a 'surrogate' – that is a technical version of the substitute (Adúriz-Bravo, 2013). To add to the complexity of the use of models in experimental design is that when scientists are thinking about how to best represent the target, they can select aspects of this target when deciding the focus of their research. Consequently their selection and adaptation of a model is particularized for each research situation for one that best provides a bridge between theory and phenomena (Oh & Oh, 2011).

With all of these variations in model use and purpose it could be pertinent to explore the possibility of assessing scientists' models and modeling competence (MC) alongside the comprehensive description of model use by scientists (Van der Valk, Van Driel, & De Vos, 2007). Although these researchers do provide a descriptive framework, their description does not allow space for an analysis of scientists' MC. Closer to this focus on competence is an identification of students' understanding of models and modeling by Upmeier zu Belzen and Krüger (2010) where they present a matrix that identifies not only five aspects of models but propose three levels of student understanding.

Nevertheless, it would be impertinent and unproductive to equate even these high levels of student understanding with an assessment of scientists' MC. Because scientists are immersed in model development and use within their research, and not as onlookers as students are, attempting to assess scientists' MC would be equivalent to assessing their 'knowledge about science' that is the Nature of Science rather than assessing how they have created 'scientific knowledge'. Furthermore it could be asserted that the complexity of model use by scientists would be difficult to unravel, let alone assess. Rather than focusing on assessment it could be instructive to explore the way in which scientists talk about models and their purpose as a way of interpreting their competence. Consequently the focus of this chapter is to recount with some comment scientists' stories of how they use models in their research because this approach was considered to be more enlightening for the reader than a direct assessment of competence. It is proposed that such an account would illuminate how a scientist viewed the nature of models, their purpose, and how they tested and evaluated the models they used in their scientific practice. Furthermore when appropriate an interpretation of an example of these scientist's modeling competence will be explained with reference to the framework for modeling competence (FMC; Chap. 1) and modeling-based Learning Framework (MLF; Chap. 3) that could be given some attention.

It is proposed that following indicative questions about the character and use of models could provide access to scientists' thinking and provide a framework for this analysis. These indicative questions are derived from the author's interpretation of Van der Valk et al.'s (2007) description of the nature and functions of models (pp. 471–472).

It could be presumed that scientists' awareness of the potential and limitation of models in their quest to discover answers to 'why' and 'how' questions about phenomena would provide some enlightenment about the complexity of the scientific process. In the following account these areas will be related to the story of how each scientist uses models in their research and, (rather than assessment), illustrative examples of their competence will be provided. The following questions will provide a framework for these scientists' accounts and provide illustrative examples of the theoretical FMC as discussed in Chap. 1 and how an understanding of the Nature of Science can deepen this analysis (Chap. 4).

- How do scientists perceive models? What are their understandings of the nature and function of models in research?
- What are their understandings of the relationship between model use and knowledge development? How are these epistemological relationships expressed by scientists?
- How do scientists develop models? What is their awareness between the relationship between questions asked and data generated?
- What do scientists say about the limitations of models? How are their epistemological understandings expressed?

5.3 Using Models in Research Science: Scientists' Stories

In order to describe and analyse how scientists viewed and used models, two scientists were identified via a snowball sample – that is a non-probablility sample in which the researcher makes initial contact with a group of people (practising scientists) who establish contact with others who can respond to the researcher's request (Bryman, 2004, p. 544). These scientists were interviewed to find out how they used models in the process of knowledge development. The following questions guided the direction of these conversations:

- Tell me about a research that you have been engaged in that uses models?
- What models do you use in your research?
- How do you use models in your research?
- Do you use more than one model in your research to represent the same explanation/data/prediction?

As part of the ethics process the chapter was returned to each scientist in order for them to check their contributions so that their quotations provided an accurate scientific account of their research. Because it is not possible to provide anonymity for these two scientists as they and their research are well known in New Zealand and at Auckland University, it was mutually decided by participants (scientists) and the researcher that both scientists would be identified, and an article that best represented this aspect of their research would be included in the literature review. Siouxsie a microbiologist investigating the evolution and transmission of as well as drug testing on pathogenic bacteria and Laura a perinatal systems physiologist who uses sheep models to provide clinical data – have agreed to allow their quotes to be acknowledged by name. They have provided papers that give more detail and context to the research they are discussing (Dalton et al., 2017; Bennet, 2017).

As this research is conducted within an interpretativist research paradigm a narrative was constructed from the scientists' quotes that best answered the questions posed (France, 2010). Because the focus of narrative enquiry is to provide an opportunity to create further meaning for the reader (Connelly & Clandinin, 1990) the narrative was constructed under the headings that provided links to these scientists ontological view of models, their epistemological beliefs, the procedural knowledge that underpinned the way they used models in their research design – that is their generative data capacity; as well showing that these scientists have a keen understanding of the limitations of models when interpreting data.

During the process of constructing the quotes that provided illustrative data about these scientists' perception and use of models the following components were paid attention. These were:

- The establishment of a collaborative relationship between the researcher and scientist that enabled the construction of these illustrative excerpts. For example the invitation was accompanied with an explanation of the expected outcome – i.e. the book proposal
- Using a process that enabled participants and scientists to make sense of the data and developing story (Bryman, 2004). For example the chapter was returned to the scientist to ensure the excerpts best reflected the explanation and interpretation of how they used their models.
- Establishing that the completed narrative reflected the complexity of the underpinning material that demonstrated the apparency and verisimilitude of the outcome (Connelly & Clandinin, 1990; Clandinin & Connelly, 2000). For example scientists were asked to provide a published research paper which exemplified how they used models in their research. These papers were included in the reference list.

In summary the illustrative narrative excerpts were constructed so that the reader could become independently aware of these scientists' theoretical perspectives of model use and limitations as well as the constructs that underpinned their procedural and epistemic knowledge. It was expected that these illustrative narratives would ensure that the scientist's competence when using models in their science practice would be illustrated (France, 2010).

5.4 Results

5.4.1 Siouxsie: A Microbiologist

Siouxise is a microbiologist who combines her passion for exploring the phenomenon of bioluminescence with her research on infectious diseases in order to understand not only how pathogenic bacteria adapt and evolve, but also to develop new antibiotics to kill them. She summarises this research focus by saying that she and her team make nasty bacteria glow in the dark in order to find new medicines.

Because one of the bacteria under research is *Mycobacterium tuberculosis*, which is very difficult to kill, spreads through the air and deadly (it is the cause of tuberculosis) – Siouxsie requires highly specialist containment laboratories in which to carry out her research. This bacterium, as well as being restricted to laboratories with specialist people who can work with the bacterium, there are limited surrogate hosts that provide source data for scientists when studying this disease and its control.

Consequently, models are central to this research group's experimental research. The substitution of models takes into account the danger and cost of working with the tuberculosis bacterium as well as needing to use a surrogate host instead of humans. This research team works with *M. tuberculosis* and a variety of closely related bacteria to replicate its growth in vitro as well as in vivo in model animal hosts such as mice. They often choose alternative microbes from the same family as *M. tuberculosis* that are less dangerous but provide important source information about its physiology and infection patterns. Furthermore, the substitutions of host and micro-organism can be studied as a model system where the surrogate microbe and surrogate host can also provide source knowledge about the target.

Siouxsie's comments about the range of models used by her team demonstrate that she has a high level of MC in that the model and model system she selects depends on the nature of the question she is investigating.

> My research uses lots of different models... and model systems. We have model hosts for the different infectious diseases. We would use different model systems depending on the organism we are asking questions about and the type of question we are asking. For example, we use different models to simulate the way the bacteria might be behaving in the human body.

Her understanding of the nature and role of the models used in her research is apparent in the following comments. Siouxsie's view is that a model is a research tool that is used to obtain information about the target which itself cannot be easily observed or measured directly. Consequently she needs to model these microbes' growth in the human or animal host – hence creating an environment to best replicate the human or animal host.

> Our model is a thing. Our models are the bugs we grow in a particular way to model what is happening in the human or animal host.

She comments that the purpose of this model is to:

> Try and replicate in the best possible way … with all the limitations we have (ethical constraints, financial constraints) … what might be happening in reality. It is a living model and more complex.

However this replication was more complex than attempting to recreate an idealized representation of the original because she and her team needed to take into consideration other issues than just the replication of growing conditions. Consequently the factors that influenced the development of this model involved deciding if the use of an infectious dangerous microbe was ethical and worth the expense of using specialized expensive staff in a specialized containment laboratory. During this stage of the researchers were investigating 'how can this microbe be grown that best replicates the host organism environment within the limitations of this research design?' New knowledge was being developed about the best growing conditions rather than scientists making an adjustment to the environmental conditions.

But her view of a model is more even complex as she notes that the substitute organisms and host systems are various. For example:

> We have the tuberculosis bug that we can use on its own and we can infect mice. And then we have several relatives of tuberculosis – the main one that we use is *Mycobacterium marinum.* We use it alone and we use it [to infect] zebra fish embryos and now we have put it into caterpillars.

It is significant that the use of model systems is a focus for Siouxsie's team. In order to replicate the way this bacterium grows in the host, decisions need to be made not only on the choice of bacteria but about how the organism will be grown – that is in vitro or in a surrogate host which could be a zebra fish embryo or a mouse. All of these decisions are focused on providing source data material that can supply pertinent information to predict how bacteria might behave in the target (human body). She states:

> We have model systems that are about the way you grow the bacteria. For example, bacteria can form communities and grow biofilms.

Making choices about how to model the micro-organism's physiology, method of infection, and ultimately its evolution into different pathogenic strains is central to these scientists' thinking. It is apparent that they are equating within their experimental design the analogous relationship between source and target all the while using this situation to predict what could happen in the human host – this is an example of sophisticated thinking about how to test the original hypothesis.

The following account of how Siouxsie's team developed the zebra fish embryo model for drug testing demonstrates that a model will always be the result of the compromise between the demand to best represent the target and a simpler model that can be managed in the laboratory. These developmental decisions reflect the nature of the research problem, the facilities available to more experienced members of the research team as well as providing an opportunity for researchers who are not qualified to work with such dangerous organisms. Consequently time, money and expertise issues as well as the personal preference of the researcher will influence what models and model systems are developed.

Siouxsie's account of how her team developed the zebra fish embryo model for drug testing also demonstrates how collaboration and dialogue between different research teams with different research agendas can influence the way in which models can develop.

The overall focus of Siouxsie's team was to find out how to "speed up drug discovery for tuberculosis by letting it be applied in wider labs in an easier way". The group were seeking the development of a consensus model that was not only quicker and involved more humane methods but enabled access to laboratories who might not have the facilities for drug testing on infectious human diseases.

> Doing this drug testing in a more humane way that would be quicker and could also be used by other labs - maybe labs that don't do any animal work or don't do any mammal work. So maybe they have a zebra fish facility and if they could do some of this drug testing in their labs.

Her group were aware that zebra fish embryos are a widely used model for research – for example on the immune system. These embryos are used at about 3 days after fertilization, that is before they have developed systems for sensing pain. Their bodies are transparent so it is possible to observe them under the microscope and the location of invading bacteria. She observes that what makes these embryo hosts so valuable is that they are "easy to genetically modify and there are versions of these zebra fish that have different types of cells that are labelled in different colours".

At this stage of planning for this model development Siouxsie's team had genetically engineered, for another research group, a relative of the tuberculosis bacterium that glowed in the dark when alive.

Siouxsie's group realised that this bacterium could provide a model for their own research. But the problem for Siouxsie's research team was to find a way to infect the embryo that did not involve yet another high level of expertise and equipment in addition to that occurring in their microbial research. Normally zebra fish embryos are immobilised and then bacteria are injected directly into different parts of the body which involves a complicated technique and requires a high level of skill. Siouxsie wanted to develop a model system of infection that didn't require such a level of skill from laboratory personnel. She used bacteria that were similar to tuberculosis bacteria but not dangerous but the problem was to find a method of infecting zebra fish embryos that was less expensive and did not require a highly skilled level of personnel to infect the embryos. "Something that could maybe be applied more widely and didn't require such skill and could maybe be automated."

Siouxsie was able to draw on her post-doctoral experience where she worked on a model system that compared the infecting of organisms using gastric gavage (where the dose of bacteria is introduced into the animal's stomach with a blunt needle) with a method of infection that more closely mimicked normal conditions. She drew on her past experience and knowledge to reflect that these embryos would be naturally infected by substances in water and developed a system to infect them naturally by adding the bacteria to the water in which these embryos were swimming.

Consequently, Siouxsie's team were able to adapt this model system so that less dangerous bacteria could be used to infect zebra fish embryos that were more accessible to other research teams testing drugs.

This alternative method of infection added to the range of the potential model systems available to Siouxsie's research team. As well as testing different drugs on the tuberculosis bacterium growing in the flask in vitro, her team have developed model systems that provide different infection routes, that use identified relatives of the tuberculosis as well as alternative surrogate hosts such as mice and zebra fish embryos. This variety of bacteria, surrogate hosts and transmission systems increased the potential for providing model systems for drug testing. Her comments indicate that she has not only provided multiple models but also new pathways for future research:

> More people can use this model system to progress the research further to get more people involved who can then say right we have now developed all these drugs can some now try them against the tuberculosis bacterium. Now we have more confidence that they will work.

However, Siouxsie is pragmatic, realising that not all models provide the source information that is expected. She described an evolution experiment that she planned which would use versions of bacteria that were coloured green and red. She had anticipated showing by colour which population of bacteria became more dominant in a culture or an infected host. However, the research team were unsuccessful in developing these differently coloured bacteria for these experimental conditions so they reverted to a previous developed method of using different varieties of bacteria that glow or not. As Siouxsie comments.

> We tried to adapt this model so we can do our really cool experiments in a really cool way. With the colour difference we could look at how they changed. For example, in an animal you could see that one is better than the other because the infection would become predominantly green or predominantly red. It didn't work so we are doing to these experiments anyway. Instead we competed the bacteria with a [with]one that glowed in the dark. We infected mice and then asked the question – do they end up with a tummy full of glowing bugs or a tummy of non-glowing bugs?

Siouxsie's description of the changes of direction that this research took illustrates that she is aware that living models do not always provide data that is expected and her expertise is demonstrated in that she was able to make the most of this failure of research design and revert to an earlier technique that provided data that could test the hypothesis. Such adaptability of the researcher as part of the aspect changing model in the FMC could be considered in future iterations of the FMC more explicitly.

5.5 Summary

Siouxsie's description of model use in her experimental research shows her level of expertise and that modeling took centre stage. For example in order to carry out this research there was a need to develop surrogate model systems that provided information about the source organisms that could enable drugs to be tested. This source data was used to identify potential drugs that have potential to kill the target bacteria – that is *Mycobacterium tuberculosis*. Because of Siouxsie's knowledge of source microbes and hosts and her awareness of the limitations of developing drug testing on human hosts, she designed this research using a variety of model systems. The implementation of these model systems shows a deep understanding of the potential and limitations of what source information can be produced that can support a theoretical prediction about the phenomenon in question. For example not only did she use microbial models to test drugs, but also she employed a variety of host organisms in which the microbes were grown – that is mice, zebra fish and caterpillars. Furthermore Siouxsie extended the capacity of the model system to enable other researchers with less expertise and without high performance laboratory facilities to take part in this adapted experimental design.

Such a capacity to produce this variability of research design demonstrates a deep and sophisticated level of understanding of the role of models and model systems in her experimental research.

5.5.1 Laura: The Perinatal Systems Integrated Physiologist

Laura leads a team researching perinatal physiology within the Faculty of Medical and Health Sciences. She describes herself as a perinatal systems integrated physiologist.

> [My research is] in fetuses and newborns, so, babies and I'm particularly interested in babies before they're born, particularly pre-term babies and how they grow and develop. How they cope with adverse events in their environment, what causes injury and what we can do to: (a) detect it; and (b) to try and ameliorate that kind of injury or prevent it. We are in the business of understanding basic physiology with a clinical translation or medical component in it. We [research] about what we can do to help these babies and newborn.

Laura reflects that collecting data from the fetus is problematic but crucial in order to establish developmental patterns from detailed physiology information on blood pressure, blood flow into organs, heart rate and brain activity. More importantly, because fetal injuries evolve over a long period of time, this physiological information needs to be recorded throughout the pregnancy in order to provide long-term information. She observes that once these developmental patterns of data are established, there is the capacity to identify environmental challenges to the fetus such as inflammation and hypoxia (oxygen deprivation). Because it is impossible to carry out such an interrogation of a human fetus, a chronically instrumented fetal sheep has been developed for an animal model. The sheep model has been selected by this research team because there are similar parameters to a human fetus in that it is a similar size, and at certain periods of its development it equates to the physiology of the human. These animal models are prepared by inserting recording devices into the fetus during an anesthetized caesarean section. After the fetus has been replaced and the host animal has recovered, data are collected while the mother sheep is exhibiting her normal behavior.

Laura comments on this model which she describes as "a paddock to bedside to cot model".

> This is a very powerful systems physiology model – that is we are looking at various systems together. We are looking at the physiology of the fetus itself. We can look at its blood pressure and its brain activity and its body movements that make the physiology of the animal and mediate injury. There are bi-markers that we can take to clinic to say – if you see this pattern of activity of body movements or heart rate then we know the baby is in a state and is doing this. And once you are understanding the physiology of what is going on you can begin to layer in potential treatments.

The focus of this model was to provide information about how to treat newborn injury that can result from a difficult birth – for example brain tissue inflammation and oxygen deprivation. It has been found that cooling the fetal brain can improve outcomes and the data from the fetal sheep model has provided information for this procedure – for example when to start, for how long, when it is too late, its effectiveness, timing and dose. This information provided data for international clinical trials of the cooling-cap treatment which nowadays is a standard of care that when there is a clinical diagnosis of damage to the newborn.

Laura's view of models and modeling shows a sophisticated view of its capacity to represent the target under research. Her level of sophistication is apparent in the way in which she reflects on the role of models in scientific research where she uses models as a tool for collecting data.

> It all comes down to the question that you are asking. Models are a tool and not the solution … because you can do a technique – that is not science. The science is the question. The question is 'what do I want to know?' and then 'how do I then get the data that is going to support the question that I am asking or the hypothesis I'm generating?'

As well as demonstrating her epistemological awareness of the role of models and their function she expresses the thinking that must occur when designing a model that can provide a valid representation of the phenomena under examination. Furthermore, she observes that the researcher needs to be very aware of the differences between the source (animal model) and the target.

> So constantly you are asking yourself - what is the clinical scenario? What is it that I'm physiologically monitoring? It's always about the question. And therefore, you adapt it to the scenario that relates to the question. And you layer it [the model] up depending on what information you need to show … either physiological changes or molecular changes or chemical changes … whatever you need to do to add to your recipe for that experiment.

> So the hypothesis is the key scientific element of an experiment. Then you have to look at how you will address that in terms of the model you might use. We need to be very careful

about any animal model ... for example in terms of equating it to certain comparisons to a human because sheep stand up when they are born, and feed and run around. Humans don't, so we have to be careful that we have got the right timing.

It is very apparent that Laura and her research team not only have searched for a close analogy that provides data for the target but they are constantly critiquing the model to provide the best clinical data that can be used for treatment.

Because there is a clinical focus for this experimentally obtained source data, a lot of time is spent developing the model. As Laura comments.

It is not something you can pluck off a shelf and say 'ok I am going to do this model today'. You actually have to set in and experiment around the parameters of what timings you are using, what ages you are using, what might cause inflammation. There is a lot of experimentation around just establishing a model let alone using it.

This aspect of MC is not able to be assessed using the model proposed in Chap. 1 as there is no space given to an analysis of how the thinking underpinning how models are developed. It appears that the testing of a model's category does not seem to encompass the deeper level of analysis that is required when the model parameters are identified that can affect the research design.

It is apparent that Laura has a deep understanding of the nature of models and that more than one model may be required to provide different data sources. Laura notes that one model isn't necessarily a fit for all purposes. She makes the observation that a model may be developed to provide specific source information and comments that often questions needing experimental data to answer them are set in the model, for example when a model is designed to provide source information about what is happening in the fetal brain during labour. She comments.

For example, I am using a model where I might just give a single squeeze of the umbilical cord to [provide] a period of time of low oxygen to the fetus – which often happens at birth. But actually in labour you have repeated squeezes because that is what contractions are about. That is a different model and you have to develop that as a separate model. So the questions are set in your model.

An awareness of the epistemic demands of this source data means that the experimental design must reflect this model's capacity to provide pertinent data. Consequently, Laura and her team pay attention to the statistical representation of the data.

You need to apply your statistics before you even start your experiment. That is - the right group-size number, group-size setting well before you start your experiment so you've got power over the statistics.

Laura states that the function of her models are the testing and revision of scientific theories. She notes that "a good model should be robust". She reflects that her model provides predictable data when trauma is applied to the sheep fetus. It is significant to acknowledge that her claims for robustness are set strongly within the framework of the experimental design and indicates her awareness of the nature and function of this model that will provide data from which scientific theories about cause and recovery theory in the human fetus and pre-terms can be developed.

> My model for pre-terms is now around the world because I tested it robustly and it took a long time to get 'yes' under these situations. I'm not claiming it to be anything else but this condition under this situation – this is what this model is producing and if you do it this way it will produce this pattern of brain injury.

Because she is acutely aware that the fundamental principle of science is not to be biased towards one's original hypothesis, her commentary also acknowledges the limitations and caveats of the model:

> We are using sheep. We don't know necessarily about the sheep differences. By definition it is no longer a normal in utero environment because you stuck a lot of tubes and things in it.

But Laura and her research team are not just restricted to experimental modeling – that is using a model as a surrogate research animal. She is also developing theoretical models when dealing with the data that can inform further model development within an experiment.

> A theoretical model can use known experimental variables and then put to see patterns from which should be able to predict. What you are trying to do is present a whole lot of information and then mathematically develop it in a predictive algorithm, for example heart rate monitoring. We develop predictive algorithms by knowing physiology and putting it into a mathematical database that allows us to look at predictive outcomes … [We arrive at] a mathematical model then how can we model that in an experiment and that is a real challenge.

The following account demonstrates Laura's depth of understanding of how her work with experimental models as well as her critique of the data on which she had based her predictions to rethink her original premise that the fetus and pre-terms would have similar responses to oxygen deprivation. Laura told the story about her discovery that led to a paradigm shift about the resilience of the fetus to oxygen deprivation. This story illustrates how Laura constantly uses the experimental data to interrogate the robustness of the experimental models on which she is working.

> Our hypothesis was based on that there was no difference between the reaction to oxygen deprivation between the fetus and the pre-terms. For a long time we didn't even know if the fetus had the capacity to detect oxygen deprivation and could respond. It was about a paradigm shift that comes back to our perception. If you look at a pre-term baby - newborn baby born at 25–26 weeks instead of 40 weeks. They are so small (500–600 grams) and you look at them in hospital and all you see is fragility and vulnerability and immaturity … so they are not able to cope with any challenge because they are so fragile.

Consequently, she assumed that when she developed an experimental model to monitor the reaction of the sheep fetus to oxygen deprivation, she would need to keep the period of insult short. To her surprise she found that she had to push out the length of time to 30 min in the sheep fetus before she saw the clinical patterns of injury that would be seen with term fetuses after 10 min. Suddenly she realised the paradigm of trauma from oxygen deprivation was different for the fetus as it is living in a very unique environment that is not the same as a pre-term baby. This account of how she discovered this unexpected data when manipulating the model meant that she was able to show the existence of a new information about the the vulnerability of fetuses that could not be equated with a pre-term baby. Such a revision to model design is also a reflection of this scientist's high level of MC because adjustments to the model were based on a paradigm shift of thinking about the vulnerability of the fetus to oxygen deprivation.

Because she looks at the model as an experimental tool that provides knowledge to inform the model to identify conditions for fetal damage in human fetus, this attention and adaption of the model is central to her research.

> You really do have to be constantly looking at your data and what does it tell you and what kind of adjustments do you make to your models. What is the information coming in from the clinical or other experimental models that tell you we should be looking at something … and it is all additive. It is the integration of knowledge

5.6 Summary

When attempting to assess the MC of Laura and her team it is evident that they have a profound understanding of the purpose of these experimental models (that is chronically instrumented fetal sheep) to provide data that can be used for clinical diagnosis of trauma in the human fetus and subsequent treatment of preterm babies.

Although the model fetal sheep provides a physical model upon which experiments can be carried out, she also sees the nature of her models as predictive sets of data that provide guideposts to detection of trauma in the human fetus and pre-term baby. She has a deep appreciation of the role of the model as a tool that can provide data for developing these clinical models. Her understanding of the epistemic issues involved in testing a hypothesis on a fetal model sheep is shown with her discussion about the need to think about the statistical representation of the data she intended to collect and analyse. This research group continually question the validity of their experimental models in representing the human fetus and in the development of a research design that will provide data to fully represent the theories they wish to support. This account of Laura and her research team's thinking about how models provide data demonstrate that modeling was centre stage when these scientists were designing their research. In fact, modeling directed their thinking and reasoning through their forays along new experimental pathways.

5.7 Discussion

5.7.1 *How Important Is it to Learn from Scientists' Modeling Competence?*

At the beginning of this chapter the argument was put forward that an assessment of MC of scientists was at worst impertinent and at best impractical. Furthermore, the author questioned how fruitful would this assessment be to develop the pedagogy to understand the role of models when science is practised.

Instead I would argue that rather than assessing a scientist's MC, providing examples of scientists' model use where the components of the model – that is source and target, their purpose and how they are developed and critiqued – could contribute to students' developing critical scientific literacy about model use by scientists. As Krell, Upmeier zu Belzen, and Krüger (2014) opine, in order to develop students' deep understanding of the nature and role of models in science, they need to be able to recognise examples of model use. These illustrations could show how models are used when building theoretical explanatory constructions or when they are used as research tools in science. It is proposed that these scientists' stories of model use could help students develop a stronger conceptual understanding of modeling as well as enable them to recognise the prevalence of modeling in scientific practice.

Finally stories about scientists developing and using models provide yet another opportunity to show the messiness, creativity and complexity inherent in how science is practised and supports this push for a deeper understanding of the culture of science that was alluded to in Latour and Woolgar's account of *Laboratory Life* (1986).

These scientists have no need for assessment of their model use - instead we can marvel at their expertise as they find their way to the 'truth of the matter'. These stories illuminate the thinking of these scientists and demonstrate the creativity, complexity and deep understanding of how models are used as they practise science.

5.7.2 *Illuminating Scientist's Modeling Competence for the Classroom*

At this point the question needs to be asked – How can a teacher get valid information about scientists' thinking as they use models in their research? Because educational researchers assert that in order to understand the central role of models in science knowledge development it is essential to access how experts use this tool.

Furthermore, if knowing about models and modeling are key features of science then it is important to make accessible some aspect of scientist's business (Coll, France, & Taylor, 2005). A presumption of their business is not just how they use models when doing science but also the social and intellectual circumstances that determine the direction of scientific research (Chap. 4).

But this understanding does not come by creating situations for students to carry out scientific research and expecting them to pick up this expertise explicitly. As Hodson (2014) opines it is better to teach about the Nature of Science explicitly rather than expecting students to pick up how scientists develop scientific knowledge by conducting their own scientific investigations. There is a compelling argument that in order to teach about model use in knowledge development it is important to access the community of scientists so that teachers can provide an entry into the subculture of science.

In this chapter these scientists' narratives have provided a glimpse of the private language and personal experience of science knowledge development that is so different to the public language of science (Hodson, 2014). Grosslight, Unger, and Jay (1991) note that the expertise of scientists are evident when they provide pragmatic responses to issues of model development and implementation. This was very evident in Siouxsie's decision to revert to bioluminescent bacteria when the coloured model bacteria did not provide the contrasting data that she had anticipated. This change in model parameters was a pragmatic solution to an experimental problem because of her expertise in setting up and adapting models that would provide experimental data but would not have appeared in the publication of her work.

Access to scientists is difficult and an ideal situation is for students to be mentored by scientific experts but although such situations are desirable it is not always practical to be part of scientists' decisions during the development and adaptation of their models. Even though it is possible for students to interview scientists about their view of models, their purpose, the types of models they use and if they ever changed their models (Grosslight, 1991), it would be less likely they would be able to pose questions that gave an indication of a scientist's MC let alone develop a scoring system. Instead it could be more informative if students were provided with scientists' examples of level III competence (Chap. 1) with an accompanying explanation.

The need to interact with scientists was realised in New Zealand with the development of the on line Science Learning Hub where scientists talked about their research which was video recorded. In each case a transcript was provided. Examples of these resources are provided as follows:

- A scientist describes the predictive capacity of models. This video allows students to observe a range of scientific models used as research tools[1].
- The building of a climate model was explained using the parallel analogy of mine craft to demonstrate the strength of a model that can be measured by the data base on which it is formed[2].
- A video entitled 'New Zealand's next top model' tells scientists' stories involved in building a more dependable climate change model. This video provides a

[1] https://www.sciencelearn.org.nz/videos/844-models-in-science

[2] https://www.sciencelearn.org.nz/resources/2232-climate-models

critique of the model and could provide data for student critique and discussion about climate change. These examples of level III competence could be used to show how models are developed to test a hypothesis[3].

What is important for students is to have access to scientists' voices as they describe how they develop, critique and adapt models as they develop some understanding of phenomena. Such narratives provide some insight into an expert's thinking or MC but more importantly they provide an example of the private language of personal experience as they build knowledge rather than the factual but flavourless public language of science (Hodson, 2014).

References

Adúriz-Bravo, A. (2013). A 'semantic' view of scientific models for science education. *Science & Education, 22*, 1593–1611. https://doi.org/10.1007/s11191-011-9431-7

Bennet, L. (2017). Sex, drugs and rock and roll: Tales from preterm fetal life. *Journal of Physiology, 595*(6), 1865–1881.

Bryman, A. (2004). *Social research methods* (2nd ed.). Oxford, U.K.: Oxford University Press.

Clandinin, D. J., & Connelly, F. M. (2000). *Narrative inquiry: Experience and story in qualitiative research*. San Francisco, CA: Jossey Bass.

Coll, R. K., France, B., & Taylor, I. (2005). The role of models/and analogies in science education: Implications from research. *International Journal of Science Education, 27*(2), 183–198.

Connelly, F. M., & Clandinin, D. J. (1990). Stories of experience and narrative inqury. *Educational Researcher, 19*(5), 2–14.

Dalton, J. P., Uy, B., Okuda, K. S., Hall, C. J., Denny, W. A., Crosier, P. S., et al. (2017). Screening of anti-mycobacterial compounds in a naturally infected zebrafish larvae model. *Journal Antimicrobial Chemotherapy, 72*(2), 421–427.

France, B. (2010). Narrative interrogation. Constructing parallel stories. In S. Rodrigues (Ed.), *Using analytical frameworks for classroom research* (pp. 90–108). Abingdon, Oxon: Routledge. OX14 4RN.

Gilbert, J. K., Boulter, C. J., & Elmer, R. (2000). Positioning models in science education. In J. K. Gilbert & C. K. Boulter (Eds.), *Developing models in science education* (pp. 3–17). Dordrecht, The Netherlands: Kluwer.

Grosslight, L., Unger, C., & Jay, E. (1991). Understanding models and their use in science: Conceptions of middle and high school students and experts. *Journal of Research in Science Teaching, 28*(9), 799–822.

Hodson, D. (2014). Learning science, learning about science, doing science: Different goals demand different learning methods. *International Journal of Science Education, 36*(15), 2534–2553.

Kind, P., & Osborne, J. (2017). Styles of scientific reasoning: A cultural rationale for science education? *Science Education, 101*(1), 8–31.

Krell, M., Upmeier zu Belzen, A., & Krüger, D. (2014). Students' level of understanding models and modeling in biology: Global or aspect-dependent. *Research in Science Education, 44*, 109–132.

Latour, B., & Woolgar, S. (1986). *Laboratory life: The construction of scientific facts*. Princeton, NJ: Princeton University Press.

[3] https://www.nzgeo.com/stories/esm

Oh, P. S., & Oh, S. J. (2011). What teachers of science need to know about models: An overview. *International Journal of Science Education, 33*(8), 1109–1130.
Olsson, E. J. (2015). Goal, rationality in science and technology. An epistemological perspective. In S. O. Hansson (Ed.), *The role of technology in science: Philosophical perspectives,* Philosophy of Egnineeering and technology, 18. https://doi.org/10.1007/978-94-017-9762-7_9
Rosenbluet, A., & Weiner, N. (1945). The role of models in science. *Philosophy of Science, 12*(4), 316–321. http://www.jstor/stable/184253
Schwartz, R., & Lederman, N. (2008). What scientists say: Scientists' views of nature of science and relation to science context. *International Journal of Science Education, 30*(6), 727–771.
Upmeier zu Belzen, A., & Krüger, D. (2010). Modellkompetenz im Biologieunterricht. Struktur und Entwicklung. *Zeitschrift für Didaktik der Naturwissenschaften, 16*, 41–57. http://www.ipn.uni-kiel.de/zfdn/pdf/16_Upmeier.pdf
Van der Valk, T., Van Driel, J. H., & De Vos, W. (2007). Common characteristics of models in present-day scientific practice. *Research in Science Education, 37*, 469–488.

Chapter 6
Combining Visual and Verbal Data to Diagnose and Assess Modeling Competence

Inga Ubben, Sara L. Salisbury, and Kristy L. Daniel

6.1 Introduction

The potential for verbal data to provide insights into modeling competence became obvious in the pathbreaking interview study by Grosslight, Unger, and Jay (1991). Placing their focus on students' and experts' conceptions of models and the use of models in science, they conducted clinical interviews with open-ended questions. Twenty eight years and more than 900 citations later (Google Scholar), their results continue to form the basis for many other studies. No matter whether younger and more and more specific studies partially reproduce results (e.g., Grünkorn, Upmeier zu Belzen, & Krüger, 2014; Trier, Krüger, & Upmeier zu Belzen, 2014) or combine them with other theoretical perspectives such as modeling-based teaching (e.g., Gilbert & Justi, 2016), the predominant role of the underlying theory has remained the same. On the one hand, theoretical considerations deductively lead to the use of the interview method and in addition determine interview questions. On the other hand, the results of interview studies again inductively underpin or expand theory.

Nevertheless, interview studies as one type of qualitative study are often the subject of discussions regarding reproducibility, soundness of interpretation, or subjectiveness (outlined by Edwards & Holland, 2013). One way to overcome these points of criticism is to triangulate several data sources (e.g., Crawford & Cullin, 2005; Orsenne, 2015). The Standards for Educational and Psychological Testing

I. Ubben (✉)
Humboldt-Universität zu Berlin, Berlin, Germany
e-mail: inga.ubben@biologie.hu-berlin.de

S. L. Salisbury
Middle Tennessee State University, Murfreesboro, TN, USA

K. L. Daniel
Texas State University, San Marcos, TX, USA

A. Upmeier zu Belzen et al. (eds.), *Towards a Competence-Based View on Models and Modeling in Science Education*, Models and Modeling in Science Education 12, https://doi.org/10.1007/978-3-030-30255-9_6

(AERA, APA, & NCME, 2014) have even accentuated the importance of verbal data by presenting sources of validity. These are the relation between test content and construct, response processes where think-aloud methods can be used to elicit participants' cognitive processes, the internal structure of a test, relations to other variables to demonstrate convergent or discriminant evidence, and consequences of testing such as the interpretation and use of test scores. As a result, methodological decisions such as to collect verbal data as a reaction to modeling competence tasks can serve to ensure validity.

The power of think-aloud methods to investigate the cognitive processes behind the responses elicited during task performance (Ericsson & Simon, 1993; van Someren, Barnard, & Sandberg, 1994) was applied, inter alia, by Terzer, Patzke, and Upmeier zu Belzen (2012) and Mathesius, Upmeier zu Belzen, and Krüger (2018). Both studies were aimed at investigating the validity of the solving processes involved in closed tasks such as multiple-choice questions, whereas the latter study even went one step further by triangulating the think-aloud method with eye-tracking technology. This method allows eye movement data to be collected as a response to visual stimuli and thus provides insight into the cognitive processes that occur when a person interacts with a stimulus (Bojko, 2013). A combination of verbal and visual data such as gaze patterns enables the reconstruction of response processes to several stimuli such as tasks, different representations of models, and modeling processes in videos or even live (Sects. 6.2.2 and 6.3). In science education, eye-tracking technology is becoming more and more relevant, for example, for evaluating teaching materials and strategies or for diagnosing differences between individuals such as in their conceptual development (reviewed by Lai et al., 2013). In contrast to a growing number of studies on the visual perception of representations (e.g., Stieff, Hegarty, & Deslongchamps, 2011; Daniel & Leone, 2017), models and modeling with regard to modeling competence are still rarely subjected to eye-tracking research (Ubben, Nitz, & Upmeier zu Belzen, 2017).

Given the broad use of verbal methods in modeling competence research and also the inseparability of eye tracking and verbal methods, we will first present the verbal approaches that are routinely implemented in research and give examples of application. This will be followed by an introduction to eye-tracking technology to show how insights into cognitive processes can be extended in this field. This overview of several methods is meant to show the diversity of the application of verbal and visual methods and the stimuli they can be combined with. Highlighting the benefits and also mentioning other aspects to bear in mind when using a certain method are meant to act as a guide for readers' methodological decisions about research designs. Finally, ideas and suggestions for new approaches for diagnosing and assessing modeling competence by triangulating eye tracking with think-aloud methods are given to open up a new field of research.

6.2 Principles of Interviewing and Think-Aloud Methods

Science education demands several skills that go far beyond the knowledge of facts from all involved individuals such as students and teachers. Here – depending on the research question – interviews can be applied to evaluate the complexity of students' conceptions about scientific content, amongst others (Mayring, 2016). Furthermore, interviewing provides insights into *why* people act as they do in a certain situation by relating individuals' behavior to a certain context and understanding what this behavior means to individuals' personal experiences (Seidman, 2008).

Depending on the theoretical background of a research question, three different types of interviews can be applied: structured, semi-structured, and unstructured (Brinkmann, 2014), forming a continuum that ranges from quantitative to qualitative and standardized to flexibly adaptable approaches (Edwards & Holland, 2013). On one side of this continuum, quantitative interviews are strongly based on theory and are used to confirm hypotheses, whereas on the other side, unstructured interviews are mostly applied to generate hypotheses in an exploratory way when theory is not available. In order to decide which form of interviewing is most suitable for the theoretical basis of a study, the purpose of each form, their characteristics, benefits, and drawbacks have to be considered carefully. For this purpose, we compiled these criteria and present them for every form in Fig. 6.1.[1]

Cognitive processes such as solving problems or thinking in an abstract way (Gerrig & Zimbardo, 2002) can be uncovered by asking participants to think aloud (Ericsson & Simon, 1993; Olson, Duffy, & Mack, 1984; van Someren et al., 1994). In contrast to interviewing, the researcher does not ask questions, but the participant is presented with a stimulus (e.g., a task to solve) and describes his or her solution process while performing the task. This concurrent think-aloud (CTA) minimizes effects of participants' interpretation and researchers' influence on verbal data (van Someren et al., 1994). Two other forms of thinking aloud after participants solved a task - retrospective think-aloud (RTA; Ericsson & Simon, 1993) and cued RTA - can be used to clarify facts from CTA or to lower cognitive load (van Gog, Paas, van Merriënboer, & Witte, 2005), respectively. All three forms of thinking aloud are described in Fig. 6.1.

In contrast to the approach used in interviewing, the researcher should avoid speaking to the participant during CTA after giving the instructions (unless the participant is no longer verbalizing) because interactions between researcher and participant are detrimental to data collection (Ericsson & Simon, 1993; van Someren et al., 1994).[2] As with other verbal methods, data may be influenced by how well participants can express their actions. To reduce this effect, all types of thinking

[1] For detailed information about how to plan and conduct interviews, see, for example, Edwards and Holland (2013), Kvale and Brinkmann (2009), and Mason (2002).

[2] Other methods of triangulation such as post-CTA interviews or recall have been discussed, for example, by Charters (2003).

Method		Purpose	Characteristics	Benefits	Drawbacks	Example
Interview	Unstructured	Elicit participants' perspectives on topics guided by their own concepts[2]	Only a few questions, at times only one introductory question[2,4,7]	High flexibility[4,7] Participants develop their own structure[2]	No comparability	
	Unstructured + structured		Interaction between researcher and participant[7]	Different information than written answers[7,9] Reconstruction of information instead of reporting facts[7]	Time consuming, expensive[9]	
	Semi-structured	Elicit participants' perspectives on topics guided by researcher[2]	Preset questions in a flexible order[2,4,7]	Response to participants' reactions[4,7] and comparability between individuals[4]		Chap. 6.2.1
	Structured	Quantitative analysis of the research question[2]	Preset order of standardized questions[2]	High standardization and comparability[2]	No flexibility[2]	
Think-aloud	Concurrent[5] (CTA)	Problem solving[8] Higher thinking processes[5]	Verbal data from working memory[13] No interaction between participant and researcher[13]	No interpretation by participants[13] No loss of facts[13] Information on actions[12] Task performance may be affected[1]	Verbalizing abstract thoughts might lead to a loss of information[3] Action advances too fast to describe it[13] High cognitive load when task is complex[5,13]	Triangulation with eye tracking (Stieff et al., 2011)
	Retrospective[5] (RTA)	Task solving strategies and conditions[11]	Retrieval of verbal data from working memory and long-term memory[5]	Clarification of information from CTA[13]	Loss of facts through forgetting[12] Information that were not part of the actual task[12]	Comparison of CTA, RTA, and cued RTA (van Gog et al., 2005)
	Cued retrospective[9] (cued RTA)	Problem solving[12] Metacognitive processes[12]	Retrospective replay of eye movements on stimulus by superimposition of gaze paths[12]	Information on actions[12] Detailed statements of good quality[12] Task performance is not affected[1]	Loss of information[6] Participants fabricate information[1] Time consuming[1]	Triangulation with eye tracking (Jarodzka et al., 2010)
Think-aloud + Interview	Multimethod interview[10] (MMI)	Metacognition in problem solving[10]	Triangulation of think-aloud, interview, and video recording[10]	Extensive characterization of cognitive processes during problem solving[10]		Orsenne (2015)

Fig. 6.1 Types of interviews and think-aloud methods. Examples of interview studies are restricted to studies on modeling competence where semi-structured interviews are predominant. Examples of think-aloud methods refer to eye-tracking studies. [1]Bojko (2013), [2]Brinkmann (2014), [3]Charters (2003), [4]Edwards and Holland (2013), [5]Ericsson and Simon (1993), [6]Holmqvist et al. (2011), [7]Mason (2002), [8]Olson et al. (1984), [9]Seidman (2008), [10]Randhawa (1994), Wilson and Clarke (2004), [11]Taylor and Dionne (2000), [12] van Gog et al. (2005), [13]van Someren et al. (1994)

aloud have to be practiced in a previous training phase long enough to familiarize participants with the ability to simultaneously work on a task and "[…] say out loud what comes to [their] mind" (van Someren et al., 1994, p. 42). An approach that can be applied to overcome these drawbacks is van Gog et al.' (2005) method of cued RTA, which is based on the assumption that attention lies where an individual is looking (Just & Carpenter, 1980). Participants' eye movements were tracked while they worked on a problem-solving task on a screen. Subsequently, eye movements were superimposed on the task in the form of red crosses indicating loci participants looked at and lines to visualize eye movements between these fixations. It was found that CTA and cued RTA outperformed RTA in terms of number of codes and that cued RTA provided information that was similar to the information provided by CTA (van Gog et al., 2005). Cued RTA often plays a role as a triangulation method especially in eye-tracking studies (e.g., Jarodzka, Scheiter, Gerjets, & van Gog, 2010) as explained later in Sect. 6.3.

All of these methods have in common that they produce large amounts of verbal data that need to be handled. Here, qualitative content analysis comes into play with category systems designed inductively from material or deductively from underlying theory depending on the study design. Transcription makes verbal data accessible for analysis and can range from transcribing each utterance to summaries or the selection of certain parts following strict protocols (Mayring, 2014, 2016).[3] The most important points and consequently the strengths of qualitative content analysis are that transcription and coding can be transferred to other contents and can be performed by others with the same outcome (Mayring, 2014, 2016), thus providing another source of validity (AERA et al., 2014).

In sum, interviews and several kinds of think-aloud methods differ in their outcomes due to the degree of guidance offered by the researcher and whether the approaches are performed during or after an action. This addresses different cognitive processes, and hence, methods have to be chosen carefully depending on what kind of information is needed to address a research question. The next section provides insights into how the presented methods have been applied in studies on modeling competence to date.

6.2.1 Application of Interviews and Think-Aloud Methods to Assess and Diagnose Modeling Competence

Because science cannot be learned or taught without models (Harrison & Treagust, 2000), modeling competence plays a crucial role in science education (Chap. 1). This concerns students as well as (pre-service) teachers because they are directly involved in applying and fostering modeling competence. However, experts' use

[3] For theory and details on the procedure of qualitative content analysis, please see Mayring (2014, 2016).

and conceptions of models also have to be considered because they can serve as a guide in class to teach a profound understanding of models. In the next section, studies that examined the modeling competence of students, pre-service teachers, teachers, and scientists will be presented to illustrate how broad methods such as interviews and think-aloud methods can be applied and which outcomes they can produce.

In their pioneering study for research on the understanding of models, Grosslight, Unger, and Jay (1991) interviewed students and scientific model experts to elicit their conceptions of models and their role in epistemology. In a semi-structured interview study, they compared American 7th and 11th-graders' answers to questions about their general ideas about models and how they think models are applied in science. Moreover, students had to decide and explain whether four presented objects were models or not and finally developed scientific models on their own before they were asked about modeling, multiple models, and changing models in science. To compare students' answers to sophisticated model use, four experts were interviewed, too. Using a scoring system, their answers were assigned to five different aspects of models: *kinds of models, purpose of models, designing and creating models, changing a model, and multiple models*. Three levels of understanding that were proposed a priori on the basis of theory (Carey, Evans, Honda, Jay, & Unger, 1989) could be captured: On Level 1, participants see models as copies of real objects; on Level 2, participants understand that a model is constructed for a certain purpose; and on Level 3, participants are aware that models can be used to generate and evaluate ideas, can be adapted for a certain purpose, and can be tested and changed. The majority of 7th-graders were on Level 1 (67%), whereas Level 1/2 (18%) and Level 2 (12%) were represented less. By contrast, Level 1 was found in only 23% of 11th-graders who were predominantly on Levels 1/2 and 2 (36% each). Nevertheless, Level 3 was reached only by experts (Grosslight et al., 1991). They concluded "[…] that students need more experience using models as intellectual tools, more experience with models that provide contrasting conceptual views of phenomena, and more discussions of the roles of models in the service of scientific inquiry" (Grosslight et al., 1991, p. 799).

These differences between students (German 10th to 13th-graders) and experts were analyzed in detail by Trier et al. (2014) with semi-structured interviews for the purpose of developing interventions for model use in biology education. They analyzed students' conceptions of all five aspects of modeling competence according to Grosslight et al. (1991) and their influence on model use in interviews and compared them to scientists' views extracted from the literature. First, students were asked to develop their own models and to talk about their understanding of these models. Subsequently, different objects from biology class were shown to students (e.g., real organisms, pictures of organisms, microscopic pictures), and they were asked to decide and explain whether these were models or not. Interestingly, some students were found to have quite sophisticated ideas about models in science but did not transfer them to models they encountered at school. This was aligned with the findings by Treagust, Chittleborough, and Mamiala (2002) who postulated that students do not have enough opportunity at school to use models, and thus, their abstract knowledge about models is not transferred to real contexts. Additionally, most students – unlike scientists – do not see the tentativeness of research processes and do not view changing a model as an opportunity to gain new insights (Trier et al., 2014).

Orsenne (2015) investigated the constructing, testing, and changing of models using a modeling-centered scientific inquiry in German 10th-graders. She conducted a multi-method interview study (MMI; Randhawa, 1994; Wilson & Clarke, 2004; Fig. 6.1) that was aimed at investigating the influence of hands-on modeling tasks in biology on student conceptions of models as inquiry tools. The MMI approach triangulates interviews, CTA, and videography, resulting in a broad characterization of cognitive processes (Randhawa, 1994). Hands-on tasks based on the framework for modeling competence (FMC; Grünkorn et al., 2014; Upmeier zu Belzen & Krüger, 2010), the model of modeling (Justi & Gilbert, 2002), and empirical studies (cf. Orsenne, 2015) have been developed to trigger different aspects of modeling and to let students use models to investigate their own hypotheses. The semi-structured interviews were based on Grosslight et al. (1991), Trier et al. (2014), and aspects of modeling competence (Grünkorn et al., 2014; Upmeier zu Belzen & Krüger, 2010). In an introductory semi-structured interview, students were asked about their general conceptions of models and which models they were familiar with. Second, students were given their first hands-on task and were asked to think aloud concurrently (CTA). Third, they were interviewed about the aspects of modeling competence: *purpose of models, testing models,* and *changing models*. Fourth, another hands-on task was carried out in combination with CTA. Fifth, an interview about all five aspects of modeling competence was conducted. As a sixth step, alternative models were presented to the students, and subsequently, an interview about *multiple models* and scientific modeling was conducted. In a final step, an example of scientific modeling was shown. With the help of these large verbal data sets from interviews and CTA, the author of the study was able to show that modeling per se triggers medial conceptions about models (Levels I and II of FMC) because the focus lies on the model object (Legrenzi, Girotto, & Johnson-Laird, 1993). Only in the aspect *purpose of models* students express more elaborated ideas on a methodological level. However, reflecting on alternative models could lead to more elaborated and epistemic ideas about models (Orsenne, 2015).

Focusing on pre-service teachers from STEM subjects with so-called *dynamic context-rich assessments* (Crawford & Cullin, 2005, p. 321), the three levels of the understanding of modeling found by Grosslight et al. (1991) were expanded to four *(limited, pre-scientific, emerging scientific, scientific*; Crawford & Cullin, 2005, p. 314). Besides an open-ended questionnaire and a process map, these assessments included a pre- and a post-modeling semi-structured interview about the understanding of modeling. Furthermore, verbal data from video recordings were analyzed. The authors concluded from their work with different assessment methods that pre-service teachers' understanding of modeling benefited from the modeling process itself. Furthermore, they emphasized the comprehensive view on the understanding of modeling gained through *dynamic context-rich assessments* (Crawford & Cullin, 2005).

In a semi-structured interview study (Justi & Gilbert, 2003), hands-on model objects, written or drawn models, and models demonstrated by the interviewer were shown to 39 Brazilian teachers, pre-service teachers, and university teachers. This investigation of their notions of models identified seven aspects: *nature, use, entities, uniqueness, time span, status for predictions*, and *accreditation of models*. These aspects again can arise in different so-called *categories of meaning* (Justi & Gilbert, 2003, p. 1374). In contrast to Grosslight et al. (1991), no levels could be identified for teachers. Interestingly, biology teachers' notions of models were less elaborated than chemistry and physics teachers' notions. This was explained by the more frequent appearance of models in chemistry and physics than in biology (Justi & Gilbert, 2003).

How and why experts from different fields use models was elucidated by France, Compton, and Gilbert (2011) by interviewing a biotechnologist and a bioinformatics scientist and using publications of the latter to create stories (Chap. 5). The stories were meant to give students and teachers insights into how different models can be used. The biotechnologist, who is supposed to work more like an engineer than a scientist, indicated that he uses models to predict the medical effect and the economic impact of a product such as a vaccine his enterprise might be developing. Hence, his reasoning about models is functional and is needed to determine whether a product can be realized but also practical so that the company can successfully sell this product. In contrast to this, the scientist used phylogenetic trees as models on the one hand to communicate his findings and on the other hand to predict evolutionary relationships. This inductive approach allowed him to create a model of evolutionary relationships and test and change it deductively (France et al., 2011).

The presented studies on modeling competence used semi-structured interviews to assess participants' understanding of models. On the one hand, this offered the opportunity to follow up on certain points that participants mentioned in contrast to more structured approaches. On the other hand, there was still a certain degree of comparability (Edwards & Holland, 2013). Mainly combined with other verbal or written methods, this can lead to comprehensive insights into modeling competence in different settings and groups. Many of the presented studies have in common that models are shown to participants or that participants even handle them in hands-on tasks. This implies a strong interaction with the model, including visual perception of the model object. The next section will introduce eye tracking as a method for investigating these visual processes with the potential to elicit the cognitive processes involved in modeling competence.

6.2.2 Application of Eye Tracking in Science Education

To date, eye tracking has been widely utilized for data collection and analysis across a variety of disciplines and topics such as mathematics (Hegarty, Mayer, & Green, 1992), reading comprehension (Rayner, Chace, Slattery, & Ashby, 2006; Rayner, 2009), information processing (Rayner, 1998), and problem solving (Tai et al.,

2006; Tsai, Hou, Lai, Liu, & Yang, 2012). Whereas many studies have addressed and/or utilized eye tracking, very few have specifically addressed science education and even fewer have addressed modeling competence (Ubben et al., 2017, 2019). Such studies within the discipline of science education have, to date, typically addressed topics in chemistry (Chien, Tsai, Chen, Chang, & Chen, 2015), physics (Chen et al., 2014; Mason, Pluchino, Tornatora, & Ariasi, 2013; Mason, Tornatora, & Pluchino, 2013), and the natural sciences (Jarodzka et al., 2010; She & Chen, 2009; Slykhuis, Wiebe, & Annetta, 2005; Tsai et al., 2012).

An exploratory investigation used eye tracking to capture biometric data on how students visually access phylogenetic tree diagrams (Daniel & Leone, 2017). Dwell times on specified phylogenetic tree components and gaze paths were identified and compared with verbal data (i.e., CTA) to compare students' retrospective semi-structured interview responses to biometric evidence in search of consistency.

Furthermore, eye tracking has been used to quantify level of cognition and affective engagement (Kinner et al., 2017; Miller, 2015). There is some potential for this technology to offer insights into representational competence (Ubben, Nitz, Daniel, & Upmeier zu Belzen, 2018). Still, few studies have explored biology students' biometrics related to the use of scientific visualizations (Daniel & Leone, 2017; Jarodzka et al., 2010; Novick et al., 2012).

Additional eye-tracking studies conducted in science education disciplines have primarily focused on the text- and photo-based materials that are used in textbooks (Hannus & Hyönä, 1999), PowerPoint™ slides (Slykhuis et al., 2005), text segments (Mason, Pluchino, Tornatora, & Ariasi, 2013; Mason, Tornatora, & Pluchino, 2013), problem questions (Tsai et al., 2012), educational presentations (She & Chen, 2009), and tests/assessments (Tai et al., 2006). In relation to these materials, numerous variables (e.g., fixation duration, gaze patterns, information processing, learning achievements) were analyzed across a wide variety of age groups.

Fixation duration is of great significance across many of these studies. An increase in the duration of fixation on science materials is associated with increased attention (Chen et al., 2014; Cheng, 2014; Chien et al., 2015; She & Chen, 2009), deeper cognitive processes (Chien et al., 2015; She & Chen, 2009), and identifying and focusing on relevant information (Hannus & Hyönä, 1999; Jarodzka et al., 2010; Tsai et al., 2012). However, an increase in the duration of fixation may also indicate that a person is struggling to understand or process the material (Mason, Pluchino, Tornatora, & Ariasi, 2013; Mason, Tornatora, & Pluchino, 2013), and/or may result in more successful learning outcomes (Chen et al., 2014; Cheng, 2014; Mason, Pluchino, Tornatora, & Ariasi, 2013; Mason, Tornatora, & Pluchino, 2013). In terms of expert-novice fixation durations, the expertise of a subject was not always found to impact fixation durations (Tai et al., 2006).

Regarding gaze patterns and visual information processing, previous studies found that incorporating images alongside scientific text impacted participants' gaze patterns (Mason, Tornatora, & Pluchino, 2013; Slykhuis et al., 2005).

Using eye tracking to track the success of multimedia learning has also grown in popularity recently. She and Chen (2009) explored the relation between learning outcomes and eye movement patterns while participants viewed meiosis and mitosis multimedia materials. The authors found that different styles of instruction during learning experiences affected learning outcomes. Jarodzka et al. (2010) also utilized multimedia to explore expertise differences when observing and interpreting science-based videos and developing instructional methods. The authors found differences in foci, viewing times, and processing speeds between those identified as novices versus experts.

6.3 Principles of Eye Tracking

Eye tracking is a method that can be used to measure how a person visually interacts with a representation (Miller, 2015). Data from eye tracking provide a dynamic view of human focus and attention (Lai et al., 2013). The procedure involves the use of infrared technology to capture biometric pupil responses as a person visually accesses images (Holmqvist et al., 2011), obtaining measures such as fixations, gaze path, dwell time, fixation counts, saccades, and pupil dilation (Holmqvist et al., 2011; Morimoto & Mimica, 2005). A fixation is defined as when the eye stays focused on one location or on one subject or object for a predetermined length of time (Duchowski, 2017), such as at least 200 milliseconds (Slykhuis et al., 2005). For analysis, these locations are often theory-driven and are defined as areas of interest (AOIs) or designated spaces inside specified borders focused around explicit features of the visualization that are being investigated. AOIs can be dynamic or stationary in nature and can vary depending on project needs and the theoretical basis (Holmqvist et al., 2011). Gaze paths refer to the order in which the participant accesses a part of the visualization in relation to the AOI. Dwell time represents the length of time a participant fixates on an AOI. Fixation counts represent how many different times the participant fixates on a given AOI. Saccades refer to the rapid movement of the eye as the gaze path shifts between fixations. Pupil dilation indices reflect the number of unusual increases in pupil size per second, which can indicate the presence of cognitive (Marshall, 2007) and/or emotional mental processes (Kinner et al., 2017). Two major types of cameras can be used to capture visual data: a bar style tracking device designed to capture data presented on a computer screen and a wearable eyeglass-like tracker designed to capture natural viewing behaviors in more real-world-style environments (Mele & Federici, 2012). Bar cameras are best suited for investigations wherein the researcher wants to capture data regarding how a participant interacts with a static or dynamic image, video, or model that can be shown on a computer screen. Eyeglass-like trackers are particularly beneficial for exploring understanding with physical manipulatives and other non-virtual models.

Visual data can be grouped on the basis of methodological and theoretical frameworks depending on the needs of a study. A common framework separates data into two categories including "types of eye movement" and "scales of measurement" (Lai et al., 2013). Types of eye movement are qualitative in nature and typically focus on fixations, saccades, and pupillary dilation. Scales of measurement are quantitative in nature and typically fall into three major categories for addressing temporal, spatial, and count measures. Temporal measures indicate at which time point and for which time span a person views something, spatial measures indicate where a person is looking, and count measures indicate how often an individual looks at a specific AOI (Lai et al., 2013). Together, these data (i.e., "types of eye movement" and "scales of measurement") provide a holistic view of a person's physical and cognitive experience during the collection of eye-tracking data. From a theoretical perspective, the collection and analysis of eye-tracking data typically rely on one or more assumptions (Gegenfurtner, Lehtinen, & Säljö, 2011; Lai et al., 2013). One of the earliest and most widely accepted assumptions regarding eye-tracking data is the eye-mind assumption (Guan, Lee, Cuddihy, & Ramey, 2006; Just & Carpenter, 1980), which suggests a link between a visual focus and paying attention and/or cognitive workload (Just & Carpenter, 1980). Theoretically, the longer a person pauses or looks at something, the larger the processing load is during that time or the deeper the level of cognitive engagement (Cheng, 2014; Just & Carpenter, 1980).

In contrast to well-tried methods in science education research, such as interviews and think-aloud methods, eye-tracking data reflect cognitive processes such as perception (Gerrig & Zimbardo, 2002). They do so online and without filtering processes such as verbalization. This unique feature of eye tracking might open up new possibilities in the field of assessing and diagnosing modeling competence. However, applying eye tracking does not exclude verbal methods but – quite the contrary – necessitates triangulation. In a study on chemical representations, data from CTA were found to be correlated with participants' eye movements so that missing or unclear data from one method could be complemented with the other. Additionally, two data sources can offer a more comprehensive view on how people deal with representations (Stieff et al., 2011). Besides CTA, van Gog et al. (2005) suggested cued RTA for eye-tracking triangulation, offering nearly the same information as CTA. One advantage of this cued RTA might be the lower cognitive load during the task than found in CTA (Bojko, 2013) and also technical issues such as participants moving their heads away from the infrared cameras when getting a prompt. Hence, choosing CTA or cued RTA to triangulate with eye tracking requires the weighing of several factors such as the characteristics of the task and the eye-tracking device used.

6.4 Implications of Eye Tracking for Modeling Competence

We cognitively process what we are looking at. This simplified aggregation of the eye-mind assumption (Just & Carpenter, 1980) is at least valid for complex tasks (Rayner, 1998). If we assume that most students are at quite a low level of modeling competence (e.g., Grünkorn et al., 2014), tasks including models should comply with this requirement of complexity. Furthermore, nearly all model objects trigger visual interaction with superficial features such as different colors, materials, lines, or written words to name just a few. This superficial approach to the model object might literally at first glance correspond with a low level of modeling competence. However, if an individual is able to make sense of superficial features and to recognize the meaning of lines in a diagram or words and text, this might stand for at least an intermediate level (e.g., Level II in the FMC by Upmeier zu Belzen & Krüger, 2010). Here, the importance of triangulation comes into play because eye-tracking data per se do not explain why people fixate on a certain AOI. Hence, in order to distinguish levels of modeling competence, verbal data are indispensable. Conversely, this also means that – as mentioned by Stieff et al. (2011) – triangulation sheds light on the cognitive processes involved in modeling competence from different perspectives.

To distinguish different levels of modeling competence, a first step is to investigate the eye-movement patterns of individuals on different levels of modeling competence.[4] For this purpose, open-ended questionnaires (Chap. 7) can initially be used to assess modeling competence. Subsequently, participants work on several tasks that involve models so that eye-movement and verbal data can be collected. Knowing participants' level of modeling competence from pre-tests and combining them with eye tracking and verbal data might allow for the categorizing of eye movements and reasoning patterns corresponding with modeling competence. In turn, this would be the basis for diagnosing modeling competence using eye-movement data when performing tasks on a model. Especially in the case of hands-on tasks (e.g., Orsenne, 2015), eye-tracking glasses are a promising tool for investigating the handling of a 3-D model object. In contrast to remote eye-tracking devices, data can be obtained during the direct manipulation of a model and during modeling processes.

One way to foster modeling competence might be eye movement modeling examples (EMMEs; e.g., Jarodzka, van Gog, Dorr, Scheiter, & Gerjets, 2013; van Gog, Jarodzka, Scheiter, Gerjets, & Paas, 2009) where expert eye movements are displayed to novices so they can learn how to visually interact with a model. For example, these EMMEs could help students learn how to read phylogenetic trees as models of and for evolution. Ubben, Nitz, and Upmeier zu Belzen (2017, 2019) triangulated eye tracking with CTA in order to compare tree reading skills between pre-service teachers (novices) and experts depending on the use of phylogenetic trees as the medium or method. Here, the relevance for education about evolution and hence for biology education in general becomes obvious because insights into complex cognitive processes can lead to the development of new ways to foster modeling competence.

[4] Ubben, Nitz, Daniel, and Upmeier zu Belzen (2018) proposed a detailed approach for how to assess levels of representational competence using the example of phylogenetic trees.

6.5 Conclusion

As shown in this chapter, verbal methods such as interviewing and think-aloud methods offer a rich amount of data and can be performed simultaneously with modeling processes or the handling of a model. Even though recording and analyzing verbal data is time consuming, and information might get lost due to methodological drawbacks, every method – if chosen adequately – offers its benefits for a certain research goal such as eliciting participants' perspectives on models and modeling in interviews or getting insights into actions performed on models via think-aloud methods without influencing the participant. Combined with eye-tracking technology, aspects of modeling competence can be investigated even more comprehensively and from even more perspectives: Where do participants look when they handle a certain model object? Which features attract attention and – probably most important – which conclusions do participants draw from superficial features, and which reasoning processes are connected to noticing specific parts of a model? Of course, this method produces a large amount of extra data, and researchers should not underestimate the possibility that confounding variables such as the graphical arrangement of the representation of a model make the design of eye-tracking studies one of the most important points. Here, researchers should carefully scrutinize which model objects are chosen, how they are represented, and which measures are chosen for analysis. Besides these challenges, eye tracking makes it possible to directly compare aspects of modeling competence between groups such as novices and experts, students from different age groups, and pre-service and experienced teachers, to name just a few. The next step is to use these detailed insights to develop eye-tracking tools to diagnose and assess modeling competence or to foster it with EMMEs from the expert handling of models – always in combination with verbal methods. Even though eye tracking is still in its fledgling stages in science education, it should be seen as a chance to enrich knowledge about modeling competence by combining well-established methods with comparatively new ones.

References

American Educational Research Association, American Psychological Association, & National Council on Measurement in Education. (2014). *Standards for educational and psychological testing*. Washington, DC: American Educational Research Association.

Bojko, A. (2013). *Eye tracking the user experience: A practical guide to research*. Brooklyn, NY: Rosenfeld Media.

Brinkmann, S. (2014). Interview. In *Encyclopedia of critical psychology* (pp. 1008–1010). New York: Springer.

Carey, S., Evans, R., Honda, M., Jay, E., & Unger, C. (1989). An experiment is when you try it and see if it works: A study of grade 7 students' understanding of the construction of scientific knowledge. *International Journal of Science Education, 11*(5).

Charters, E. (2003). The use of think-aloud methods in qualitative research: An introduction to think-aloud methods. *Brock Education Journal, 12*(2).

Chen, S. C., She, H. C., Chuang, M. H., Wu, J. Y., Tsai, J. L., & Jung, T. P. (2014). Eye movements predict students' computer-based assessment performance of physics concepts in different presentation modalities. *Computers & Education, 74*, 61–72.

Cheng, W. K. (2014). *Relationship between visual attention and flow experience in a serious educational game: An eye tracking analysis*. Doctoral dissertation, George Mason University.

Chien, K. P., Tsai, C. Y., Chen, H. L., Chang, W. H., & Chen, S. (2015). Learning differences and eye fixation patterns in virtual and physical science laboratories. *Computers & Education, 82*, 191–201.

Crawford, B., & Cullin, M. (2005). Dynamic assessments of preservice teachers' knowledge of models and modelling. In K. Boersma, M. Goedhart, O. de Jong, & H. Eijkelhof (Eds.), *Research and the quality of science education* (pp. 309–323). Dordrecht, the Netherlands: Springer.

Daniel, K.L. & Leone, E.A. (2017, June). Student visual interpretations of a cladogram. Poster presented at the AP Biology Posters on the Prairie symposium, Kansas City, MO, USA.

Duchowski, A. (2017). *Eye tracking methodology* (3rd ed.). Cham, Switzerland: Springer.

Edwards, R., & Holland, J. (2013). *What is qualitative interviewing? What is? Research methods series*. London/New Delhi, India: Bloomsbury.

Ericsson, K. A., & Simon, H. A. (1993). Verbal reports as data. *Psychological Review, 87*(3), 215–251.

France, B., Compton, V. J., & Gilbert, J. K. (2011). Understanding modelling in technology and science: The potential of stories from the field. *International Journal of Technology and Design Education, 21*(3), 381–394.

Gegenfurtner, A., Lehtinen, E., & Säljö, R. (2011). Expertise differences in the comprehension of visualizations: A meta-analysis of eye-tracking research in professional domains. *Educational Psychology Review, 23*(4), 523–552.

Gerrig, R. J., & Zimbardo, P. G. (2002). *Psychology and life* (16th ed.). Boston, MA: Allyn and Bacon.

Gilbert, J., & Justi, R. (2016). Approaches to modelling-based teaching. In *Modelling-based teaching in science education. Models and modeling in science education: Vol. 9*. (pp. 57–80). Cham, Switzerland: Springer.

Grosslight, L., Unger, C., & Jay, E. (1991). Understanding models and their use in science: Conceptions of middle and high school students and experts. *Journal of Research in Science Teaching, 28*(9), 799–822.

Grünkorn, J., Upmeier zu Belzen, A., & Krüger, D. (2014). Assessing students' understandings of biological models and their use in science to evaluate a theoretical framework. *International Journal of Science Education, 36*(10), 1651–1684.

Guan, Z., Lee, S., Cuddihy, E., & Ramey, J. (2006, April). The validity of the stimulated retrospective thinkaloud method as measured by eye tracking. In CHI '06: Proceedings of the SIGCHI conference on human factors in computing systems (pp. 1253–1262). New York: ACM Press.

Hannus, M., & Hyönä, J. (1999). Utilization of illustrations during learning of science textbook passages among low-and high-ability children. *Contemporary Educational Psychology, 24*(2), 95–123.

Harrison, A. G., & Treagust, D. F. (2000). A typology of school science models. *International Journal of Science Education, 22*(9), 1011–1026.

Hegarty, M., Mayer, R. E., & Green, C. E. (1992). Comprehension of arithmetic word problems: Evidence from students' eye fixations. *Journal of Educational Psychology, 84*(1), 76.

Holmqvist, K., Nyström, M., Andersson, R., Dewhurst, R., Jarodzka, H., & van de Weijer, J. (2011). *Eye tracking: A comprehensive guide to methods and measures*. Oxford, UK/New York: Oxford University Press.

Jarodzka, H., Scheiter, K., Gerjets, P., & van Gog, T. (2010). In the eyes of the beholder: How experts and novices interpret dynamic stimuli. *Learning and Instruction, 20*(2), 146–154.

Jarodzka, H., van Gog, T., Dorr, M., Scheiter, K., & Gerjets, P. (2013). Learning to see: Guiding students' attention via a Model's eye movements fosters learning. *Learning and Instruction, 25*, 62–70.

Just, M. A., & Carpenter, P. A. (1980). A theory of reading: From eye fixations to comprehension. *Psychological Review, 87*, 329–354.

Justi, R., & Gilbert, J. (2003). Teachers' views on the nature of models. *International Journal of Science Education, 25*(11), 1369–1386.

Justi, R. S., & Gilbert, J. K. (2002). Modelling, teachers' views on the nature of modelling, and implications for the education of modellers. *International Journal of Science Education, 24*(4), 369–387.

Kinner, V. L., Kuchinke, L., Dierolf, A. M., Merz, C. J., Otto, T., & Wolf, O. T. (2017). What our eyes tell us about feelings: Tracking pupillary responses during emotion regulation processes. *Psychophysiology, 54*(4), 508–518.

Kvale, S., & Brinkmann, S. (2009). *Learning the craft of qualitative research interviewing*. Thousand Oaks, CA: SAGE Publications.

Lai, M.-L., Tsai, M.-J., Yang, F.-Y., Hsu, C.-Y., Liu, T.-C., Lee, S. W.-Y., et al. (2013). A review of using eye-tracking technology in exploring learning from 2000 to 2012. *Educational Research Review, 10*, 90–115.

Legrenzi, P., Girotto, V., & Johnson-Laird, P. N. (1993). Focussing in reasoning and decision making. *Cognition, 49*(1), 37–66.

Marshall, S. P. (2007). Identifying cognitive state from eye metrics. *Aviation, Space, and Environmental Medicine, 78*(5), 165–175.

Mason, J. (2002). *Qualitative researching* (2nd ed.). London/Thousand Oaks, CA: SAGE Publications.

Mason, L., Pluchino, P., Tornatora, M. C., & Ariasi, N. (2013). An eye-tracking study of learning from science text with concrete and abstract illustrations. *The Journal of Experimental Education, 81*(3), 356–384.

Mason, L., Tornatora, M. C., & Pluchino, P. (2013). Do fourth graders integrate text and picture in processing and learning from an illustrated science text? Evidence from eye-movement patterns. *Computers & Education, 60*, 95–109.

Mathesius, S., Upmeier zu Belzen, A., & Krüger, D. (2018). Eyetracking als Methode zur Untersuchung von Lösungsprozessen bei Multiple-Choice-Aufgaben zum wissenschaftlichen Denken [Eye tracking as a method for investigating the solving processes involved in multiple-choice tasks in scientific reasoning]. In M. Hammann & M. Lindner (Eds.), *Lehr- und Lernforschung in der Biologiedidaktik*. Innsbruck, Germany: Studienverlag.

Mayring, P. (2014). *Qualitative content analysis: theoretical foundation, basic procedures and software solution*. Klagenfurt, Austria.

Mayring, P. (2016). *Einführung in die qualitative Sozialforschung: Eine Anleitung zu qualitativem Denken* [Introduction to qualitative social research] (6th ed.). Weinheim, Germany/Basel, Switzerland: Beltz Verlag.

Mele, M. L., & Federici, S. (2012). Gaze and eye-tracking solutions for psychological research. *Cognitive Processing, 13*(1), 261–265.

Miller, B. W. (2015). Using reading times and eye-movements to measure cognitive engagement. *Educational Psychologist, 50*(1), 31–42.

Morimoto, C. H., & Mimica, M. R. (2005). Eye gaze tracking techniques for interactive applications. *Computer Vision and Image Understanding, 98*(1), 4–24.

Novick, L. R., Stull, A. T., & Catley, K. M. (2012). Reading phylogenetic trees: The effects of tree orientation and text processing on comprehension. *Bioscience, 62*(8), 757–764.

Olson, G. M., Duffy, S. A., & Mack, R. L. (1984). Thinking-out-loud as a method for studying real-time comprehension processes. In D. E. Kieras & M. A. Just (Eds.), *New methods in reading comprehension research* (pp. 253–286). Hillsdale, NJ: Erlbaum.

Orsenne, J. (2015). *Aktivierung von Schülervorstellungen zu Modellen durch praktische Tätigkeiten der Modellbildung* [Activating students' conceptions of models with hands-on modeling tasks]. Dissertation, Humboldt-Universität zu Berlin, Berlin, Germany.

Randhawa, B. (1994). Theory, research, and assessment of mathematical problem solving. *The Alberta Journal of Educational Research, 40*(2), 213–231.

Rayner, K. (1998). Eye movements in reading and information processing. 20 years of research. *Psychological Bulletin, 124*(3), 372–422.

Rayner, K. (2009). The 35th Sir Frederick Bartlett Lecture: Eye movements and attention in reading, scene perception, and visual search. *Quarterly Journal of Experimental Psychology, 62*(8), 1457–1506.

Rayner, K., Chace, K. H., Slattery, T. J., & Ashby, J. (2006). Eye movements as reflections of comprehension processes in reading. *Scientific Studies of Reading, 10*(3), 241–255.

Seidman, I. (2008). *Interviewing as qualitative research: A guide for researchers in education and the social sciences* (3rd ed.). New York/London: Teachers College Press.

She, H. C., & Chen, Y. Z. (2009). The impact of multimedia effect on science learning: Evidence from eye movements. *Computers & Education, 53*, 1297–1307.

Slykhuis, D. A., Wiebe, E. N., & Annetta, L. A. (2005). Eye-tracking students' attention to PowerPoint photographs in a science education setting. *Journal of Science Education and Technology, 14*(5), 509–520.

Stieff, M., Hegarty, M., & Deslongchamps, G. (2011). Identifying representational competence with multi-representational displays. *Cognition and Instruction, 29*(1), 123–145.

Tai, R. H., Loehr, J. F., & Brigham, F. J. (2006). An exploration of the use of eye-gaze tracking to study problem-solving on standardized science assessments. *International Journal of Research & Method in Education, 29*(2), 185–208.

Taylor, K. L., & Dionne, J. P. (2000). Accessing problem-solving strategy knowledge: The complementary use of concurrent verbal protocols and retrospective debriefing. *Journal of Educational Psychology, 92*(3), 413.

Terzer, E., Patzke, C., & Upmeier zu Belzen, A. (2012). Validierung von multiple-choice items zur Modellkompetenz durch lautes Denken [Validation of multiple choice items on modeling competence through think aloud]. In U. Harms & F. X. Bogner (Eds.), *Lehr- und Lernforschung in der Biologiedidaktik*. Innsbruck, Austria: Studienverlag.

Treagust, D. F., Chittleborough, G., & Mamiala, T. L. (2002). Students' understanding of the role of scientific models in learning science. *International Journal of Science Education, 24*(4), 357–368.

Trier, U., Krüger, D., & Upmeier zu Belzen, A. (2014). Students' versus scientists' conceptions of models and modelling. In D. Krüger & M. Ekborg (Eds.), *Research in biological education* (pp. 103–115). Brunswick, Germany: Westermann.

Tsai, M. J., Hou, H. T., Lai, M. L., Liu, W. Y., & Yang, F. Y. (2012). Visual attention for solving multiple-choice science problem: An eye-tracking analysis. *Computers & Education, 58*, 375–385.

Ubben, I., Nitz, S., Daniel, K. L., & Upmeier zu Belzen, A. (2018). Assessing representational competence with eye tracking technology. In K. L. Daniel (Ed.), *Towards a framework for representational competence in science education*. New York: Springer.

Ubben, I., Nitz, S., & Upmeier zu Belzen, A. (2017, August). *Visual perception of phylogenetic trees as models of and for evolution*. Poster at ESERA 2017 Conference, Ireland.

Ubben, I., Nitz, S., & Upmeier zu Belzen, A. (2019, August). *Influence of experience on tree Reading performance*. Italy: Poster at ESERA 2019 Conference.

Upmeier zu Belzen, A., & Krüger, D. (2010). Modellkompetenz im Biologieunterricht [Modeling competence in biology education]. *ZfDN, 16*, 41–57.

van Gog, T., Jarodzka, H., Scheiter, K., Gerjets, P., & Paas, F. (2009). Attention guidance during example study via the model's eye movements. *Computers in Human Behavior, 25*(3), 785–791.

van Gog, T., Paas, F., van Merriënboer, J. J. G., & Witte, P. (2005). Uncovering the problem-solving process: Cued retrospective reporting versus concurrent and retrospective reporting. *Journal of Experimental Psychology. Applied, 11*(4), 237–244.

van Someren, M. W., Barnard, Y. F., & Sandberg, J. A. C. (1994). *The think aloud method: A practical guide to modelling cognitive processes*. London: Academic.

Wilson, J., & Clarke, D. (2004). Towards the modelling of mathematical metacognition. *Mathematics Education Research Journal, 16*(2), 25–48.

Chapter 7
Assessing Modeling Competence with Questionnaires

Sabrina Mathesius and Moritz Krell

7.1 Introduction

The development of abilities related to models and modeling is one goal of science education on different educational levels in various countries all over the world (e.g. Australia: VCAA, 2016; Germany: KMK, 2005; USA: NGSS Lead States, 2013). Consequently, the development and evaluation of assessment instruments focusing on the different aspects of the framework for modeling competence (FMC; Chap. 1) are one important goal of science education research (cf. Nicolaou & Constantinou, 2014). Here, different methodological approaches have been applied, ranging from performance-assessment to closed-ended tasks. This chapter aims to provide an overview of studies that have employed instruments with either open-ended tasks or closed-ended tasks as a way to elicit individuals' abilities with respect to models and modeling. The aim of the chapter is to provide researchers in science education with a summary of instruments that have been proposed for the assessment of modeling competence and to discuss the advantages and limitations of each instrument on the basis of current standards for educational assessment (cf. AERA, APA, & NCME, 2014; Kane, 2013; Shavelson, 2013).

S. Mathesius (✉) · M. Krell
Freie Universität Berlin, Berlin, Germany
e-mail: Sabrina.Mathesius@fu-berlin.de

A. Upmeier zu Belzen et al. (eds.), *Towards a Competence-Based View on Models and Modeling in Science Education*, Models and Modeling in Science Education 12, https://doi.org/10.1007/978-3-030-30255-9_7

7.2 Questionnaires as Tools for Assessing Modeling Competence

Taking into account the FMC, which includes aspects and levels as possible parts of the progression of learning, there is a need for appropriate instruments for assessing individuals' abilities with respect to models and the modeling process in science. Using such instruments as diagnostic tools can help teachers improve students' learning opportunities and makes individual support possible (cf. Oh & Oh, 2011). The development and rigorous evaluation of assessment instruments with respect to competencies as highlighted in standard documents is critically important because of the possible consequences of testing for the participants but also because it was found that teachers tend to focus on "competencies specific to assessment and testing procedures" (Osborne, 2013, p. 267) in their lessons.

Shavelson (2013) proposed an approach for assessing competencies and evaluating the quality of test scores. This approach is in line with current standards for educational assessment (cf. AERA et al., 2014) and will therefore be used to illustrate crucial aspects of the assessment of modeling competence. Shavelson (2013) conceptualized competence assessment as a triangle with the construct, observation, and interpretation as its vertices. In relation to modeling competence, this means that a clear definition of this competence (*the construct*), a thorough understanding of the nature of the data gathered with an instrument (*observation*), and legitimate inferences based on these data (*interpretation*) are necessary.

The construct: By definition, competencies are complex and latent constructs that are not directly observable; an inference from an observable performance to an individual's competence has to be made (Shavelson, 2013). Modeling competence in science education is understood as a multidimensional construct (Nicolaou & Constantinou, 2014), comprising abilities to engage in modeling practices as well as knowledge about models and the modeling process in science ("meta-modeling knowledge"). Some definitions additionally include motivational aspects (e.g. Upmeier zu Belzen & Krüger, 2010). Furthermore, meta-modeling knowledge is usually subdivided into different aspects, each including hierarchical levels of understanding (cf. Krell, Upmeier zu Belzen, & Krüger, 2014a). Typically, the following aspects are considered: Describing the extent to which a model looks like the corresponding original, explaining reasons for multiple models, judging the purpose of a model, explaining how one can test a model, and demonstrating the reasons to change a model (cf. Nicolaou & Constantinou, 2014; Gilbert & Justi, 2016; Krell, Upmeier zu Belzen, & Krüger, 2016). Consequently, researchers have to define precisely which aspect of this complex construct is to be assessed.

Observation: Observation means an individual's performance on a set of tasks, where the "universe of possible tasks and responses for observing performance, […] logically follows from the definition of the construct" (Shavelson, 2013, p. 78). However, in relation to the assessment of modeling competence, there is still a large universe of possible tasks, containing, for example, different test formats (e.g. performance-assessment, open-ended tasks, or closed-ended tasks) and different task contexts, both of which can influence the cognitive demands of a task and, consequently, the nature of the observed performance (cf. Krell, Upmeier zu Belzen, & Krüger, 2014b; Martinez, 1999).

Interpretation: Interpretation refers to the question of the extent to which valid inferences from observed performance to (the level of) an individual's competence can be drawn (Shavelson, 2013). The interpretation of test scores, especially in relation to complex constructs such as modeling competence, means generalizing from some scores to an individual's competence. For this generalization to be valid, the tasks have to be representative of "the entire universe of tasks" that are suitable for assessing the targeted construct (Shavelson, 2013, p. 79). This is important, for example, for the operationalization of the construct: The interpretation of test scores on the basis of tasks that have been developed for assessing meta-modeling knowledge as indicators of individuals' modeling competence may be questioned because modeling competence is not only comprised of meta-modeling but also the ability to engage in modeling practices and, depending on the definition, motivational aspects (Nicolaou & Constantinou, 2014; Upmeier zu Belzen & Krüger, 2010). Hence, the evaluation of the validity of the proposed interpretation of test scores is critical and complex, and different sources of evidence are usually needed to support the claim that the proposed inferences from test scores to an individual's competence are valid (e.g. evidence based on test content, response processes, relations to other variables, or internal structure; AERA et al., 2014). This is why "the evidence required for validation is the evidence needed to evaluate the claims being made" (Kane, 2015, p. 64). Gathering evidence based on *test content* hereby means analyzing the relation between the construct and observed performance, which is often a starting point for constructing questionnaires. Sources of evidence based on *test content* often consist of expert judgments. With respect to the assessment of modeling competence, it is necessary, for example, to ask why specific test formats and task contexts have been chosen and to what extent these decisions influence the intended interpretation of the test scores (cf. Krell et al., 2014b; Martinez, 1999). Gathering evidence on the basis of *response processes* takes into account individuals' reasoning while answering the tasks in order to evaluate the extent to which the expected skills and knowledge are de facto initiated (Leighton, 2004). The sources of this process are often interviews and think-aloud protocols. Gathering evidence based on *relations to other variables* means considering relevant external variables, for example, test scores from other assessments or categorical variables such as different subsamples (known groups). Furthermore, quality criteria such as objectivity and reliability are necessary prerequisites for the valid interpretation of test scores (AERA et al., 2014), and replication studies can contribute to consolidating validity arguments (cf. Borrmann, Reinhardt, Krell, & Krüger, 2014). The current concept of validity includes aspects of reliability and fairness in testing as part of the criteria that offer evidence of a sufficient *internal structure*.

7.2.1 Aims and Procedures for Analyzing Questionnaires Designed to Assess Modeling Competence

In the following, published instruments that are used to assess modeling competence will be analyzed and discussed on the basis of the ideas about competence assessments sketched out above. The publications under consideration were selected by using the Google scholar database to search the archives of five science education journals: *Journal of Research in Science Teaching* (2016 Impact Factor 3.179), *Science Education* (2.506), *International Journal of Science and Mathematics Education* (1.474), *Research in Science Education* (1.329), and *International Journal of Science Education* (1.240). The following word combinations were used: Questionnaire AND (model(l)ing OR meta model(l)ing knowledge OR model competence OR scientific models OR models in science OR model(l)ing processes) (cf. Campbell, Oh, Maughn, Kiriazis, & Zuwallack, 2015; Nicolaou & Constantinou, 2014). In addition, reference lists of pertinent articles were searched as well as articles from key authors in the field. Only articles that explicitly described instruments that were designed to assess (aspects of) FMC in adequate detail were considered.

7.2.2 Results of the Review, or: How Is Modeling Competence Assessed in Science Education?

In the following, the identified studies are summarized on the basis of the three aspects of *the construct* (Fig. 7.1), *observation* (task context and test format; Fig. 7.2), and *interpretation* (sources of evidence; Shavelson, 2013; Fig. 7.2). In addition, sample information is provided (Fig. 7.2).

7.2.2.1 The Construct

The assessed constructs were diverse, but some aspects of meta-modeling knowledge were considered in many studies (e.g. *nature of models*, *purpose of models*; Fig. 7.1). One reason for this partial consensus regarding the assessed construct may be that many authors (e.g. Crawford & Cullin, 2005; Treagust, Chittleborough, & Mamiala, 2002; van Driel & Verloop, 1999) explicitly referred to the study by Grosslight, Unger, Jay, and Smith (1991), which can therefore be seen as seminal for research on models and modeling in science education. Nonetheless, both the abstract de-contextualized approach (Krell et al., 2014b; Sins, Savelsbergh, van Joolingen, & van Hout Wolters, 2009) and the global levels of understanding (Crawford & Cullin, 2005; Krell et al., 2014a) proposed by Grosslight et al. (1991) have been critically discussed, leading to more differentiated theoretical frameworks (e.g. Crawford & Cullin, 2005; Krell et al., 2014a).

	Study	Construct					
		NAT	ALT	PUR	TES	CHA	Others
1	Crawford & Cullin (2004)	X	X	X	X	X	
2	Crawford & Cullin (2005)	X	X	X	X	X	
3	Everett et al. (2009)	X	X	X	X	X	Understanding of models in science
4	Grünkorn et al. (2014)	X	X	X	X	X	
5	Justi & Gilbert (2005)	X	X	X	X	X	
6	Justi & van Driel (2005)	X	X	X	X	X	
7	Krell (2012)	X	X	X	X	X	
8	Krell et al. (2014a)	X	X	X	X	X	
9	Krell et al. (2014b)	X	X	X	X	X	
10	Krell et al. (2015)	X	X	X	X	X	
11	Krell & Krüger (2016)	X	X	X	X	X	
12	Lin (2014)	X	X	X	X	X	Knowledge of model functions and modeling process
13	Schwarz & White (2005)	X	X	X	X	X	Meta-modeling knowledge
14	Terzer (2013)	X	X	X	X	X	
15	Treagust et al. (2004)	X	X	X	X	X	Views of models and modeling in science
16	van der Valk et al. (2007)	X	X	X	X	X	Creativity
17	Cheng & Lin (2015)	X	X	X		X	Understanding of models in science
18	Derman & Kayacan (2017)	X	X	X		X	Understanding of models in science
19	Gobert et al. (2011)	X	X	X		X	Understanding of models in science
20	Lee (2017)	X	X	X		X	Types of representations of models
21	Treagust et al. (2002)	X	X	X		X	Understanding of models in science
22	Wei et al. (2014)	X	X	X		X	Understanding of models in science
23	Lee et al. (2017)	X	X	X			Types of representations of models
24	Chittleborough et al. (2005)	X	X		X	X	Role of models in science and learning
25	Borrmann et al. (2014)	X		X	X	X	Types of representations of models
26	Gogolin et al. (2017)	X		X	X	X	
27	Sins et al. (2009)	X		X	X	X	Epistemological understanding of models and of modeling processes
28	van Driel & Verloop (1999)	X		X	X	X	Types of representations of models
29	Danusso et al. (2010)	X		X			Knowledge about scientific models and modeling
30	Gogolin (2017)	X		X			Meta-modeling knowledge
31	Krell & Krüger (2017)			X	X	X	Meta-modeling knowledge
32	Patzke et al. (2015)			X	X	X	
33	Krell et al. (2012)			X			
34	Al-Balushi (2011)						Epistemologies about the credibility of scientific models
35	Bamberger & Davis (2013)						Modeling performances

Fig. 7.1 Constructs assessed in the reviewed studies. *NAT* Nature of models; *ALT* Alternative models; *PUR* Purpose of models; *TES* Testing models; *CHA* Changing models (cf. Krell et al., 2016)

Figure 7.1 also shows that many researchers called their construct *meta-modeling knowledge* (or similar), referring to the seminal study by Schwarz and White (2005) and highlighting the procedural role of modeling as a scientific practice (e.g. Crawford & Cullin, 2005). Others emphasized the role of models as types of scientific knowledge and called their construct, for example, an *understanding of scientific models* (e.g. Treagust et al., 2002). Some researchers included both, resulting in

Study	Observation: Task context Contextualized + De-contextualized	Observation: Test format (no. of tasks)	Interpretation: Sources of evidence	Sample
1	De-contextualized	O (8)	Response processes, other variables	14 P
2	De-contextualized	O (8)	Response processes, other variables, replication study (1)	17 P
3	Various models + De-contextualized	O (6), concept maps, short answer, RS (27; SUMS), yes-no tasks (9)	Interrater agreement (O: = 80 %, CM: = 97.5%), internal structure, other variables, replication study (1, 21)	>200 P
4	Biological models	O (15)	Test content, interrater agreement (.81 ≤ κ ≤ .90), internal structure	1177 S
5	Chemical models, historical models + De-contextualized	RS (74; VOMM A), O (not reported; VOMM C)	Test content, response processes, other variables	19 P (RS), 74 T (RS), 63 P (O)
6	Teaching and historical models + De-contextualized	O (not reported; VOMM C)	Response processes, internal structure, other variables, replication study (5)	5 T
7	Biological models	FC (30)	Internal structure, EAP/PV reliability = .69	901 S
8	Biological models	FC (30)	Test content, internal structure, other variables	1180 S
9	Biological models + De-contextualized	RS (35)	Test content, internal structure, EAP/PV reliability = .26-.34	1349 S
10	Biological models, chemical models, physical models + De-contextualized	FC (20), O (20, short answer)	Test content, response processes, other variables	617 S (FC), 115 S (O)
11	De-contextualized	O (5)	Interrater agreement (.63 ≤ κ ≤ .87), other variables	148 T
12	De-contextualized	RS (16); web-based	Test content, response processes, internal structure, Cronbach's α = .92, other variables	187 T
13	Various models + De-contextualized	MC (18), true-false questions, categorization task	Response processes, internal structure, Cronbach's α = .19-.80, other variables	>72 S
14	Biological models	MC (40)	Test content, response processes, internal structure, EAP/PV reliability = .28-.38, other variables	1136 S
15	Molecular representations + De-contextualized	RS (44), MC + justification (6; VOMMS)	Test content, internal structure, Cronbach's α = .69-.85 (Molecular representations), Cronbach's α = .87 (VOMMS), other variables	36 S
16	Statements about common features of models	correct-incorrect + explanation (10)	Test content	77 E
17	De-contextualized	RS (27; SUMS)	Internal structure, Cronbach's α = .72-.81, other variables, replication study (21)	402 S
18	Model examples + De-contextualized	RS (26+4; SUMS)	Internal structure, Cronbach's α = .74, other variables, replication study (21)	76 P
19	De-contextualized	RS (26; SUMS)	Internal structure, Cronbach's α = .56-.86, other variables, replication study (21)	736 S
20	Biological models + De-contextualized	RS (36); web-based	Test content, internal structure, person reliability = .84, other variables	983 S
21	De-contextualized	RS (27; SUMS)	Test content, internal structure, Cronbach's α = .71-.84	228 S
22	De-contextualized	RS (27; SUMS)	Internal structure, EAP/PV reliability = .59-.68, replication study (21)	629 S
23	Biological models	O (27); web-based, true-false-questions	Response processes, interrater agreement (.80-.83), other variables	189 S
24	De-contextualized	MC + justification (5; VOMMS)	Test content, Cronbach's α = .87, other variables, replication study (15)	275 S
25	De-contextualized	RS (32)	Internal structure, Cronbach's α = .53-.68, other variables, replication study (28)	226 T
26	Biological models	FC (36)	Response processes, other variables	107 S

Fig. 7.2 Observation (task context and test format), interpretation (sources of evidence), and sample information from the reviewed studies (*Note*: Test format: *O* open-ended tasks; *MC* multiple-choice tasks; *FC* forced-choice tasks; *RS* rating scale tasks; Sample: *S* students from different school grade levels; *P* pre-service science teachers; *T* in-service science teachers; *U* university students; *E* experts; for replication studies, the replicated study is provided in parentheses. ∗ sample size mentioned in abstract $N = 1207$)

27	De-contextualized	O (10); web-based	Interrater agreement (κ = .70), other variables	26 S
28	De-contextualized	O (7), RS (32)	Test content, internal structure, Cronbach's α = .64-.75, other variables	15 T (O), 71 T (RS)
29	Real phenomena + De-contextualized	O (3+3), MC (3)	Test content, response processes, other variables, replication study (Pintó & Gutierrez, 2005)	180+115+ 93 P
30	Biological models	FC (12), O (2)	Test content, response processes, internal structure, EAP/PV reliability$_{FC}$ = .51, EAP/PV reliability$_{O}$ = .55, other variables	382 S
31	Respondents' subject of study, one chosen model known in this subject	O (6)	Interrater agreement (.64 ≤ κ ≤ .92), other variables	184 U
32	Biological models	MC (25), O (9)	Test content, internal structure, EAP/PV = .58-.75, other variables, replication study (4, 14)	514 S
33	Biological models + De-contextualized	FC (7)	Test content	1209 S*
34	Natural entities and phenomena that are located at different points along the concrete-abstract continuum	MC (19-30); specific versions for each grade level	Test content, other variables	845 S, 108 P
35	Models (smell, evaporation, friction)	O drawing (3)	Test content, interrater agreement (>.80), other variables	65 S

Fig. 7.2 (continued)

constructs such as *views of models and modeling in science* (e.g. Treagust, Chittleborough, & Mamiala, 2004). However, a closer look at the respective studies revealed that, independent of the name of the construct, most researchers included aspects related to both modeling as a practice and models as types of knowledge in their frameworks (e.g. Crawford & Cullin, 2005; Treagust et al., 2002). Therefore, if researchers want to refer to other studies, it is critically important not to rely on the given label of the construct but to precisely examine the operationalization in terms of the assessment instrument.

It is evident that the vast majority of studies included in Fig. 7.1 are related to meta-knowledge (about models, modeling, or both) but that the elements of the practice have largely been neglected (cf. Nicolaou & Constantinou, 2014). However, this neglect may be a result of the focus of this article on written assessments with questionnaires (Chap. 6).

7.2.2.2 Observation

As one aspect of observation, the abovementioned criticism of the abstract de-contextualized approach by Grosslight et al. (1991) resulted in contextualized assessments that explicitly referred to specific models or situations (e.g. Grünkorn, Upmeier zu Belzen, & Krüger, 2014). Studies have shown that the assessment context may significantly affect respondents' answers (e.g. Al-Balushi, 2011; Krell, Upmeier zu Belzen, & Krüger, 2012). These findings suggest that it is not valid to generalize observations that are based on assessments as indicators of respondents' overall meta-modeling knowledge (or similarly named constructs; see above) as long as the effect of the included contexts is not fully understood and considered (cf. Shavelson, 2013).

As another aspect of observation, the chosen task format should be considered because it can influence the cognitive demands of an assessment (Martinez, 1999). In the studies included in Fig. 7.2, open-ended task formats were chosen most often ($n = 16$), followed by rating scales ($n = 13$), multiple-choice tasks ($n = 7$), and forced-choice tasks ($n = 6$). Some researchers combined different formats, especially open-ended and rating scale tasks. The prevalence of task formats corresponds with the popularity of established instruments. For example, many researchers adopted the "Students' Understanding of Models in Science" (SUMS) questionnaire developed by Treagust et al. (2002), which uses rating scale tasks (e.g. Gobert et al., 2011).

7.2.2.3 Interpretation

The evaluation of the validity of inferences being made is a necessary prerequisite for the interpretation of assessment observations (Shavelson, 2013), and different sources of evidence have been proposed for this reason (AERA et al., 2014; Kane, 2015). In the studies shown in Fig. 7.2, evidence based on *test content* was considered most often ($n = 19$), for example, by conducting expert reviews of the developed instruments and judging whether the tasks adequately represent the construct (e.g. Chittleborough, Treagust, Mamiala, & Mocerino, 2005; Lin, 2014; van der Valk, van Driel, & de Vos, 2007). In addition, it should be noted that all questionnaires of the reviewed studies are based on a theoretical framework. Evidence based on *response processes* was considered in $n = 12$ studies, for example, by conducting concurrent (e.g. "thinking aloud"; Gogolin et al., 2017) or retrospective interviews (Justi & Gilbert, 2005; Lin, 2014). Reliability estimates (as evidence based on *internal structure*) were provided in many studies, for example, for all proposed rating scale instruments (e.g. van Driel & Verloop, 1999). Although not always explicitly treated in this way, evidence of validity based on *relations to other variables* was provided in some studies. For example, Cheng and Lin (2015) compared students' results on the SUMS questionnaire (Treagust et al., 2002) with their science learning performance and found significant positive correlations, which can be interpreted as validity evidence because it is assumed that an epistemological understanding supports the learning of science concepts (Schwarz & White, 2005).

Another important source of evidence is the implementation of replication studies (cf. Borrmann et al., 2014). Fig. 7.2 proposes that there are four instruments that have been subjected to replication studies so far: The SUMS questionnaire (Treagust et al., 2002; replicated by, e.g. Gobert et al., 2011), the questionnaire about "Models and Modeling in Science" (van Driel & Verloop, 1999; replicated by Borrmann et al., 2014), the "My Views of Models and Modeling in Science" (VOMMS) questionnaire (Treagust et al., 2004; replicated by Chittleborough et al., 2005), and the "Views on Models and Modeling C" (VOMM C) questionnaire (Justi & Gilbert, 2005; replicated by Justi & van Driel, 2005). However, only one instrument, the SUMS questionnaire, seems to be established because it has been used in several studies so far (Fig. 7.2).

7.3 Conclusion and Discussion

As stated above, validity is a fundamental requirement for the interpretation of assessment observations (Shavelson, 2013; Kane, 2013), and it "refers to the degree to which evidence and theory support the interpretations of test scores for proposed uses of tests" (AERA et al., 2014, p. 11). Kane (2013) further argued that researchers have to critically demonstrate the validity of test interpretations on the basis of a variety of evidence, especially by considering the evidence that potentially threatens the intended interpretation (cf. *falsificationism*). On the basis of the present review, it can be concluded that there are hardly any questionnaires for the assessment of modeling competence (or selected aspects) that meet these requirements (cf. Nicolaou & Constantinou, 2014). This conclusion is in line with Osborne (2013), who offered the criticism that there is a lack of evidence supporting the validity of questionnaires for assessing scientific reasoning competencies. Thus, the community needs to put more effort into the systematic evaluation of questionnaires. Two exceptional studies can be highlighted here: The SUMS questionnaire (Treagust et al., 2002) was adopted and evaluated by different researchers, resulting in validity evidence based on samples with different educational and cultural backgrounds (Cheng & Lin, 2015; Derman & Kayacan, 2017; Everett, Otto, & Luera, 2009; Gobert et al., 2011; Treagust et al., 2002; Wei et al. 2014). Furthermore, Gogolin (2017) systematically evaluated her instrument in line with the AERA et al. (2014) standards, resulting in a forced-choice questionnaire suitable for assessing 11th- to 12th-graders' meta-modeling knowledge. However, even this instrument does not take into account the influence of different task contexts on students' responses.

As discussed above, modeling competence is conceptualized as comprising abilities to engage in modeling practices, as well as knowledge about models and the modeling process in science ("meta-modeling knowledge"). Many instruments included in this review focus on single aspects of FMC, especially on the knowledge dimension of competence, and have been developed to assess, for example, students' understanding of models in science (Treagust et al., 2002) or students' meta-modeling knowledge (Gogolin, 2017). As mentioned above, the interpretation of such test scores on the basis of such tasks as indicators of individuals' modeling competence may be questioned because modeling competence not only comprises meta-modeling knowledge but also abilities to engage in modeling practices and, depending on the definition, motivational aspects (Nicolaou & Constantinou, 2014; Upmeier zu Belzen & Krüger, 2010). Therefore, the interpretation of such test scores as indicators of individuals' modeling competence would require a powerful argument for validity about, for example, meta-modeling knowledge strongly contributing to or being a prerequisite for engaging in modeling practices. This assumption has been made in the science education literature (e.g. Schwarz & White, 2005), but the empirical evidence has shown that there might not be a coherent relation between students' meta-modeling knowledge and the quality of their modeling practices (Chap. 9). Hence, depending on the goals of research, scholars have to be cautious about which instrument they choose.

One crucial aspect that is not yet understood by the research community is the influence of different task contexts on observed test performance (Al-Balushi, 2011; Krell et al., 2014b). This fundamentally calls into question the validity of existing questionnaires because the interpretation of test scores as indicators of respondents' competence levels means generalizing from "a person's performance on a small sample of tasks [...] the level of competence in the full domain" (Shavelson, 2013, p. 80). As Shavelson (2013) further emphasized, this generalization requires that the tasks on an instrument are representative of the whole universe of tasks that are suitable for assessing the targeted construct. Therefore, as long as the research community only knows that there is an effect of task contexts on test performance but is not able to explain or predict this effect, we will not be able to claim representativity, and thus, we will not be able to make valid generalizations from test scores (Krell et al., 2014b).

Another crucial aspect that directly concerns the focus of this review on written assessments is the chosen task format. In line with the argument of test score interpretation as a generalization (Shavelson, 2013), the task format is important, too. Following the established conceptualization of modeling competence as a multidimensional construct, comprising abilities to engage in modeling practices, as well as knowledge about models and the modeling process in science, the aspect of meta-modeling knowledge seems to be "over-evaluated" (Nicolaou & Constantinou, 2014, p. 72), and it makes sense to ask: To what extent is it valid to argue that modeling competence can be assessed with questionnaires at all? Hence, Nicolaou and Constantinou (2014) concluded that there is a need "for a more explicit and more coherent theoretical framework for assessing knowledge, practices and processes related to the modeling competence" (p. 72).

Finally, it is important to mention that many studies included in this review were conducted before the argument-based approach for validation had been established in science education research (AERA et al., 2014; Kane, 2013; Shavelson, 2013). Most of the scholars involved in these studies did excellent work that was in line with the current standards of test development at the time. However, from a contemporary point of view, more research is clearly necessary for developing and evaluating scales and questionnaires for the assessment of the different aspects of the FMC.

References

AERA, APA, & NCME (Eds.). (2014). *Standards for educational and psychological testing*. Washington, DC: American Educational Research Association.

Al-Balushi, S. M. (2011). Students' evaluation of the credibility of scientific models that represent natural entities and phenomena. *International Journal of Science and Mathematics Education, 9*, 571–601. https://doi.org/10.1007/s10763-010-9209-4

Bamberger, Y. M., & Davis, E. A. (2013). Middle-school science students' scientific modelling performances across content areas and within a learning progression. *International Journal of Science Education, 35*, 213–238. https://doi.org/10.1080/09500693.2011.624133

Borrmann, J. R., Reinhardt, N., Krell, M., & Krüger, D. (2014). Perspektiven von Lehrkräften über Modelle in den Naturwissenschaften: Eine generalisierende Replikationsstudie [Teachers perspectives' on models in science: A generalizing replication study]. *Erkenntnisweg Biologiedidaktik, 13*, 57–72.

Campbell, T., Oh, P. S., Maughn, M., Kiriazis, N., & Zuwallack, R. (2015). A review of modeling pedagogies: pedagogical functions, discursive acts, and technology in modeling instruction. *EURASIA Journal of Mathematics, Science & Technology Education, 11*, 159-176. https://doi.org/10.12973/eurasia.2015.1314a

Cheng, M.-F., & Lin, J.-L. (2015). Investigating the relationship between students' views of scientific models and their development of models. *International Journal of Science Education, 37*, 2453–2475. https://doi.org/10.1080/09500693.2015.1082671

Chittleborough, G. D., Treagust, D. F., Mamiala, T. L., & Mocerino, M. (2005). Students' perceptions of the role of models in the process of science and in the process of learning. *Research in Science & Technological Education, 23*, 195–212. https://doi.org/10.1080/02635140500266484

Crawford, B. A., & Cullin, M. J. (2004). Supporting prospective teachers' conceptions of modelling in science. *International Journal of Science Education, 26*, 1379–1401. https://doi.org/10.1080/09500690410001673775

Crawford, B. A., & Cullin, M. J. (2005). Dynamic assessments of preservice teachers' knowledge of models and modelling. In K. Boersma, M. Goedhart, O. de Jong, & H. Eijkelhof (Eds.), *Research and the quality of science education* (pp. 309–323). Dordrecht, The Netherlands, Springer. https://doi.org/10.1007/1-4020-3673-6_25

Danusso, L., Testa, I., & Vicentini, M. (2010). Improving prospective teachers' knowledge about scientific models and modelling: Design and evaluation of a teacher education intervention. *International Journal of Science Education, 32*, 871–905. https://doi.org/10.1080/09500690902833221

Derman, A., & Kayacan, K. (2017). Investigation of the relationship between the views of the prospective science teachers on the nature of scientific models and their achievement on the topic of atom. *European Journal of Education Studies, 3*, 541–559.

Everett, S. A., Otto, C. A., & Luera, G. R. (2009). Preservice elementary teachers' growth in knowledge of models in a science capstone course. *International Journal of Science and Mathematics Education, 7*, 1201–1225. https://doi.org/10.1007/s10763-009-9158-y

Gilbert, J. K., & Justi, R. (2016). *Modelling-based teaching in science education*. Cham, Switzerland: Springer.

Gobert, J. D., O'Dwyer, L., Horwitz, P., Buckley, B. C., Levy, S. T., & Wilensky, U. (2011). Examining the relationship between students' understanding of the nature of models and conceptual learning in biology, physics, and chemistry. *International Journal of Science Education, 33*, 653–684. https://doi.org/10.1080/09500691003720671

Gogolin, S. (2017). *Diagnosing students' meta-modelling knowledge: Gathering validity evidence during test development.* Doctoral dissertation. Freie Universität, Berlin, Germany. Retrieved from http://www.diss.fu-berlin.de/diss/receive/FUDISS_thesis_000000105919

Gogolin, S., Krell, M., Lange-Schubert, K., Hartinger, A., Upmeier Zu Belzen, A., & Krüger, D. (2017). Erfassung von Modellkompetenz bei Grundschüler/innen [Assessment of elementary students' modeling competence]. In H. Giest, A. Hartinger, & S. Tänzer (Eds.), *Vielperspektivität im Sachunterricht* (pp. 108–115). Bad Heilbrunn, Germany: Klinkhardt.

Grosslight, L., Unger, C., Jay, E., & Smith, C. L. (1991). Understanding models and their use in science: Conceptions of middle and high school students and experts. *Journal of Research in Science Teaching, 28*, 799–822. https://doi.org/10.1002/tea.3660280907

Grünkorn, J., Upmeier zu Belzen, A., & Krüger, D. (2014). Assessing students' understandings of biological models and their use in science to evaluate a theoretical framework. *International Journal of Science Education, 36*, 1651–1684. https://doi.org/10.1080/09500693.2013.873155

Justi, R., & Gilbert, J. K. (2005). Investigating teachers' ideas about models and modelling: Some issues of authenticity. In K. Boersma, M. Goedhart, O. de Jong, & H. Eijkelhof (Eds.), *Research and the quality of science education* (pp. 325–335). Dordrecht, The Netherlands: Springer.

Justi, R., & van Driel, J. (2005). The development of science teachers' knowledge on models and modelling: Promoting, characterizing, and understanding the process. *International Journal of Science Education, 27*, 549–573. https://doi.org/10.1080/0950069042000323773

Kane, M. T. (2013). Validating the interpretations and uses of test scores. *Journal of Educational Measurement, 50*, 1–73. https://doi.org/10.1111/jedm.2013.50.issue-1

Kane, M. (2015). Validation strategies. Delineating and validating proposed interpretations and uses of test scores. In M. Raymond, S. Lane, & T. Haladyna (Eds.), *Handbook of Test Development* (pp. 64-80). New York: Routledge.

KMK (Ed.). (2005). *Bildungsstandards im Fach Biologie für den Mittleren Schulabschluss [Biology education standards for the Mittlere Schulabschluss]*. München, Neuwied: Wolters Kluwer.

Krell, M. (2012). Using polytomous IRT models to evaluate theoretical levels of understanding models and modeling in biology education. *Science Education Review Letters. Theoretical Letters 2012*, 1–5.

Krell, M., & Krüger, D. (2016). Testing models: A key aspect to promote teaching activities related to models and modelling in biology lessons? *Journal of Biological Education, 50*, 160–173. https://doi.org/10.1080/00219266.2015.1028570

Krell, M., & Krüger, D. (2017). University students' meta-modelling knowledge. *Research in Science & Technological Education, 35*, 261–273. https://doi.org/10.1080/02635143.2016.1274724

Krell, M., Reinisch, B., & Krüger, D. (2015). Analyzing students' understanding of models and modeling referring to the disciplines biology, chemistry, and physics. *Research in Science Education, 45*, 367–393. https://doi.org/10.1007/s11165-014-9427-9

Krell, M., Upmeier zu Belzen, A., & Krüger, D. (2012). Students' understanding of the purpose of models in different biological contexts. *International Journal of Biology Education, 2*, 1–34. Retrieved from http://www.ijobed.com/2_2/Moritz-2012.pdf

Krell, M., Upmeier zu Belzen, A., & Krüger, D. (2014a). Students' levels of understanding models and modelling in biology: Global or aspect-dependent? *Research in Science Education, 44*, 109–132. https://doi.org/10.1007/s11165-013-9365-y

Krell, M., Upmeier zu Belzen, A., & Krüger, D. (2014b). Context-specificities in students' understanding of models and modelling: An issue of critical importance for both assessment and teaching. In C. Constantinou, N. Papadouris, & A. Hadjigeorgiou (Eds.), *E-book proceedings of the ESERA 2013 conference: Science education research for evidence-based teaching and coherence in learning* (pp. 1024–1035). Nicosia, Cyprus: European Science Education Research Association. Retrieved from https://www.esera.org/publications/esera-conference-proceedings/esera-2013

Krell, M., Upmeier zu Belzen, A., & Krüger, D. (2016). Modellkompetenz im Biologieunterricht [Modeling competence in biology education]. In A. Sandmann, & P. Schmiemann (Eds.), *Biologiedidaktische Forschung: Schwerpunkte und Forschungsstände* (pp. 83–102). Berlin, Germany: Logos.

Lee, S. W.-Y. (2017). Identifying the item hierarchy and charting the progression across grade levels: Surveying Taiwanese students' understanding of scientific models and modeling. *International Journal of Science and Mathematics Education*. https://doi.org/10.1007/s10763-017-9854-y

Lee, S. W.-Y., Chang, H.-Y., & Wu, H.-K. (2017). Students' views of scientific models and modeling: Do representational characteristics of models and students' educational levels matter? *Research in Science Education, 47*, 305–328. https://doi.org/10.1007/s11165-015-9502-x

Leighton, J. P. (2004). Avoiding misconception, misuse, and missed opportunities. The collection of verbal reports in educational achievement testing. *Educational Measurement: Issues and Practice, 23*, 6–15. https://doi.org/10.1111/j.1745-3992.2004.tb00164.x

Lin, J.-W. (2014). Elementary school teachers' knowledge of model functions and modeling processes: A comparison of science and non-science majors. *International Journal of Science and Mathematics Education, 12*, 1197–1220. https://doi.org/10.1007/s10763-013-9446-4

Martinez, M. E. (1999). Cognition and the question of test item format. *Educational Psychologist, 34*, 207–218.

NGSS Lead States (Eds.). (2013). *Next generation science standards: For states, by states*. Washington, DC: The National Academies Press.

Nicolaou, C. T., & Constantinou, C. P. (2014). Assessment of the modeling competence: A systematic review and synthesis of empirical research. *Educational Research Review, 13*, 52–73. https://doi.org/10.1016/j.edurev.2014.10.001

Oh, P. S., & Oh, S. J. (2011). What teachers of science need to know about models. An overview. *International Journal of Science Education, 33*, 1109–1130. https://doi.org/10.1080/09500693.2010.502191

Osborne, J. (2013). The 21st century challenge for science education: Assessing scientific reasoning. *Thinking Skills and Creativity, 10*, 265–279. https://doi.org/10.1016/j.tsc.2013.07.006

Patzke, C., Krüger, D., & Upmeier zu Belzen, A. (2015). Entwicklung von Modellkompetenz im Längsschnitt [Development of modeling competence in a longitudinal study]. In M. Hammann, J. Mayer, & N. Wellnitz (Eds.), *Lehr- und Lernforschung in der Biologiedidaktik: Vol. 6. Theorie, Empirie & Praxis: Internationale Tagung der Fachsektion Didaktik der Biologie im VBIO, Kassel 2013* (pp. 43–58). Innsbruck, Austria: Studienverlag.

Pintó, R., & Gutierrez, R. (2005). Teachers' conceptions of scientific models. In R. Pinto & D. Couso (Eds.), *Proceedings of the 5th international ESERA conference on contributions of research to enhancing students' interest in learning science* (pp. 866–868). Barcelona, Spain: ESERA.

Schwarz, C. V., & White, B. Y. (2005). Metamodeling knowledge: Developing students' understanding of scientific modeling. *Cognition and Instruction, 23*, 165–205. https://doi.org/10.1207/s1532690xci2302_1

Shavelson, R. J. (2013). On an approach to testing and modeling competence. *Educational Psychologist, 48*, 73–86. https://doi.org/10.1080/00461520.2013.779483

Sins, P. H. M., Savelsbergh, E. R., van Joolingen, W. R., & van Hout Wolters, B. (2009). The relation between students' epistemological understanding of computer models and their cognitive processing on a modelling task. *International Journal of Science Education, 31*, 1205–1229. https://doi.org/10.1080/09500690802192181.

Terzer, E. (2013). *Modellkompetenz im Kontext Biologieunterricht: Empirische Beschreibung von Modellkompetenz mithilfe von Multiple-Choice Items* [Modeling competence in the context of biology education]. Doctoral dissertation. Humboldt-Universität, Berlin, Germany. Retrieved from: http://edoc.hu-berlin.de/dissertationen/terzer-eva-2012-12-19/PDF/terzer.pdf

Treagust, D. F., Chittleborough, G., & Mamiala, T. L. (2002). Students' understanding of the role of scientific models in learning science. *International Journal of Science Education, 24*, 357–368. https://doi.org/10.1080/09500690110066485

Treagust, D. F., Chittleborough, G., & Mamiala, T. L. (2004). Students' understanding of the descriptive and predictive nature of teaching models in organic chemistry. *Research in Science Education, 34*, 1–20. https://doi.org/10.1023/B:RISE.0000020885.41497.ed

Upmeier zu Belzen, A., & Krüger, D. (2010). Modellkompetenz im Biologieunterricht [Model competence in biology education]. *Zeitschrift für Didaktik der Naturwissenschaften, 16*, 41–57.

Van Der Valk, T., Van Driel, J. H., & De Vos, W. (2007). Common characteristics of models in present-day scientific practice. *Research in Science Education, 37*, 469–488. https://doi.org/10.1007/s11165-006-9036-3

Van Driel, J. H., & Verloop, N. (1999). Teachers' knowledge of models and modelling in science. *International Journal of Science Education, 21*, 1141–1153. https://doi.org/10.1080/095006999290110

VCAA. (2016). *Victorian certificate of education. Biology*. Melbourne, VIC: VCAA.

Wei, S., Liu, X., & Jia, Y. (2014). Using Rasch measurement to validate the instrument of students' understanding of models in science (SUMS). *International Journal of Science and Mathematics Education, 12*, 1067–1082. https://doi.org/10.1007/s10763-013-9459-z

Chapter 8
Drawing-Based Modeling in Teaching Elementary Biology as a Diagnostic Tool

Wouter R. van Joolingen, Juliette Schouten, and Frank Leenaars

8.1 Introduction and Theoretical Background

An important goal of science education is to acquaint students with the goals and methods of science, including the roles and functions of models. According to several authors (DeBoer, 2000; Longbottom & Butler, 1998), not only must students learn the facts of science, but they must also learn *about* science. Students should adopt scientific thinking characteristics such as open-minded and critical thinking, problem solving, an understanding of the relation between theory and evidence, and hence, an understanding of the nature of scientific knowledge (Kuhn & Pearsall, 2000; Longbottom & Butler, 1998). As discussed in section A in this book and in other literature, modeling is an important element of such scientific reasoning, and modeling competence is the basis of science education (Louca & Zacharia, 2012; Magnani, Nersessian, & Thagard, 2012; Windschitl, Thompson, & Braaten, 2008).

When modeling is brought to the classroom, students can develop a scientific view on the world and engage in scientific thinking (Zimmerman, 2007). Science education should explicitly address the acquisition of these higher order skills so that students can develop the scientific literacy needed to be able to function as citizens in modern society. Such skills and literacy are generally considered to be part of the set of "21st century skills," even though there are many versions of what these skills entail (e.g. McComas, 2014).

Models play the role of objects that represent the relation between reasoning and reality. In science education, models can be used to explain scientific phenomena (e.g. as pictures or animations to display how enzymes work), as an object of study (e.g. using a computer simulation), and as objects that can be constructed and

W. R. van Joolingen (✉) · J. Schouten · F. Leenaars
Freudenthal Institute, Utrecht University, Utrecht, The Netherlands
e-mail: w.r.vanjoolingen@uu.nl

A. Upmeier zu Belzen et al. (eds.), *Towards a Competence-Based View on Models and Modeling in Science Education*, Models and Modeling in Science Education 12, https://doi.org/10.1007/978-3-030-30255-9_8

modified by students (de Jong & van Joolingen, 2008; Grosslight, Unger, Jay, & Smith, 1991).

Modeling in science education is often implemented as computer-based modeling in which students create executable models by writing programming code (Blikstein, Abrahamson, & Wilensky, 2005; Brady, Holbert, Soylu, Novak, & Wilensky, 2015) or differential or difference equations (Neves, Neves, & Teodoro, 2013; Teodoro & Neves, 2011), which are often supported by graphical representations that are based on system dynamics (Doerr, 1996; Milrad, 2002). For younger children who lack the language needed to understand code or equations, such modeling tools are out of reach, but other ways to specify computational models are possible. In the approach described in this chapter, *annotated drawings* are used to specify the behavior of the model. Drawings have been used to express and communicate knowledge in science education (Ainsworth, Prain, & Tytler, 2011), and in our context, drawings are used as the basis for the modeling program SimSketch (Bollen & van Joolingen, 2013; van Joolingen, Aukes, Gijlers, & Bollen, 2015). In this approach, students create drawings of scientific phenomena and, using the language of icons representing system behavior, convert these drawings into computational models. Furthermore, drawing-based modeling is very useful for early science education because drawing enables children to turn their spontaneous thoughts into more scientific concepts (Brooks, 2009).

In the current chapter, we investigate how teachers can integrate drawing-based modeling into their classroom practice and how the development of students' scientific reasoning can be diagnosed as a result of having them engage in drawing-based modeling with SimSketch. We provide a short introduction to scientific and model-based reasoning before discussing the method and results of our study.

8.1.1 Scientific Thinking and Modeling Competence

Scientific thinking skills are required to link empirical evidence to theoretical considerations (Kuhn, Amsel, & O'Loughlin, 1988). Zimmerman (2007) reviewed many studies that have been conducted on this subject. He then explained that scientific thinking is:

> *...the application of the methods or principles of scientific inquiry to reasoning or problem-solving situations, and involves the skills implicated in generating, testing and revising theories, and in the case of fully developed skills, to reflect on the process of knowledge acquisition and change (p. 173).*

Modeling involves these aspects of generating, testing, and revising theories because models can be seen as theoretical representations of phenomena. Scientific knowledge is a complex and dynamic network of models. Models are used to test hypotheses and describe scientific phenomena. Learning goals for modeling are related to the subject matter taught, to learning to model, and to the role of models in science. In modeling, students learn to discuss and criticize their thoughts about their model and to reflect on their model (Louca, Zacharia, & Constantinou, 2011). Modeling in science classes often takes the form of *computer-based* modeling. Using software tools such as Co-Lab (Van Joolingen et al., 2005) or NetLogo (Wilensky & Reisman, 2006), students create computer models of the phenomena they are investigating. If students learn to model at a young age, using relatively simple phenomena, we assume that this will be beneficial for their modeling education in later years, involving more complex scientific phenomena. Earlier research and reviews about learning by modeling have identified two reasons for why modeling has not gained ground in early education: the lack of tools and educational materials and teachers' lack of experience with using models for learning (Louca & Zacharia, 2012, 2014; Louca et al., 2011).

Drawing-based modeling is aimed at addressing these reasons by providing an easily accessible tool, *SimSketch* (Fig. 8.1), for teachers and students, without the need for more advanced modeling skills (Bollen & van Joolingen, 2013). In earlier studies, it was shown that learners from the age of 10 years old were capable of creating computational models in the domains of astronomy (van Joolingen, Aukes, Gijlers, & Bollen, 2015) and evolution (Heijnes, van Joolingen, & Leenaars, 2018). In the current study, we integrated this tool into a series of lessons and investigated how it functions in building modeling competence. In the lessons, students learn to create their own representation of a scientific concept, while finding out which resources will be relevant and reliable when used in their models. Drawing allows them to deal with different representations of the same scientific concept, which ensures creative reasoning (Ainsworth et al., 2011). Modeling with SimSketch takes the form of assigning behaviors to the elements of a drawing. For instance, in a model of evolution, the behavior of "reproducing" can be assigned to a drawn animal, resulting in offspring with a slightly mutated color. Predators can hunt and eat

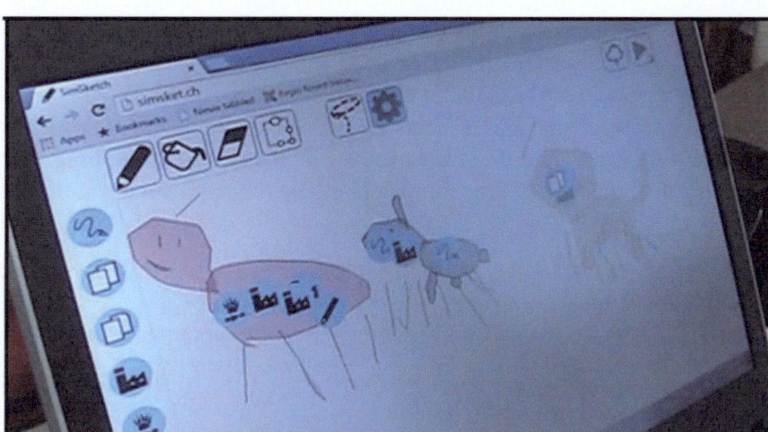

Fig. 8.1 Drawings in SimSketch of Dyad A during the practice lesson. In the online modeling and drawing tool SimSketch, different behaviors (e.g. reproduction, mutation, and hunting behavior) can be given to the objects

animals when they have been assigned the "hunting" behavior. Creating this representation is modeling in itself, but SimSketch adds the executable nature of a model, allowing for a quick cycle of constructing, evaluating, and revising activities that are closely connected to scientific thinking and the aspects of modeling competence. In this way, drawing-based modeling is a real scientific activity that may contribute to scientific thinking.

8.1.2 Assessing the Understanding of Models

Earlier research found that modeling can contribute to a better understanding of the nature of science (Louca et al., 2011; Sins, Savelsbergh, van Joolingen, & van Hout-Wolters, 2009b; Chap. 4). This in turn may lead to more proficient scientific thinking because modeling is an important part of scientists' work. When students learn more about modeling, they also gain more insights into the approach that scientists use. In this study, the level of understanding of models is used as an indicator of a shift in students' thoughts about the nature of science. Grünkorn, Upmeier zu Belzen, and Krüger (2014) developed a framework for assessing students' understanding of scientific models (Chap. 1). They used five categories: *nature of models*, *multiple models*, *purpose of models*, *testing models,* and *changing models*, reflecting both the epistemological aspects of models as representing scientific knowledge and the roles of models in science, including their relations with empirical evidence. Both aspects are important as the roles link models with scientific reasoning, and the epistemological aspects stress epistemological understanding, which influences students' cognitive processing on a modeling task (Sins et al., 2009b).

The aim of this study is to gain insight into the opportunities for using drawing-based modeling for teaching, learning, and assessment in science education, integrated in a classroom environment. The main research question is:

How can drawing-based modeling be used in a science education classroom to develop and assess students' modeling competence?

We focused on two specific aspects of modeling competence: (a) students' understanding of models and (b) the ways in which students' reasoning processes can be indicators of their modeling competence. The main question was divided into three partial questions:

1. How can we extract students' modeling competence from their behavior in creating SimSketch models?
2. How does drawing-based modeling change students' understanding of models?
3. How is student reasoning about drawing-based models related to their modeling competence?

As part of a design-based study design, we developed a series of lessons with the topic of the evolution of snails, with a target audience of ninth-grade students (14–15 year olds) in general secondary education. In order to address the link between empirical data and theory, we established a link with a large European science project. In four lessons, students learned to use the modeling tool, collected data, created models for explaining the data, and reflected on the modeling process. We expected this lesson series to result in students' deeper understanding of scientific reasoning and the nature and use of models.

8.2 Method

In the *design* phase of this study, learning goals were elaborated, and the lessons were designed in collaboration with teachers and domain experts. In the *implementation* phase, the lessons were taught in seven ninth-grade classes. Process and outcome measures were collected in order to assess students' modeling competence.

8.2.1 Design

The students learned to understand modeling in the context of the evolution of snails. Prior to the lesson series, students had acquired knowledge about evolution as part of their regular teaching program. The lesson series focused on the construction of models showing the dynamics of evolutionary processes.

The lesson series was developed in cooperation with Naturalis Biodiversity Centre in Leiden, the Netherlands. Naturalis participates in "The Evolution MegaLab," a European citizen science project designed by the Open University in the United Kingdom in 2009 (Worthington et al., 2012). On the website of the Evolution MegaLab, the color polymorphism of the snails is explained, and observations of the shell color, banding, and environment can be studied. Participants can also collect their own data by finding snails in their environment and can add their data to the database. An expert group was formed to discuss the contents of the lesson series. This expert group consisted of two experienced employees of Naturalis from the department "Educational development" as well as the first two authors of this chapter.

The lesson series was designed to achieve four learning goals related to the development of modeling competence. These goals were supposed to be met at the end of the lesson series:

1. The students are able to relate a model to the real situation (understanding the nature of models).
2. The students are able to evaluate the model they created (testing and changing models).
3. The students are aware of the fact that a model is not a copy of reality (understanding the nature of models as well as testing and changing them).

4. The students recognize similarities between their own method of working and the way scientists use models in their work (understanding the nature and purpose of models).

To summarize, these learning goals are related to modeling competence as described in Chap. 1, especially to understanding the *nature* of modeling (on levels II and III, learning goals 1 and 3) and to *testing* and *changing* models in relation to their purpose (on levels II and III, learning goals 2 and 3).

The lesson series consisted of four lessons, and separate learning goals were formulated for each lesson. The general construction of the lesson series was:

- Lesson 1: Introduction to modeling and practicing with SimSketch
- Lesson 2: Collecting data in the field about the evolution of the snail
- Lesson 3: Modeling the evolution of the snail with SimSketch
- Lesson 4: Reflecting on models and their connection to science

The lessons were adjusted to the test school, where each lesson took 45 min. The first versions of two of the four lessons (lessons 1 and 3) were given in a grade 10 class with 21 students. These two lessons included the most innovative aspect of the series: using SimSketch to create computational models. After the pilot, the lesson series was adapted on the basis of the students' and teacher's experiences.

8.2.2 Participants

Seven ninth-grade classes from one secondary school with a total of 204 students participated in this study. The classes belonged to two levels of general secondary education, four classes were part of "higher General education" (marked G below), and three classes were "Preparatory higher education" (marked P below). These classes were taught by three different teachers. An overview can be found in Fig. 8.2. Participation was obligatory for all students, and the exercises in their student man-

Teacher	Class	Number of students	Complete sets of modeling questions
1	G1	32	20
	G2	32	25
2	G3	29	10
	G4	29	10
	P1	28	17
	P2	28	8
3	P3	26	20
Total:	7 classes	204 students	110 sets

Fig. 8.2 Overview of the teachers, classes, number of students, complete sets of modeling questions on pre- and post-tests, and the videotaped dyads

ual were graded. Unfortunately, due to practical circumstances, only about half of the data sets were completed (Fig. 8.2). None of the students had previous experience with drawing-based modeling or with any other kind of computer-based modeling. We deleted the pre- and post-tests of students who did not finish, did not fill in the answers seriously, or submitted unreadable answers due to bad handwriting.

8.2.3 Conditions

In all lessons, the students worked in pairs or triplets. The conditions of the fieldwork differed between the classes. Five worked during a week with cold nights. Two classes went to the field a week later because it was damper and warmer outside, and there were probably more snails. In the final lesson, a Skype session with a scientist from Naturalis was planned to help students make the connection between models and their use in science. Four classes actually engaged in this session. The students in the other three classes were able to ask questions of the first author of this chapter, who was present at every lesson.

8.3 Data Collection and Analysis

8.3.1 Change in Students' Understanding of Models

We examined the effect of the lesson series on students' understanding of models using a pre- and post-test design. Eight open questions on the tests asked about students' understanding of models, thus modeling competence. These questions were used in a questionnaire from a previous study about the relation between students' epistemological understanding of computer models and their cognitive processing on a modeling task (Sins, Savelsbergh, van Joolingen, & van Hout-Wolters, 2009a). There were two questions each concerning the aspects of modeling competence (Fig. 1.3) except changing models. Questions addressed general modeling features such as "What is a model in your opinion? (nature)" and "Why do scientists use models? (purpose)" and also included reactions to statements such as "Scientists need to test their models (testing models)" and "It is impossible to determine which model is the best (multiple models)."

We developed a scoring system for these eight questions on the basis of the revised framework for students' understanding of models and their use in science, including levels of complexity and their categories (Grünkorn et al., 2014). In general, a level 1 understanding implies that students see models as simple copies of reality. At level 2, students realize that there are specific choices that they need to make to arrive at a suitable scientific model and that a model is a possible variant of reality. At level 3, students understand that models can be used to test hypotheses and that the modeler plays an active role in the modeling process (Grosslight et al.,

1991; Grünkorn et al., 2014; Sins et al., 2009a). We examined whether there was a shift in students' understanding between the pre-test and post-test. Indications of this shift would be a shift from modeling competence levels 1 and 2 to level 3, which is a shift from seeing models as media to seeing models as a research tool.

8.3.2 *Scientific Reasoning*

In addition to the quantitative data, we videotaped two pairs of students during the two lessons in which they worked with SimSketch in order to gain insights into their reasoning with the models and modeling tool. Video transcripts were analyzed for statements about scientific reasoning, and the SimSketch drawings students made were used to support our findings from the evaluation of students' scientific reasoning processes.

8.4 Results

In this section, we combine quantitative data from the pre- and post-tests with a qualitative analysis of the statements students made, their answers on the exercises, and the models they drew.

8.4.1 *Understanding of Models*

A change in students' understanding of models was tested by a statistical analysis of the pre- and post-tests the students ($N = 110$) had to fill out. Descriptive statistics of the test scores of all students can be found in Fig. 8.3, such as the test scores of subgroups consisting of G and P students.

			Total test score		
	Students	N	M (SD)	Median	95 % CI
Pre	All	110	6.92 (2.15)	7.0	[6.51, 7.32]
Post	All	110	8.32 (2.38)	8.5	[7.87, 8.77]
Pre	G	65	6.56 (2.34)	6.0	[5.85, 7.26]
Post	G	65	7.53 (2.32)	8.0	[6.84, 8.23]
Pre	P	45	7.47 (1.77)	8.0	[6.94, 8.00]
Post	P	45	9.42 (2.20)	9.0	[8.76, 10.08]

Fig. 8.3 Means, standard deviations, and confidence intervals for the total test scores of all students together, the G students, and the P students (*Note.* The maximum score of a test is 24)

8.4.2 The Progress of All Students

Because students' test scores were not normally distributed, a Wilcoxon signed-ranks test was used. The output indicated that the median post-test scores were statistically significantly higher than the median pre-test scores ($Z = 5.43$, $p < 0.001$). Cohen's d, a measure of effect size, ($d = 0.62$) suggested a medium effect, although the final level could not be considered very high.

8.4.3 Difference Between G and P Students

As expected, Mann Whitney U tests showed higher pre-test scores for P students than for G students ($U = 1022.00$, $p = 0.007$). However, the effect size ($d = 0.45$) suggested a small effect. On the post-test, a similar difference was found: ($U = 797.50$, $p < 0.001$). In this case, the effect size ($d = 0.83$) suggested a large effect. Separate Wilcoxon signed-ranks tests for the P and G groups showed a small gain for G students ($Z = 3.16$, $p = 0.002$, Cohen's $d = 0.44$) and a large effect for P students ($Z = 4.55$, $p < 0.001$, $d = 0.99$).

8.4.4 Progress per Aspects of Modeling Competence

The answers to the questions on the pre- and post-test with an open format were qualitatively categorized into the four aspects of the framework for modeling competence (Fig. 8.4).

Wilcoxon signed-ranks tests show that students scored significantly higher on questions on the post-test on the nature of models ($d = 0.74$) and multiple models ($d = 0.28$) than they did on the pre-test. No significant differences were found for the purpose of models and model testing.

				Wilcoxon signed-ranks test	
Aspect	n	Mdn pre-test	Mdn post-test	Z	p
Natureof models	110	2.00	3.00	5.47	< .001
Multiple models	110	1.00	2.00	2.28	.023
Purposeof models	110	2.00	2.00	1.76	.079
Testing models	110	2.00	2.00	.88	.380

Fig. 8.4 Medians for the test scores for the different aspects and output from the Wilcoxon signed-ranks test

8.4.5 Scientific Reasoning Process

Two pairs of students were videotaped during the first and the third lessons. During the first lesson, we could see how they practiced with SimSketch, and during the third lesson, we could see what elements of the model they drew, what they discussed about the model, and the amount of help they needed. In this section, we report the results of the analyses about statements, drawings, and exercises for these two dyads on the different levels of scientific reasoning (Fig. 8.4).

8.4.5.1 Dyad A: Class G1

In the first lesson, this dyad tried all the buttons available in SimSketch. For example, they drew a rabbit and a lion and gave them all kind of behaviors (Fig. 8.1).

In the third lesson, student 1 of this dyad took the lead. A few statements made by this student in this discussion can be read below. Student 2 did not respond and was distracted most of the time. All quotes were translated from Dutch into English.

Student 1:	"How can the snails suddenly have other colors?" (1.26 min.) […]
Student 1:	"All kinds of new species originated." (2.12 min.) […]
Student 1:	"They adapt to their environment, I like that." (3.35 min.) […]
Student 1:	"Miss, what should we do now?" (5.44 min.)
Teacher:	"You need to create a background, like a big colored surface, which is the forest." […]
Student 1:	"We need to draw a yellow and a brown or pink snail." (8.38 min.)
Student 1:	"Wow! New species originated." (13.26 min.) […]
Teacher:	"It looks good, but what is missing from your drawing?" (14.50 min.)
Student 1:	"A background and a bird."

The student's and teacher's discussion consisted of four subtopics. The student made two observations and had one idea of his own. The student did not elaborate on his observations. The student said that the snails adapted to their environment, even though he had not yet drawn an environment. Furthermore, the student incorrectly inferred that a new species had emerged. Help from the teacher was needed to get the student to think about other possible objects in his model, such as a background and a thrush. No further explanations or logical connections were made by this student. From this point of view, the student did not engage in higher levels of scientific reasoning. No improvement in modeling competence could be gleaned directly from the conversation with the teacher either.

The student drew a yellow and a brown snail, and he assigned behavior to the snails so they could move and split and change colors. A background (the surroundings) and the bird were missing from his drawing. In the preparatory scheme on the worksheet, they indicated that there should be two different kinds of backgrounds, so he understood all the aspects he was supposed to draw. This exercise was meant to get the student to first think about possible elements and behaviors in his model.

In general, after each simulation, this dyad as well as other dyads needed to change the behaviors or drawings in order to match their model with their own conclusions. Many students did not go beyond simply drawing some elements.

8.4.5.2 Dyad B: Class G3

This dyad practiced with SimSketch by drawing a rabbit and giving it all kinds of possible behaviors to see what happened. After some confusion about the exercises in the third lesson, they focused on modeling the evolution of the snail.

Student 1:	"You can see that green conceals better on green than green on orange." (21.42 min.) […]
Student 2:	"We have to make it complete." (26.35 min.) […]
Student 1:	"In fact a snail cannot really evolve." (29.29 min.) […]
Student 1:	"Now we are going to make a background." (30.34 min.) […]
Student 1:	"It must be a very simple model." (31.05 min.) […]
Student 1:	"The snail does not need to split right?" (33.00 min.)
Teacher:	"Splitting, or in this case reproduction, seems to be useful for natural selection. You need to make a background in which the bird sees the snail or not. The snail with the best camouflage colour stays alive and can reproduce himself." […]
Student 1:	"We need to make the background green, in order that the green snail can conceal better than the red snail." (36.10 min.) […]
Student 1:	"The bird eats the red snail and the green snail can split now." (41.25 min.)
Student 2:	"Actually the bird must split and eats snails. But the birds die eventually because there are no snails anymore."
Student 1:	"They now adapt to the background."

These students had some ideas about modeling, such as keeping it simple and making it as complete as possible. Student 1 in particular elaborated on his thoughts about the evolution of the snail. Student 1's statement that a snail cannot really evolve came out of the blue. Furthermore, they had some trouble with the behaviors they should use in SimSketch, such as splitting. The students thought about the environment they wanted to draw but again needed help from the teacher to really draw the background in SimSketch. Compared with the first dyad, this dyad engaged in higher levels of scientific reasoning, especially because they justified their ideas at the end of the lesson.

If we evaluate the exercises they completed on their worksheet, we can conclude that they gained a deeper understanding about how to model the evolution of the snail. This can be seen in Fig. 8.5. The students explained what kinds of changes they made in their model to arrive at a more realistic view. They understood that snails who adapted to the environment had a higher probability of surviving and reproducing, thus leading to more adapted snails.

Simulation round	Changes I made	Effects of the changes
2	Addition of a background	The snail is better camouflaged
3	Let the bird hunt	The bird eats the least camouflaged snail
4	The snail adapts	The bird does not see the snail anymore
5	Camouflaged snail repro-duces	More camouflaged snails appear
6	The snail mutates	More adapted snails appear,and they are quite safe from the birds

Fig. 8.5 The notes on changes to the model made by dyad B in the student guide

Despite the fact that some of the changes were formulated in an incorrect way (e.g. in the simulations, individual snails do not mutate, but a snail's offspring can have a slightly mutated color), students showed progress by developing increasingly adequate models.

8.5 Conclusion

The aim of the present study was to gain insight into the role of drawing-based modeling in supporting scientific thinking and to obtain an impression of students' modeling competence by inspecting their reasoning with the models. We did this by investigating the effects of the lesson series on students' understanding of models and their scientific reasoning processes, which in turn reflected students' level of understanding. From the two cases that we presented, we could see that this was not trivial. Only in the second of these cases did we see a clear reference to the purpose of the model (it should be simple) and the relationship to reality: "a snail cannot evolve." In contrast to this, the student from the first group was very task-directed when working on the 11 modeling task and showed no meta-modeling knowledge. In such a way, by inspecting the students' statements, we obtained information not only about their reasoning about the domain but also on the extent to which they understood models.

8.5.1 Understanding of Models and the Role of Models in Science

The lesson series contributed to a slight increase in students' understanding of models. Effects were significant but small. Students specifically scored higher on the questions about the nature of models and multiple models. This seemed strange because, during the building process, students would be expected to better understand the need to test models because they had to test the different versions of the models while building them. A possible cause may be that they do not see that kind of testing as part of the modeling process but only as a technical procedure that needs to be followed to reach a certain goal. This idea was partly confirmed by the students' logs and statements, which tended to focus on technical matters. The reflection lesson involving a question and answer session with the researcher focused more on the aspects on which the learners improved.

There was a clear difference between the scores achieved by the general education G group and the preparatory higher education P group. The students in the P group scored higher on both the pre-test and post-test than the G students. Especially on the post-test, the P students scored higher, which was confirmed by the large effect size. This conclusion was expected because P students were expected to have a deeper understanding. Furthermore, the P students' drawings and discussions were of higher quality on average. In general, they drew more elements related to the evolution of the snail in their models, and they made more adaptations to their models. Despite these findings, it is important to point out that the G students also achieved a significant shift in scores from their pre-test to post-test even though the scores were lower than the P students' scores. Although the G group showed only a small effect, drawing-based modeling can still be a useful learning method for G students as well.

Overall, the progress in modeling competence was small but measurable. Although this may look disappointing at first glance, it should be kept in mind that this progress occurred after only four lessons. This means that this small increase is encouraging as a basis for further research involving more extensive modeling activities. What is important is that measurable progress was made, and it could be measured with a relatively simple measurement instrument.

8.5.2 *Tracing Reasoning with Models*

The use of drawing-based modeling in a lesson series can be used to support students' model-based reasoning. Moreover, in combination with assessments of modeling competence (Grünkorn et al., 2014), we can get an indication of students' modeling competence by having them work with the modeling tool, by tracing their model, and by observing the changes and predictions they make about the effects of these changes.

However, the modeling activities that students' recorded on their worksheets sometimes showed that the teacher intervened in the modeling process by scaffolding the learning process. A teacher's support has an influence on the modeling task and the scientific reasoning processes of the students. Although help is needed to reach higher levels of scientific reasoning, as can be seen in the videotapes of the dyads, if given in a spontaneous way, it may blur insights into how students develop their own reasoning. Part of the support the teacher had to give was related to the changes students had to make in their models in order to get them up and running.

Although the students practiced with SimSketch for at least 20 min in the first lesson, it took a while before they understood the principle of modeling in the third lesson. This principle implies that students first drew a model with some basic elements and behaviors. Scaffolding and probably more time to complete the task potentially led to learners expressing more ideas, explanations, justifications, and elaborations on their scientific reasoning processes.

Overall, SimSketch proved to be a modeling tool that can be used to foster and study modeling processes for students in lower secondary education. Drawing-based modeling provides a way for students to create computational models before they are able to program or process mathematical equations. Students were able to create reasonable models and reason about them even though a large amount of time still had to be devoted to technical issues with the tool, and scaffolding was required. To use the environment as a way to assess students' modeling competence, it is important to take into account the number and type of scaffolds given by the teacher. Despite this fact, students' models and reasoning logs in SimSketch will provide teachers and researchers with valuable insight into the development of students' modeling competence.

References

Ainsworth, S., Prain, V., & Tytler, R. (2011). Drawing to learn in science. *Science, 333*(6046), 1096–1097.

Blikstein, P., Abrahamson, D., & Wilensky, U. (2005). NetLogo: Where we are, where we're going. In M. Eisenberg & A. Eisenberg (Eds.), *Paper presented at the annual meeting of Interaction Design and Children*. Boulder, Colorado.

Bollen, L., & van Joolingen, W. (2013). SimSketch: Multi-agent simulations based on learner-created sketches for early science education. *IEEE Transaction on Learning Technologies, 6*(3), 208–216.

Brady, C., Holbert, N., Soylu, F., Novak, M., & Wilensky, U. (2015). Sandboxes for model-based inquiry. *Journal of Science Education and Technology, 24*(2–3), 265–286. https://doi.org/10.1007/s10956-014-9506-8

Brooks, M. (2009). Drawing, visualisation and young Children's exploration of "big ideas". *International Journal of Science Education, 31*(3), 319–341.

de Jong, T., & van Joolingen, W. (2008). Model-facilitated learning BT – Model-facilitated learning. In M. Spector (Ed.), *Model-facilitated learning* (pp. 457–468). Handbook of research on educational communications and technology.

DeBoer, G. E. (2000). Scientific literacy: Another look at its historical and contemporary meanings and its relationship to science education reform. *Journal of Research in Science Teaching, 37*(6), 582–601.

Doerr, H. M. (1996). Stella ten years later: A review of the literature. *International Journal of Computers for Mathematical Learning, 1*, 201–224.

Grosslight, L., Unger, C., Jay, E., & Smith, C. L. (1991). Understanding models and their use in science: Conceptions of middle and high school students and experts. *Journal of Research in Science Teaching, 28*(9), 799–822.

Grünkorn, J., Upmeier zu Belzen, A., & Kruger, D. (2014). Assessing students' understandings of biological models and their use in science to evaluate a theoretical framework. *International Journal of Science Education, 36*(10), 1651–1684.

Heijnes, D., van Joolingen, W., & Leenaars, F. (2018). Stimulating scientific reasoning with drawing-based modeling. *Journal of Science Education and Technology, 27*(1), 45–56.

Kuhn, D., Amsel, E., & O'Loughlin, M. (1988). *The development of scientific thinking skills. Developmental psychology series*. New York: Academic.
Kuhn, D., & Pearsall, S. (2000). Developmental origins of scientific thinking. *Journal of Cognition and Development, 1*(1), 113–129.
Longbottom, J. E., & Butler, P. H. (1998). Why teach science? Setting rational goals for science education. *Science Education, 83*, 473–492.
Louca, L. T., & Zacharia, Z. C. (2012). Modeling-based learning in science education: Cognitive, metacognitive, social, material and epistemological contributions. *Educational Review, 64*(4), 471–492.
Louca, L. T., & Zacharia, Z. C. (2014). Examining learning through modeling in K-6 science education. *Journal of Science Education and Technology, 24*, 192–215.
Louca, L. T., Zacharia, Z. C., & Constantinou, C. P. (2011). In Quest of productive modeling-based learning discourse in elementary school science. *Journal of Research in Science Teaching, 48*(8), 919–951. 20435.
Magnani, L., Nersessian, N., & Thagard, P. (2012). *Model-based reasoning in scientific discovery*. Berlin/Heidelberg, Germany: Springer.
McComas, W. F. (Ed.). (2014). *The language of science education*. Rotterdam, The Netherlands: Sense Publishers. https://doi.org/10.1007/978-94-6209-497-0_1
Milrad, M. (2002). Using construction kits, modeling tools and system dynamics simulations to support collaborative learning. *Educational Technology & Society, 5*(4), 76–87.
Neves, R. G. M., Neves, M. C., & Teodoro, V. D. (2013). Modellus: Interactive computational modeling to improve teaching of physics in the geosciences. *Computers & Geosciences, 56*, 119–126.
Sins, P. H. M., Savelsbergh, E. R., van Joolingen, W. R., & van Hout-Wolters, B. H. A. M. (2009a). The relation between students' epistemological understanding of computer models and their cognitive processing on a modeling task. *International Journal of Science Education, 31*(9), 1205–1229.
Sins, P. H. M., Savelsbergh, E. R., van Joolingen, W. R., & van Hout-Wolters, B. H. A. M. (2009b). The relation between students' epistemological understanding of computer models and their cognitive processing on a modeling task. *International Journal of Science Education, 31*(9), 1205–1229.
Teodoro, V. D., & Neves, R. G. (2011). Mathematical modeling in science and mathematics education. *Computer Physics Communications, 182*(1), 8–10.
van Joolingen, W., Aukes, A. V. A., Gijlers, H., & Bollen, L. (2015). Understanding elementary astronomy by making drawing-based models. *Journal of Science Education and Technology, 24*, 256–264. Retrieved from http://link.springer.com/article/10.1007/s10956-014-9540-6/fulltext.html
van Joolingen, W., de Jong, T., Lazonder, A. W., Savelsbergh, E. R., & Manlove, S. (2005). Co-lab: Research and development of an online learning environment for collaborative scientific discovery learning. *Computers in Human Behavior, 21*(4), 671–688. Retrieved from http://www.sciencedirect.com/science/article/pii/S074756320400189X
Wilensky, U., & Reisman, K. (2006). Thinking like a wolf, a sheep, or a firefly: Learning biology through constructing and testing computational theories – An embodied modeling approach. *Cognition and Instruction, 24*(2), 171–209. Retrieved from http://search.ebscohost.com/login.aspx?direct=true&db=pbh&AN=20785410&site=ehost-live
Windschitl, M., Thompson, J., & Braaten, M. (2008). Beyond the scientific method: Model-based inquiry as a new paradigm of preference for school science investigations. *Science Education, 92*(5), 941–967.
Worthington, J. P., Silvertown, J., Cook, L., Cameron, R., Dodd, M., Greenwood, R. M., et al. (2012). Evolution MegaLab: A case study in citizen science methods. *Methods in Ecology and Evolution, 3*(2), 303–309.
Zimmerman, C. (2007). The development of scientific thinking skills in elementary and middle school. *Developmental Review, 27*(2), 172–223.

Chapter 9
The Black Box Approach: Analyzing Modeling Strategies

Moritz Krell and Susann Hergert

9.1 Introduction

As outlined in Chap. 1, modeling competence in science education is understood as a multidimensional construct, comprising abilities to perform modeling practices as well as knowledge about models and the modeling process in science ("meta-modeling knowledge"). Researchers have proposed a positive relationship between these two dimensions, suggesting that "metamodeling knowledge guides the practice [...], enabling students to more effectively plan and evaluate their investigations" (Schwarz et al., 2009, p. 635) and that engaging in modeling practices contributes to developing and deepening meta-modeling knowledge. There is some evidence that supports these ideas (e.g. Cheng & Lin, 2015; Gobert & Pallant, 2004; Jong, Chiu, & Chung, 2015; Schwarz & White, 2005). However, most studies have been correlational (e.g. Schwarz & White, 2005) and therefore do not allow causal inferences to be drawn. Recently published review articles (Louca & Zacharia, 2012; Nicolaou & Constantinou, 2014) have revealed that research on modeling competence tends to focus on the assessment of meta-modeling knowledge (e.g. Justi & Gilbert, 2003; Krell & Krüger, 2017; Schwarz & White, 2005). Furthermore, the quality of modeling processes has mostly been assessed post hoc by analyzing the appropriateness of modeling products (i.e. models) (e.g. Cheng & Lin, 2015; Jong et al., 2015). Consequently, Nicolaou and Constantinou (2014, p. 72); emphasized that "there is no completely coherent way to conceptualize or to assess modeling [processes]."

This contribution argues that the black box approach is suitable for conducting process-based analyses of modeling and for fostering modeling abilities when

M. Krell (✉) · S. Hergert
Freie Universität Berlin, Berlin, Germany
e-mail: moritz.krell@fu-berlin.de

A. Upmeier zu Belzen et al. (eds.), *Towards a Competence-Based View on Models and Modeling in Science Education*, Models and Modeling in Science Education 12, https://doi.org/10.1007/978-3-030-30255-9_9

additional guidance and opportunities for explicit reflections are provided. In the theoretical part of this contribution (Sects. 9.2 and 9.3), the appropriateness of the black box approach for diagnosing and fostering modeling abilities is explored. In the empirical part (Sects. 9.4 and 9.5), we will illustrate how the black box approach can be used to analyze pre-service science teachers' modeling strategies and to foster secondary school students' modeling competences. Whereas study 1 contributes to science education research by providing different modeling strategies and an instrument (category system) that can be used to analyze them, study 2 offers an instructional setting that can be adapted by practitioners in science education.

9.2 Modeling

In a simplified form, scientific modeling can be regarded as an iterative, cyclical process of developing and evaluating representations of phenomena with the aim of further investigating the phenomena under consideration (Clement, 1989; Giere et al., 2006; Krell, Upmeier zu Belzen, & Krüger, 2016). Model development is understood as a creative process in which analogy generation, metaphorical reasoning, thought experiments, and imagistic simulations occur (Bailer-Jones, 1999, 2009; Clement, 2009). On the basis of the modeler's knowledge and experiences, an initial model that represents selected parts or variables of the system is developed (Clement, 1989). The model then has to be evaluated with respect to its internal consistency and the extent to which it can provide an adequate representation of what was observed (Clement, 1989; Mahr, 2011). Thus, the model itself (i.e. the model object; Mahr, 2011) has to be (logically) consistent, and the model needs to be able to reproduce or to explain the phenomenon retrospectively. From this perspective, the model can be conceptualized as a medium for adequately representing selected parts of the system (*model of something*; Gouvea & Passmore, 2017; Krell et al., 2016; Mahr, 2011; Chap. 1). Furthermore, it is possible to deduce predictions about how the system should behave under certain conditions by mentally or materially manipulating the model (Giere et al., 2006; Godfrey-Smith, 2006). These predictions can be tested by conducting experiments or by making scientific observations. If the predictions turn out to be false, it is likely that the model does not fit the system (Giere et al., 2006; Godfrey-Smith, 2006). Consequently, the model has to be changed or rejected, and the evaluation of the model starts from the beginning (cyclical process). This leads to the evaluation of assumptions and to further insights about the underlying phenomenon. From this perspective, models can be conceptualized as tools for scientific reasoning (*model for something*; Gouvea & Passmore, 2017; Krell et al., 2016; Mahr, 2011; Chap. 1).

The strategy of scientific modeling can be summarized as follows:

> *The modeler's strategy is to gain an understanding of a complex real-world system via an understanding of a simpler hypothetical system that resembles it in relevant respects* (Godfrey-Smith, 2006, p. 726).

Standard documents in science education in various countries have emphasized that scientific modeling practices should be implemented in science classes (e.g. Australia: VCAA, 2016; Germany: KMK, 2005; USA: NGSS Lead States, 2013). Campbell and colleagues proposed five "modeling pedagogies" that can be applied in science classes (e.g. Campbell, Oh, Maughn, Kiriazis, & Zuwallack, 2015): *exploratory modeling* (investigating a pre-existing model), *expressive modeling* (developing a model to express ideas about a phenomenon), *experimental modeling* (deducing predictions from a model and testing them empirically), *evaluative modeling* (comparing and evaluating alternative models of/for the same original), and *cyclic modeling* (being engaged in the cyclical process of model development, evaluation, and modification). Studies have found that expressive and exploratory modeling are the most frequently used pedagogies in science education, whereas cyclic modeling is least often applied (Campbell et al., 2015; Krell & Krüger, 2016).

9.3 Modeling and the Black Box Approach in Science Education

One approach for initiating science practices – for example, modeling – in science classes is the black box approach (e.g. Koch, Krell, & Krüger, 2015; Ruebush, Sulikowski, & North, 2009). Hereby, a black box is an entity with an invisible internal system that can be investigated by manipulating the input and observing the resulting output. A generic definition of the term black box was proposed by Glanville (1982, p. 1):

> *Briefly, a black box can be characterized as: (a) being believed to be distinct, (b) having observable (and relatable) inputs and outputs, (c) being black (that is, opaque to the observer).*

Upmeier zu Belzen (2014) highlighted that a black box may be used in science education to represent elements of science and scientific practices on three different levels. On the first level, the black box represents a natural phenomenon, and the exploration of the black box represents the process of scientific discovery. On the second level, the black box and its exploration can be seen as an abstract representative of the nature of science, and reflections on the exploration of the black box provide opportunities to reflect on the nature of science (cf. Lederman & Abd-El-Khalick, 2002). On the third level, the process of exploring the black box can be regarded as a problem-solving process that is applied not only in the sciences but also in other scientific disciplines and everyday life (Upmeier zu Belzen, 2014).

Consequently, various black boxes are used in science education for different purposes. Most published approaches for using black boxes in science education have proposed that a black box can be used as a teaching/learning aid to foster conceptual knowledge (e.g. Berge, 2007; Chakrabarti et al., 2013) or knowledge about (the nature of) science (e.g. Abd-El-Khalick, 2002; Crowe, 1968; Ferstl & Schneider, 2007; Miller, 2014). Most of these articles have been related to physics education (e.g. Berge, 2007; Chakrabarti et al., 2013; Keller & Wang, 1994; Lietz, 2007). Only a few of the studies in which a black box was used for teaching/learning provided evidence for the efficacy of the approach. For example, Akerson et al. (2000) showed that a reflective, explicit, activity-based approach that included two black box activities successfully improved pre-service teachers' views of the nature of science. Other authors were successful in fostering subjects' views of models and modeling in science by means of black box activities (e.g. Cartier, 2000; Koch et al., 2015; Ruebush et al., 2009). Furthermore, some studies have provided evidence that students positively evaluate black box activities (e.g. Hildebrandt & Oliver, 2000; Kücük et al., 2011). Finally, some authors suggest that black boxes should be used for assessment/diagnostic purposes (e.g. assessment of lateral thinking skills (Arsad et al., 2012), problem solving skills (Bünder et al., 2006; Mie & Friege, 2004), or modeling strategies (Krell, Walzer, Hergert, & Krüger, 2017)). To summarize, black box approaches are used to achieve various educational goals (e.g. fostering conceptual knowledge, knowledge about science), but empirical evidence for the efficacy of the approaches is often missing.

This article focuses on the use of a water black box (MUSE, 2002) for assessing and fostering skills related to modeling competence in science education. Hence, the black box is treated as a rather abstract representation of a natural phenomenon, and the respondents are asked to explore the black box, thereby simulating the process of scientific discovery (Upmeier zu Belzen, 2014). In study 1 (Sect. 9.4), pre-service science teachers individually engage in modeling a black box without further guidance. Their activities are videotaped and analyzed using a category system. Modeling strategies are inferred by analyzing the pattern of activities. In study 2 (Sect. 9.5), pairs of secondary school students follow an instructional sequence to model the black box in given phases and subsequently reflect on their activities. The findings propose that the sequence is appropriate for fostering students' meta-modeling knowledge and for making their modeling activities explicit.

Both studies that are introduced next use a black box that is literally a black box with a funnel on top of it so it can be filled with water. As a consequence of the arrangement of the inner system of tanks and overflow pipes (two "siphons"), and depending on the input, the output flows out through a pipe at the bottom of the box. For example, when 400 ml of water is poured into the black box six times in a row, the output pattern is 0 ml, 400 ml, 600 ml, 400 ml, 0 ml, 1000 ml (Krell et al., 2017: detailed description of the black box).

9.4 Study 1: Analyzing Pre-service Science Teachers' Modeling Strategies

9.4.1 Design, Methods

The main objectives of this ongoing study are to provide a qualitative analysis of pre-service science teachers' activities in the process of scientific modeling and to infer pre-service science teachers' modeling strategies (cf. Göhner & Krell, 2018). For this purpose, pre-service biology teachers who are enrolled in bachelors (currently $n = 1$) or masters (currently $n = 5$) programs at one public university in Germany volunteered to take part in this study. To get the participants engaged in the process of scientific modeling, the abovementioned water black box was used. Participants' task was to graphically develop a model of the inner system of the black box. Thereby, it was not necessary for participants to figure out the "correct solution" because the focus of the data analysis was on the modeling process and not on the final model.

The participants worked on this study individually. In order to get insights into their reasoning processes, they were asked to think aloud (Ericsson & Simon, 1998). The activities of the participants were audio- and videotaped, and their verbalizations were fully transcribed.

The data analysis falls within the methodological framework of a qualitative content analysis (Schreier, 2012). A deductively developed and inductively refined category system was applied to identify the participants' modeling activities. The category system included the following categories (i.e. activities): *perceiving a phenomenon, exploring the system, activating analogies and experiences, developing a model, testing the model as a model of something, changing the model as a model of something, rejecting the model, confirming the model as a model of something, testing the model as a model for something, refuting hypotheses, supporting hypotheses, changing the model as a model for something* (note that most categories were further subdivided into sub-categories; cf. Krell et al., 2017). Each participant's pattern of activities was analyzed and compared with theoretical descriptions of modeling processes (e.g. Campbell et al., 2015) to infer the participants' modeling strategies.

Before the participants were introduced to the black box activity, their meta-modeling knowledge was assessed using five open-ended questions (Krell & Krüger, 2016) that were developed on the basis of the framework for modeling competence. A category system (Krell & Krüger, 2016) was used to decide whether the participants expressed meta-modeling knowledge related to level I (naïve), level II (intermediate), or level III (sophisticated).

The analyses of both data sources were independently conducted by two researchers, and Cohen's Kappa was calculated as a measure of interrater agreement. Differences in the assigned categories were resolved through discussion.

9.4.2 Findings

Cohen's Kappa was $0.60 \leq K \leq 0.80$ for the analysis of the open-ended questions, and it was $0.46 \leq K \leq 0.84$ for the analysis of the modeling activities. In the open-ended questions, the participants mainly expressed an intermediate level of meta-modeling knowledge (level II), and only two of them expressed sophisticated views on level III.

In the following, data from one case ("Julia") are presented as an example. Julia was a pre-service biology teacher with food science as a second subject, studying in the fifth semester of a bachelors program at the time of data analysis. Julia was selected because her pattern of activities exemplifies cyclic modeling (cf. Campbell et al., 2015), which is rather seldom identified in samples of (pre-service) science teachers (see Krell et al., 2017, for a detailed description of a case of expressive modeling).

In the open-ended questions, Julia expressed meta-modeling knowledge on level II. The codeline (Fig. 9.1) illustrates the pattern of Julia's modeling activities in a chronological sequence and suggested a modeling strategy. In the codeline, each circle represents a coding unit (i.e. activity). The process analysis of Julia's modeling activities revealed that she mainly operated in six phases (Fig. 9.1): (I) an exploration phase, (II) a modeling phase (model of something), (III) a modeling phase (model for something), (IV) an exploration phase, (V) a modeling phase (model of something), and (VI) a modeling phase (model for something).

Exploration Phases (I, IV): Julia mainly explored the behavior of the black box by pouring water into the black box and observing the resulting output.

Modeling Phases (Model of Something) (II, V): Julia performed a sequential development of models on the basis of her observations (i.e. model of something).

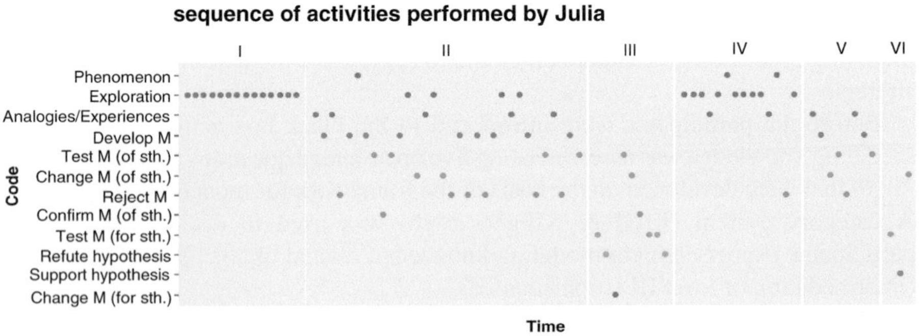

Fig. 9.1 The codeline illustrates the sequence of activities performed by Julia with time on the x axis increasing from left to right. Note that the solid circles are related to coding units (i.e. activities) and not to a standardized amount of time. See the text (paragraph 4.1) for the full names of the activities

She evaluated the models' explanatory power by retrospectively comparing her observations with the behavior she expected from the respective model. This led to the rejection of various models. During these phases, Julia activated experiences and used analogies for model development.

Modeling Phases (Model for Something) (III, VI): Julia used the models to predict the behavior of the black box and, by testing these predictions, indirectly evaluated the adequacy of the models (i.e. model for something).

The identification of the six phases in Julia's modeling process led to the interpretation of her modeling strategy as cyclic modeling (cf. Campbell et al., 2015) because she repeatedly developed, evaluated, and improved her models.

Four of the five remaining pre-service science teachers engaged in expressive modeling because they developed models of the black box on the basis of their observations but did not further evaluate their models by deducing and testing their predictions. One pre-service science teacher demonstrated a rather unsystematic method of model development because, for example, he did not consequently develop his models on the basis of the observations he made, but he instead used models to express his ideas without evaluating the ideas with respect to the data.

Julia expressed meta-modeling knowledge that would fall on level II, which means an understanding of models as models of something, but she showed a cyclical modeling strategy and used her models as models for something (Chap. 1). As in Julia's case, there was no coherent relationship between meta-modeling knowledge and modeling strategies for two other participants: They expressed an understanding on level III but did not perform cyclic modeling (but instead engaged in expressive and unsystematic modeling). The other three participants consistently expressed an understanding that fell on level II and engaged in expressive modeling.

9.4.3 Conclusion

In science education research, a positive relationship between meta-modeling knowledge and modeling processes is assumed (e.g. Schwarz et al., 2009; Schwarz & White, 2005). However, most related studies have been correlational and thus did not allow inferences to be made about causal relationships. Furthermore, modeling processes are often assessed post hoc by analyzing modeling products (e.g. Cheng & Lin, 2015; Jong et al., 2015). Consequently, researchers have emphasized that there is no coherent way to assess modeling processes (Nicolaou & Constantinou, 2014; Chap. 3) and that "one of the most pressing needs for future research is to study the relationship between [...] explicit knowledge concerning the nature of science and the process of modeling, with the ways in which students engage in model creation and revision" (Louca & Zacharia, 2012, p. 486). This study contributes to filling in these gaps in science education research by providing a category system that can be used to analyze individual pre-service science teachers' modeling strategies (Krell et al., 2017). Furthermore, the findings so far – based on a rather small sample of six pre-service biology teachers – suggest that there is not necessarily a coherent relationship between pre-service science teachers' meta-modeling knowledge and their modeling strategies. This calls into question the assumption that is quite popular in science education research that meta-modeling knowledge guides modeling practices (e.g. Schwarz et al., 2009).

9.5 Study 2: Fostering Students' Understanding of Models and Modeling

9.5.1 Design, Methods

The purpose of this study was to explore the effectiveness of an intervention concerning secondary school students' (grades 10, 11) meta-modeling knowledge (Koch et al., 2015). The intervention consisted of three parts: (1) a black box activity that provided a modeling task, (2) reflective classroom discussions about models and modeling, and (3) application tasks with biological contexts. We used a quasi-experimental design with an experimental group ($n = 89$) and a comparison group ($n = 84$) involving a pre-test and a post-test. The comparison group only participated in the pre- and post-tests. Between the two testing occasions, they took part in regular biology classes with no focus on models or meta-modeling knowledge.

9.5.1.1 Black Box Activity

The aim of the first part was to enable the students to participate in a modeling situation. The black box was the water black box described above, which was programmed as an interactive computer experiment (https://tetfolio.fu-berlin.de/web/440484). The students used tablets to examine the black box. The activity was structured around different modeling tasks that referred to the development and evaluation of models: (1) Pour 400 ml of water into the black box. (2) Draw a model of the inner mechanism of the black box. (3) Deduce a prediction about what could happen if you pour another 400 ml of water into the black box again. The purpose of this procedure was to get the participants to run through a cyclical modeling process. The participants worked in pairs in order to support communication and to offer mutual support.

9.5.1.2 Reflective Classroom Discussions

During the second part of the intervention, the students reflected on their activities. To initiate the reflection process, the students were asked to visualize the modeling process. For this purpose, the students were asked to show how predefined and

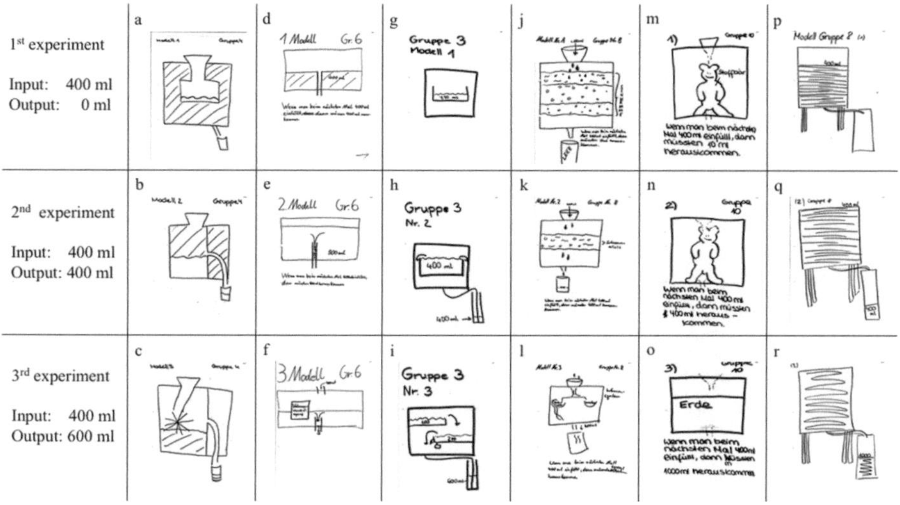

Fig. 9.2 (**a–r**) A selection of the students' models of the black box (Meaning of texts in the pictures: "If we pour 400 ml in, 400 ml (**d**, **e**, **k**, **l**, **n**)/0 ml (**j**) / 10 ml (**m**) /1000 ml (**o**) should pour out")

self-selected terms (e.g. "black box," "model 1," "model 2," "prediction 1," "prediction 2") were related to each other in a process diagram. While they were reflecting on their models, the students also presented and compared their models (Fig. 9.2). Additionally, they were asked questions that referred to the different aspects of modeling competence, for example, concerning the relationship between the black box (as an original) and the drawing of the black box (as the related model) or concerning the role of deduced predictions.

9.5.1.3 Application Tasks

The students were prompted to apply their generic meta-modeling knowledge to different biological contexts by relating the process diagrams to different biological examples (e.g. modeling DNA; cf. Giere et al., 2006).

Paper-pencil tests with the five open-ended questions described above (Sect. 9.4.1; cf. Krell & Krüger, 2016) were used to assess the students' meta-modeling knowledge. The data were analyzed as described (Sect. 9.4.1), which means that a category system (Krell & Krüger, 2016) was used by two researchers independently to decide whether the participants expressed meta-modeling knowledge related to level I (naïve), level II (intermediate), or level III (sophisticated). Cohen's Kappa was calculated as a measure of interrater agreement, and differences in the assigned categories were resolved through discussion.

In the following section, the results of the pre- and post-tests are provided to argue for the efficacy of the intervention in fostering students' meta-modeling knowledge. In addition, we present some of the models that were developed by the students.

9.5.2 *Findings*

The students developed different models of the black box. Fig. 9.2 shows examples of the students' drawings in order to illustrate the diversity of the models that were developed. The progression of each model during the black box activity is arranged in a column. Different models in a row allow for the comparison of different ideas on the basis of the same empirical data (i.e. in the same phase of the black box activity).

Some students (e.g. pictures p-r) drew only observable aspects and neglected the tasks that required them to develop a model of the presumed inner mechanism. Other groups (e.g. pictures c, f, i) considered the hypothetical structure of the black box but did not evaluate the internal consistency. It can be seen that the assumed mechanisms could not explain the observed data, especially when the output changed in different ways even when the input was the same.

Some of the students added a prediction to their drawing (e.g. picture d: "If we pour 400 ml in, 400 ml should pour out"; j: "If we pour 400 ml in, 0 ml should pour out"). Some of these predictions were based on the model as students were asked to do, but some were just guesses. Even though the task was to formulate a prediction, not all students did so.

Cohen's Kappa for the analysis of the open-ended questions was $K = 0.65$. In the experimental group, there was a significant shift in understanding in the aspects of the nature of models ($p < 0.001$, $r = 0.451$), purpose of models ($p < 0.001$, $r = 0.429$), testing models ($p < 0.001$, $r = 0.510$), and changing models ($p < 0.001$, $r = 0.412$), with mostly medium-sized effects. For the aspect of the nature of models, there were no students who expressed a sophisticated view (level III) on the pre-test, which changed to 37% on the post-test (purpose of models: from 3% to 22%; testing models: from 2% to 29%; changing models: from 0% to 13%).

Positive significant differences can also be observed in the comparison group regarding the aspects of multiple models ($p = 0.016$, $r = 0.287$) and changing models ($p = 0.021$, $r = 0.272$; small effect sizes). These differences reflected small shifts ranging from 20% to 21% (multiple models) or from 0% to 1% (changing models) of the students with a sophisticated meta-knowledge of modeling (level III).

9.5.3 *Conclusion*

On the pre-test, the participants primarily expressed naïve or intermediate views (levels I, II). This is in line with findings from other studies (e. g. Grünkorn, 2014). The occurrence of sophisticated views on the post-test indicated the efficacy of the intervention. Based on similar studies (e.g. Akerson et al., 2000; Krell, Koska, Penning, & Krüger, 2015), it can be argued that the combination of engaging in scientific practices and explicit reflections caused the positive shift in students' meta-modeling knowledge.

It can be further argued that a structured learning environment enables students to engage in the process of model development. On the basis of the available data and students' personal experiences, they developed models of the (assumed) inner mechanism of the black box. The formulation of a model-based prediction was intended to support the application of the models as *models for something* (Gouvea & Passmore, 2017). The absence of model-based predictions in some groups pointed toward difficulties for students with the cyclical process of modeling and emphasized that often guidance or scaffolding by teachers is necessary for students to run through this process (Louca & Zacharia, 2015).

9.6 Summary and Overall Conclusion

To summarize, the studies discussed in this article highlight the idea that black box activities can be used to facilitate modeling practices (Göhner & Krell, 2018). More precisely, this article contributes to science education research by providing qualitative, process-based analyses of individual modeling processes (Louca & Zacharia, 2012; Nicolaou & Constantinou, 2014; Chap. 3) and by providing empirical evidence for the efficacy of the black box approach to foster modeling competence in science education.

The category system, which was used to analyze the pre-service teachers' modeling activities is available (Krell et al., 2017) and provides a tool for process-based analyses that can be used by science education researchers. Practitioners in science education can use the black box intervention (available online, see above) in their classes to get their students engaged in modeling processes.

As emphasized above, black box approaches are widely used in science education research to reach various educational goals (e.g. fostering conceptual knowledge, knowledge about science), but empirical evidence for the efficacy of the approaches has been missing. From this point of view, this article provides evidence for the educational power of black box activities for facilitating and fostering scientific practices (Upmeier zu Belzen, 2014).

This chapter is based in part on work supported by the German Research Foundation under project number 327507949. Any opinions, findings, and conclusions or recommendations expressed in this chapter are those of the authors and do not necessarily reflect the views of the DFG.

References

Abd-El-Khalick, F. (2002). Rutherford's enlarged: A content-embedded activity to teach about nature of science. *Physics Education, 37*, 64–68.

Akerson, V. L., Abd-El-Khalick, F., & Lederman, G. (2000). Infuence of a reflective explicit activity-based approach on elementary teachers' conceptions of nature of science. *Journal of Research in Science Teaching, 37*(4), 295–317.

Arsad, N., Sanusi, H., Majid, R. A., Ali, M. M., & Husain, H. (2012). Lateral thinking through black box experiment among engineering students. *Procedia – Social and Behavioral Sciences, 60*, 14–20.

Bailer-Jones, D. (1999). Creative strategies employed in modelling: A case study. *Foundations of Science, 4*, 375–388.

Bailer-Jones, D. (2009). *Scientific models in philosophy of science*. Pittsburgh, PA: University of Pittsburgh Press.

Berge, O. E. (2007). Einstiege in das Thema "Transformator" [First steps into the theme "transformer"]. *Unterricht Physik, 102*, 18–21.

Bünder, W., Mie, K., & Friege, G. (2006). Wie lösen Experten experimentelle Fragestellungen? [How do experts solve experimental problems?]. In A. Pitton (Ed.), *Lehren und Lernen mit neuen Medien* (pp. 349–351). Berlin, Germany: LIT.

Campbell, T., Oh, P., Maughn, M., Kiriazis, N., & Zuwallack, R. (2015). A review of modeling pedagogies: Pedagogical functions, discursive acts, and technology in modeling instruction. *Eurasia Journal of Mathematics, Science & Technology Education, 11*, 159–176.

Cartier, J. (2000). Assessment of explanatory models in genetics: Insights into students´ conceptions of scientific models. *University of Wisconsin-Madison, Research Report*, No. 98–1.

Chakrabarti, B., Pathare, S., Huli, S., & Nachane, M. (2013). Experimental determination of unknown masses and their positions in a mechanical black box. *Physics Education, 48*, 477–483.

Cheng, M.-F., & Lin, J.-L. (2015). Investigating the relationship between students' views of scientific models and their development of models. *International Journal of Science Education, 37*, 2453–2475. https://doi.org/10.1080/09500693.2015.1082671

Clement, J. (1989). Learning via model construction and criticism. In J. Glover, C. Reynolds, & R. Royce (Eds.), *Handbook of creativity* (pp. 341–381). Berlin, Germany: Springer.

Clement, J. (2009). *Creative model construction in scientists and students*. Dordrecht, The Netherlands: Springer.

Crowe, M. J. (1968). Black boxes and white cans: An effective experiment for use in college science classes – And elsewhere. *The Journal of General Education, 19*, 263–268.

Ericsson, K., & Simon, H. (1998). How to study thinking in everyday life: Contrasting think-aloud protocols with descriptions and explanations of thinking. *Mind, Culture, and Activity, 5*, 178–186.

Ferstl, A., & Schneider, L. (2007). Film canister science. *The Science Teacher, 74*, 46–50.

Giere, R., Bickle, J., & Mauldin, R. (2006). *Understanding scientific reasoning*. London: Thomson Learning.

Glanville, R. (1982). Inside every white box there are two black boxes trying to get out. *Behavioral Science, 27*, 1–11.

Gobert, J., & Pallant, A. (2004). Fostering students' epistemologies of models via authentic model-based tasks. *Journal of Science Education and Technology, 13*, 7–22.

Godfrey-Smith, P. (2006). The strategy of model-based science. *Biology and Philosophy, 21*, 725–740. https://doi.org/10.1007/s10539-006-9054-6

Göhner, M., & Krell, M. (2018). Modellierungsprozesse von Lehramtsstudierenden der Biologie [Modeling processes of pre-service biology teachers]. *Erkenntnisweg Biologiedidaktik, 17*, 45–61.

Gouvea, J., & Passmore, C. (2017). 'Models of' versus 'Models for'. *Science & Education, 26*, 49–63. https://doi.org/10.1007/s11191-017-9884-4

Grünkorn, J. (2014). *Modellkompetenz im Biologieunterricht. Empirische Analyse von Modellkompetenz bei Schülerinnen und Schülern der Sekundarstufe I mit Aufgaben im offenen Antwortformat* (Doctoral dissertation). Berlin: Freie Universität Berlin. Retrieved from http://www.diss.fu-berlin.de/diss/receive/FUDISS_thesis_000000097320

Hildebrandt, C., & Oliver, J. (2000). The mind as black box: A simulation of theory building in psychology. *Teaching Psychology, 27*, 195–197.

Jong, J.-P., Chiu, M.-H., & Chung, S.-L. (2015). The use of modeling-based text to improve students' modeling competencies. *Science Education, 99*, 986–1018. https://doi.org/10.1002/sce.21164

Justi, R., & Gilbert, J. K. (2003). Teachers' views on the nature of models. *International Journal of Science Education, 25*, 1369–1386.

Keller, C., & Wang, Y. (1994). A golden oldie: A black box circuit. *The Physics Teacher, 32*, 222–223.

KMK [Sekretariat der Ständigen Konferenz der Kultusminister der Länder in der BRD]. (2005). *Bildungsstandards im Fach Biologie für den Mittleren Schulabschluss* [Biology education standards for the Mittlere Schulabschluss]. München/Neuwied, Germany: Wolters Kluwer.

Koch, S., Krell, M., & Krüger, D. (2015). Förderung von Modellkompetenz durch den Einsatz einer Black box [Fostering modeling competence using a black box]. *Erkenntnisweg Biologiedidaktik, 14*, 93–108.

Krell, M., Koska, J., Penning, F., & Krüger, D. (2015). Fostering pre-service teachers' views about nature of science: Evaluation of a new STEM curriculum. *Research in Science & Technological Education, 33*, 344–365. https://doi.org/10.1080/02635143.2015.1060411

Krell, M., & Krüger, D. (2016). Testing models: A key aspect to promote teaching-activities related to models and modelling in biology lessons? *Journal of Biological Education, 50*, 160–173. https://doi.org/10.1080/00219266.2015.1028570

Krell, M., & Krüger, D. (2017). University students' meta-modelling knowledge. *Research in Science & Technological Education, 35*, 261–273. https://doi.org/10.1080/02635143.2016.1274724

Krell, M., Upmeier zu Belzen, A., & Krüger, D. (2016). Modellkompetenz im Biologieunterricht [modeling competence in biology education]. In A. Sandmann & P. Schmiemann (Eds.), *Biologiedidaktische Forschung. Schwerpunkte und Forschungsgegenstände* (pp. 83–102). Berlin, Germany: Logos.

Krell, M., Walzer, C., Hergert, S., & Krüger, D. (2017). Development and application of a category system to describe pre-service science teachers' activities in the process of scientific modelling. *Research in Science Education*. https://doi.org/10.1007/s11165-017-9657-8

Küçük, H., Schallies, M., & Balim, A. G. (2011). An argumentation based black box activity example about transformation of energy in science and technology education. *Western Anatolia Journal of Educational Sciences, Special Issue*, 53–58.

Lederman, N., & Abd-El-Khalick, F. (2002). Avoiding de-natured science: Activities that promote understandings of the nature of science. In W. McComas (Ed.), *The nature of science in science education. Rationales and strategies* (pp. 83–126). Dordrecht, The Netherlands: Kluwer.

Lietz, M. (2007). Make a mystery circuit with a Bar light fixture. *The Physics Teacher, 45*, 244–245.

Louca, L., & Zacharia, Z. (2012). Modeling-based learning in science education: Cognitive, metacognitive, social, material and epistemological contributions. *Educational Review, 64*, 471–492. https://doi.org/10.1080/00131911.2011.628748

Louca, L. T., & Zacharia, Z. C. (2015). Examining Learning Through Modeling in K-6 Science Education. *Journal of Science Education and Technology, 24* (2–3):192–215.

Mahr, B. (2011). On the epistemology of models. In G. Abel & J. Conant (Eds.), *Rethinking epistemology* (pp. 301–352). Berlin, Germany/Boston: De Gruyter.

Mie, K., & Friege, G. (2004). Black-Box-Aufgaben im experimentellen Physikunterricht [Black-box-tasks in experimental physics classes]. In A. Pitton (Ed.), *Chemie- und physikdidaktische Forschung und naturwissenschaftliche Bildung* (pp. 66–68). Münster, Germany: Lit.

Miller, S. (2014). Modeling the nature of science with the mystery tube. *The Physics Teacher, 52*, 548–551.

MUSE. (2002). *Course material 1D: Black box*. Retrieved from http://ncisla.wceruw.org/muse/index.html

NGSS Lead States (Ed.). (2013). *Next generation science standards: For states, by states*. Washington, DC: The National Academies Press.

Nicolaou, C., & Constantinou, C. (2014). Assessment of the modeling competence: A systematic review and synthesis of empirical research. *Educational Research Review, 13*, 52–73. https://doi.org/10.1016/j.edurev.2014.10.001

Ruebush, L., Sulikowski, M., & North, S. (2009). A simple exercise reveals the way students think about scientific modeling. *Journal of College Science Teaching, 38*, 18–22.

Schreier, M. (2012). *Qualitative content analysis in practice*. Thousand Oaks, CA: Sage.

Schwarz, C., Reiser, B., Davis, E., Kenyon, L., Achér, A., Fortus, D., et al. (2009). Developing a learning progression for scientific modeling: Making scientific modeling accessible and meaningful for learners. *Journal of Research in Science Teaching, 46*, 632–654. https://doi.org/10.1002/tea.20311

Schwarz, C., & White, B. (2005). Metamodeling knowledge: Developing students' understanding of scientific modeling. *Cognition and Instruction, 23*, 165–205.

Upmeier zu Belzen, A. (2014). Black Box: Modellierung von Prozessen naturwissenschaftlicher Erkenntnisgewinnung [Black box: Modeling processes in scientific inquiry]. In D. Luwig, C. Weber, & O. Zauzig (Eds.), *Das materielle Modell. Objektgeschichten aus der wissenschaftlichen Praxis* (pp. 99–106). Paderborn, Germany: Fink.

VCAA [Victorian Curriculum and Assessment Authority]. (2016). *Victorian certificate of education. Biology. Advice for teachers*. Melbourne, Australia: VCAA.

Part III
Educating Teachers for Competence-Based Teaching of Models and Modeling

Chapter 10
Teachers' Views About Models and Modeling Competence Towards Developing Scientific Literacy in Young People

Barbara A. Crawford and Kayla P. Flanagan

10.1 Introduction

A scientifically literate public is crucial as our modern world is on the brink of environmental crisis; and solutions for global problems reside in scientific knowledge, evidence, and creativity in solving problems. The concept of scientific literacy is not clearly defined by all, and the concept has shifted over time (Deboer, 2000). Science is a way of thinking used to develop explanations of natural phenomena using evidence and logic (Crawford, 2014). Scientific literacy includes application of scientific knowledge to the situations individuals will encounter as citizens (Bybee, 2015). The current emphasis on scientific literacy connects with a citizen's view of contemporary and sometimes controversial scientific research. For our purposes, scientific literacy is the understanding of scientific concepts, in addition to understanding how scientists think and construct knowledge, including how scientists create and use models; in short, learning about inquiry/practices and nature of science (American Association for the Advancement of Science [AAAS], 1993; National Research Council [NRC], 1996, 2012). Scientific inquiry consists of the methods and systematic ways of investigating phenomena (Crawford, 2000). Nature of science relates to values and underlying assumptions intrinsic to scientific knowledge, including the human aspects of scientific work (Schwartz, Lederman, & Crawford (2004). Reforms for teaching science emphasize developing learners' epistemological views of science (NRC, 1996, 2012). One of the most important products of science is that of models. Thus, teaching about aspects of scientific models and modeling, is of high importance in classrooms, in developing scientific literacy in young people. This chapter focuses on teachers' views about models and modeling competence in the classroom, as teachers engage students in learning how

B. A. Crawford (✉) · K. P. Flanagan
The University of Georgia, Athens, GA, USA
e-mail: barbarac@uga.edu

A. Upmeier zu Belzen et al. (eds.), *Towards a Competence-Based View on Models and Modeling in Science Education*, Models and Modeling in Science Education 12, https://doi.org/10.1007/978-3-030-30255-9_10

to think and reason through inquiry and modeling, and, in turn, foster scientific literacy.

Scientific models are vital in understanding our natural world. The development and use of models by scientists leads to trustworthy scientific knowledge. Citizens encounter models in everyday life. For example, meteorologists create models of weather patterns and of storms and the possible paths of a particular hurricane over water and over land. Citizens see weather forecasters depicting changing models of weather patterns; yet many people, including youth, may not fully understand the changeable nature of scientific models; and, thus, discount the trustworthiness of the models. Many citizens attribute the changeability of models to a lack of knowledge or true understanding. Therefore, in the minds of many citizens they may mistrust science. Understanding how scientists build and use models is at the heart of what teachers need to know in order to develop in their students an understanding of how scientists use logic and evidence.

Models are powerful tools that enable scientists to generate predictions, as well as guide explanation, interpretation, understanding, and discovery (Jungck & Calley, 1985). An important element of models and modeling is that of abstraction (Chap. 17). By simplifying the complex phenomenon (abstraction), that then can be tested, building a model leads to a better understanding of the phenomenon being studied (target) to better understand the target. In this way, models are used by scientists to reconstruct the idea of a phenomenon, to better study it and generate new knowledge. As such, models are refined over time, based on new evidence or new ways of looking at the same evidence. One important aspect of modeling, is to start with what justifies conceiving of something as a model, which relates to an epistemic pattern of model-being (Mahr, 2011).

A vision of teaching science in the twenty-first century is one of teachers supporting students in understanding the nature of science, and engaging in inquiry/practices, including building and using models (i.e. NRC, 2012). While science educators are unified in this goal, promoting scientific literacy in citizens around the world has been a challenge for more than a century (Dewey, 1916). During the early years of the twentieth century, United States education focused on the basis of relevance to contemporary life and contribution to a shared understanding of the physical world by all members of society (Dewey, 1916). Although it is evident children benefit from model-based instruction, the reality is many teachers view scientific models in a limited and narrow sense (Crawford & Cullin, 2004). Practicing and prospective science teachers may view models mainly as pedagogical tools, and they often fail to attribute to models the function of idea testing or idea generating (Crawford & Cullin, 2004, Crawford & Cullin, 2005; De Jong & van Driel, 2001; Justi & Gilbert, 2003). In addition to not fully realizing the power of modeling, teachers may meet resistance from stakeholders when teaching about modeling through inquiry-based approaches and extended projects. Resistance can come from administrators, as well as their teaching colleagues, who prioritize memorization of science vocabulary; over learning about models and modeling, and development of deep understandings of science modeling (Flanagan & Crawford, 2018). Further a teacher's personal beliefs about inquiry and understandings of modeling are important (Justi & Gilbert, 2003) and can present personal barriers to teachers engaging students in modeling.

In this chapter we address teachers' views of competence related to models and modeling and nature of science (model-based science teaching). The use of model-based science teaching connects tightly to developing scientific literacy in young people (NRC, 2012). We draw upon the empirical literature and data from our work with prospective and practicing teachers. There are few published studies specifically on prospective teachers' understandings of models and views of using scientific models in classrooms (e.g. Crawford & Cullin, 2004; van Driel & Verloop, 1999). This chapter will suggest implications related to teachers' modeling competence for the future of teacher education and various kinds of teacher professional development in countries around the world (see Crawford et al., 2014).

10.2 Theoretical Background

10.2.1 Models and Modeling in Teaching Science

We align our view of models and modeling with that of Gouvea and Passmore (2017), "Models are not simply knowledge representations of the world they are epistemic tools for making sense of the world (p. 56)". Viewing models as epistemic tools, one of the important aspects of modeling in science classrooms is that of engaging students in sense making. Mahr (2011) identifies this distinction of that between models of and models for, which is assigned to models as media and models as research tools by Upmeier zu Belzen and Krüger (2010). Further, we view engaging students in modeling as a dynamic endeavor versus a static one. Static models are similar to models as medium as they are representations of a phenomenon. Dynamic models are similar to models as a method, as they represent models for understanding a particular phenomenon. A classic example of using a static model in a biology classroom involves clay or ceramic 3-D representations of the stages of cellular mitosis and meiosis. Models purchased from scientific education companies can illustrate the different phases of mitosis and meiosis. In the classroom, teachers might display these models, and students may make drawings of the different stages in their notebooks, label the parts, and memorize what scientists have already figured out. A dynamic model of mitosis may involve an animation of a human skin cell undergoing mitosis over time, depicting the various time intervals of each stage of mitosis.

10.2.2 Nature of Science

We suggest an understanding of the nature of scientific models is tightly connected with an understanding of what science is, and what science is not, and that science is a way of knowing (Lederman, 1992). We refer the reader to Chap. 4 on the nature of science in connection with models and modeling. We agree modeling competence necessitates an understanding of the nature of science and the practices of scientific inquiry (Schwartz et al., 2004). And the other way around becoming model competent involves learning the nature of science. The recent United States framework (NRC, 2012) for science education promotes teaching about aspects of the nature of science. "Epistemic knowledge is knowledge of the constructs and values that are intrinsic to science. Students need to understand what is meant, for example, by an observation, a hypothesis, an inference, a model, a theory, or a claim and be able to distinguish among them" (NRC, 2012, p 79). Aspects of the nature of science important to teach students include, science investigations use a variety of methods, scientific knowledge is based on empirical evidence, scientific knowledge is open to revision in light of new evidence, scientific models, laws, mechanisms, and theories explain natural phenomena, science is a way of knowing, scientific knowledge assumes an order and consistency in natural systems, science is a human endeavor, science addresses questions about the natural and material world.

10.2.3 Socio-cultural Perspective of Learning

From a constructivist perspective, a learner comes into a new situation already with one's own ideas (Driver, Asoko, Leach, Mortimer, & Scott, 1994). Experiences shape a learner's own ideas. Newly acquired knowledge is built upon previous knowledge. A socio-cultural perspective of learning is one that takes into account both the social and the cultural environment; and what an individual learns is culturally and socially dependent. Knowledge is developed in the context of personal experiences in association with others (Vygotsky, 1978). Taber (2013) writes about students, "what students know, think, and learn are not phenomena at all (they are not directly observed features of the world); they are conjectured theoretical entities that form parts of our explanatory schemes (p. 327)". We view teacher's learning and competences from a socio-cultural perspective, much as we view young people learning from their experiences in context, and influenced by culture and others in a society.

10.2.4 Description of Levels of Competence Development by Integration of Epistemological Views

Facilitating children in classrooms in learning how to think in ways similar to that of a scientist involves constructing mental models, as they develop understandings of complex phenomena. In contemporary science teaching a primary goal is facilitating children in developing a way of thinking. A goal of school science is not just about acquisition of concrete science concepts and principles, but in developing in children the kinds of thinking aligned with that of scientists, as they create and test and modify or discard models. Oftentimes, teachers may believe the best way to help students to learn about the world, is by efficiently transferring to their students the teacher's own knowledge of scientific facts using a direct teaching approach. However, in bypassing the opportunity for children to struggle with making sense of data and creating models, children cannot fully understand models as epistemic tools. It is important to involve children in the hard and messy work of grappling with data and using empirical data to build and test and critique models (Grünkorn, Upmeier zu Belzen, & Krüger, 2014; Upmeier zu Belzen & Krüger, 2010).

10.2.5 Teachers' Modeling Competence and Teaching

In this chapter we take a problematizing stance related to learning and teaching. What students and teachers know and think and the reasons for their decisions can only be inferred, rather than known absolutely. One cannot directly observe what a person learns about science, nor what a teacher believes about science and science teaching. Similarly, we cannot fully understand a teacher's knowledge base or his or her beliefs and intentions to teach science in reform-based ways (see NRC, 2012). We can only conjecture, based on observations; but, we can never really know for sure. This relates to the notion of competence which is defined as a latent construct getting manifest while performing, e.g. during solving a task (Chap. 1).

Our research aims to answer the following questions:

1. What is the extent of teachers' competence in teaching models and modeling, with a focus on models as epistemic tools?
2. How can we assess teachers' competence in using models and modeling in teaching science and developing scientific literacy in their young students?

10.2.6 Teachers' Views About Scientific Models and Modeling in School Classrooms

It is necessary for teachers to hold conceptions of models and modeling at a deeper level than their own students, if they are to be successful in engaging their students in the scientific practice of building and using models. Further, teachers need to understand models and modeling related to the epistemology of science. In reality, teachers may not have had the necessary experiences during their lifetimes that support deep conceptions of scientific modeling. First, previous research suggests teachers themselves may likely have limited experiences in the process of scientific modeling during their traditional teacher preparation programs (Crawford & Cullin, 2004, 2005; van Driel & Verloop, 1999). Second, teachers may not appreciate the purpose of models, or the power of cognitive activities associated with building and using models (Crawford & Cullin, 2004, 2005). Third, it is not evident that many teachers value prioritizing classroom time on engaging students in modeling and in understanding the nature of scientific modeling, versus learning definitions of scientific terms and memorizing key facts related to disciplinary core ideas. Previous studies have addressed these limitations (Crawford, 2007). Combining teachers' knowledge of models and modeling with an understanding of the nature of science, creates an important shift towards an epistemic focus on teachers' competence of using models and modeling.

Creating educational and professional development opportunities for prospective and practicing teachers can be of great benefit (Schwarz, 2009). It is important for teachers to have opportunity to reflect on and apply a framework, and to address potential roadblocks in teaching about models and modeling. The notion of a teacher's intentions to teach in a certain way, is as important as a teacher's competence, in this case, of models and modeling, and in teaching about models.

10.3 Design and Methodology

10.3.1 Study of Prospective Teachers' Views of Modeling Competence

In the following we present data from a recent study of a group of prospective secondary science teachers in the United States. These new teachers represent the future of science teaching, as they had opportunities to engage in learning about the most recent frameworks for teaching science (NRC, 2012). The study took place in a large university in the southeastern part of the United States. The university has a known reputation for admitting highly qualified students. The students in this teacher education program earn the equivalent of a major in a science discipline (chemistry, biology, physics, or earth science). The teacher education program typically spans two semesters of the final year of a university student's science teacher certification degree program. The teacher preparation program is similar to many other research-intensive university teacher education programs in the United States, in that there are two to three semesters of work related to pedagogy, including the practicum work.

10.3.2 Context of the Study

During their first semester of the teacher education program prospective science teachers completed three science teacher education courses (a technology course, a science teaching methods course, and a practicum in a local school). During the second semester, prospective science teachers engaged in a full-time student teaching internship in a local school and they participated in a 3-hour evening course that met once a week at the university. This course was titled: Reflections on Teaching Science. The course focused on critical reflection by each prospective teacher of his/her classroom teaching through written and oral analyses of both pedagogy and student learning. The course emphasized teaching science as inquiry, with a focus on scientific inquiry/practices, including that of building and using models and modeling. During the course the prospective teachers submitted weekly reflections on two different incidents that happened in their teaching that week. One incident they identified as a challenge; and the other, as a celebration. Prospective teachers read selected published articles about inquiry teaching and learning and scientific practices, including specifically the practice of building and using models (i.e. Falk & Brodsky, 2013). Articles included those published in both research and practitioner journals. Prospective teachers wrote commentary on how their own teaching connected, or did not connect, with pedagogy described in these articles. Class discussions gave opportunity for prospective teachers to publicly exchange ideas with their peers and the instructor. Further, the prospective teachers wrote two versions of a philosophy of teaching and learning statement; first as a draft early in the semester, and later, as a revised statement at the end of the program, incorporating real examples from their own teaching.

10.3.3 Participants and Data

The research participants (n = 35) included secondary science prospective teachers, from two consecutive years in the program (2016 and 2017). In the middle of the second semester of the program, participants completed open-ended surveys to demonstrate their understandings of models and modeling. The survey included questions such as, "What are scientific models and what do they do?" Participants were asked to provide examples from their own teaching experiences in support of their statements. The survey did not directly ask about each of the five aspects within the framework for modeling competence (FMC; Chap. 1) used for analysis. The survey was intended to allow for open response by participants.

10.3.4 Data Analyses

We analyzed the written responses using the FMC to determine teacher understandings of both aspects of models and complexity of understanding (Fig. 1.3). As described in Chap. 1, the FMC categorizes models and modeling into five aspects *Nature of Models, Multiple Models, Purpose of Models, Testing Models, Modifying Models* (Chap. 1). Within each of these categories there are three levels of modeling competence ranging from limited (level I) to more sophisticated understandings (level III). First, written responses, including examples from teaching, were deductively coded for the five aspects of models. Then the responses were coded for level of complexity ranging from I to III as defined by the theoretical framework. Finally, responses were inductively coded to find any emerging understandings that were not applicable to the framework.

10.4 Findings

The written surveys yielded 108 responses from the 35 total participants. *Nature of Models* coded for 40 of the responses, and *Purpose of Models* coded for 24. At least one of the two aspects (*Nature of Models and Purpose of Models*), were mentioned by nearly all 35 prospective teachers and were mentioned together by 16 of those. Many participants believed models to be a representation, or a visual copy of a natural phenomenon; and, the purpose of models was mainly to teach students about science concepts (Fig. 10.1). One participant acknowledged the use of models for prediction (level III), and two wrote about the integration of related variables (level II). However, while the majority recognized the nature and purpose of models, the prospective teachers' perceptions of modeling were generally limited to a level I understanding in both of these aspects.

The remaining three aspects of models were mentioned rarely. *Changing Models* was coded for four responses, all at level II complexity. Prospective teachers seemed to skip the level I concept of correcting errors in the model, to revising the model based on new findings. These prospective teachers detailed this process in their examples. Some prospective teachers would ask students to construct models of phenomena they were studying, and then ask students to revisit and revise their models throughout instruction, using new knowledge gained through instruction. *Testing Models* and *Multiple Models* were coded once in all 108 responses, both at level I (Fig. 10.1).

A theme that emerged from the analysis of prospective teachers' responses was that models are a limited way to view the world. When giving actual examples of how they used modeling in their own classrooms, many of these prospective teachers highlighted their communication of the limitation of models to their students (Fig. 10.1). This communication occurred, either through class critique of the usefulness of a model for representing concepts or through direct instruction.

In summary, of the 35 study participants, most prospective teachers provided explanations and examples related to only two aspects of models and modeling, and these were at a level I complexity. The aspects related to models as scientific thinking tools for learners, including that of testing a model and revising a model, were not clearly evident.

Aspects of Modeling (Grünkorn et al. 2014) Aspect	Level	Evidence from student answers	Frequency (n=35)
Nature of Models	**Level 1** *Models as replications of phenomenon*	"a representation of a scientific phenomenon" "simplified representation of observable phenomenon" "correct and accurate representation of a phenomenon or system." "replica, blown up atom, mini solar system, concept map, artifact that puts something big in perspective" "representation of a mechanism or phenomenon"	**27**
	Level 1 Models are Limited *Emerging theme*	"Represent some type of scientific phenomena in a visual way that puts a larger idea into a different perspective. But they are LIMITED" "I had students work with many different types of models and critique their helpfulness at the end" "I emphasized it is just a model and was a limited example"	**3**
Purpose of Models	**Level 1** *Models to describe phenomena* -help explain -pedagogical tools	"a representation of a system, process, or concept meant to communicate the idea/concept clearly" "A concept that is illustrated in a particular way so that it is easier understood by the viewer ". "They provide an opportunity to visual(ize) phenomena that can't be seen or to explain observed phenomena from the natural world". "something that describes a scientific idea (usually in visual format)"	**21**
	Level 2 *Models to explain relatioships between variables.* -variables interacting in a model	"an interactive water cycle or a stream table for island migration" "computer simulation of tides and corresponding moon or drawing".	**3**
	Level 3 *Models to predict connections*	"are representations of a real phenomenon or system and they are used to simplify the complexity to easily predict or explain the phenomenon or system."	**1**

Fig. 10.1 Representative examples of student responses coded for modeling competence

	between variables		
Changing Models	**Level 2** *Models revised based on new knowledge* -revise while learning	"models help to identify misconceptions. It is beneficial to have students come back to and revise models over time." "Revising models is a great practice to further understandings." "scientific models can and should be revised by students as they gain more knowledge of the different phenomena"	**4**
Multiple Models	**Level 1** *Several models for one phenomena differing in materials and dimensions*	"I showed many different models of the same molecule then allowed them to draw their own molecules using a model they liked best"	**1**
Testing Models	**Level 1** *Testing the model for functionality*	"We modeled a hand using straws, fishing line, and tape. Then we were able to test our models."	**1**

Fig. 10.1 (continued)

10.5 Discussion of the Study Findings

Despite the emphasis by the course instructors on inquiry/practices during the class, the majority of prospective teachers held the view that a model is a primarily a pedagogical tool, a medium rather than a model as a method for students to make sense of phenomena (Upmeier zu Belzen & Krüger, 2010). Participants viewed a model as a way to teach facts by describing or representing phenomena. Prospective teachers recognized the function of models as that of describing and representing phenomena, but did not view using models as mirroring scientific methods. In other words, these new teachers understood models *of* something but not models *for* something (Gouvea & Passmore, 2017). In addition to low complexity in their understanding, these prospective teachers lacked awareness of the multiple aspects of modeling.

Grünkorn et al. (2014) conducted a similar analysis of responses using the FMC. In their study, 1177 seven to tenth grade students completed a 15-question survey about models. The survey aligned three questions per each aspect in the framework. Responses were open and researchers coded for each competence aspect and level, as we did above. Higher frequencies in levels I and II than in level

III for each aspect were observed from their analysis, similar to our findings. Our prospective teachers performed similarly on modeling competence to the tenth grade students in the Grünkorn et al. (2014) study. We evaluated our participants' responses using the revised framework proposed by Grünkorn et al. (2014, p. 26), and the coded levels remained the same. It appeared our prospective teachers were products of their education experiences coming up through the various school grades, with no changes during their college level science course experiences.

Analyses of the data suggested these prospective teachers were inferring that, because models can be false or revisable, they are limited in their usefulness. This view contrasts with one of higher competence, that models can be falsifiable and adapted, and it is this aspect that makes models tools for the development of scientific knowledge. The participants' views suggest that models are static and cannot be changed which is consistent with novice perspectives (Crawford & Cullin, 2004; Grünkorn et al., 2014).

Prospective teachers' limited understanding of models seems to affect their beliefs about the usefulness of models by scientists and importance for the teaching of both science concepts and *practice*. Further these beliefs held strong while participating in the University's pedagogical model instruction. Many prospective teachers felt unsupported in their classroom placement, believing there was a disconnect between the teaching at the University and the realities of the classroom. This demonstrates the influence of a context (Vygotsky, 1978) in which models are majorly viewed as media instead of methods (Grünkorn et al., 2014). Prospective teachers, once in the classroom, will most likely pass these beliefs onto their students, therefore perpetuating this inadequate conception of the nature of models and leading to a mistrust of scientific evidence and knowledge (Crawford & Cullin, 2004, 2005). Focusing on prospective teachers' understanding of models and their ability to effectively teach them to their students should be of upmost importance for our University teacher education programs as it can have a direct influence on school culture.

In their capstone teacher preparation course, these prospective teachers had been offered opportunities to read articles on teaching about models and modeling, to have seminar discussions with their peers, to write reflectively, and were encouraged to teach their own students about models and modeling. Yet, there was limited empirical evidence most of them demonstrated modeling competence that would position their future teaching to include teaching students about all aspects of models and modeling in a robust way. It appears that prospective teachers, as well as science school students, need more authentic experiences with modeling that are aimed at a level III complexity (Chap. 1). While a model is not a perfect system for understanding a phenomenon, and by the very nature of the practice, may be limited, a model is consistent with the characteristics of scientific inquiry and, like other ways of knowing in the field of science, leads to valuable knowledge.

10.5.1 Possibilities of Professional Development for Enhancing Teachers' Modeling Competence

Given the empirical findings of the study of prospective teachers described above, it is important to consider how to support all science teachers, as they enter the teaching field and as they continue to teach. Designing teacher professional development programs should take into account how teachers learn in similar ways to their own students, using a socio-cultural perspective. In other words, teacher learning involves eliciting prior knowledge, building on one's experiences, gaining new experiences and reflecting on previous knowledge and views, in collaboration with others.

10.5.2 Example of a Successful Teacher Development Program

The Fossil Finders project (Crawford, 2012) is an example of an effective professional development program supporting teachers' views and teaching children about scientific practices. In this program we immersed teachers of 10–15 year old children in gathering, analyzing and interpreting data, and in creating models of the distant past, specifically of the Devonian, using authentic fossil data. One of the aspects of this professional development program involves earth sciences. To the best of our knowledge there are few studies in the literature using earth sciences as a context related to students' and teachers' use of models and understandings of models and modeling.

10.5.3 Context of the Program

The design of the program was based on a socio-cultural learning perspective. During 6 days in each of two summers, we immersed a total of 30 secondary science teaches in an authentic scientific investigation, in this case in the work of paleontologists, creating a model of what the past environment might have been 280 million years ago. The teachers would later engage their own students in the same scientific investigation, creating a context for teaching their students about important key science concepts and principles, scientific practices and nature of science. During the program teachers worked collaboratively with their peers, science teachers, educators, and scientists in collecting samples of rock from a road cut in upstate New York State. The samples were collected from different horizons in the road cut (along a vertical line), related to age of the rock. Teachers first found fossils in the collected samples, and then learned how to identify the fossils to the taxa level, including brachiopods, clams, crinoids, cephalopods. Teachers collected other important fossil data, including color of the rock, fragmentation of the fossils, and size, including length and width. The teachers learned how to make inferences about the past environment using all the data collected, and how to compare these data to an aggregate database. A nearby museum provided a site for teachers to study fossils from different eras. Teachers also were given opportunity through reflection, to connect the various aspects of the Fossil Finders investigation with what paleontologists do, and aspects of the nature of science, including creativity, subjectivity, and that models may change. The following year the teachers engaged their students in this authentic investigation, with support from the teacher educators and scientists. One of the features of the authentic investigation was to create a model to predict how populations of organisms in the shallow Devonian sea may have changed in response to changes in the ancient environment.

10.5.4 Results of the Professional Development Program

Data included pre-posttests of teachers' views of science. In addition, researchers collected classroom videotapes of approximately 2 weeks of lessons for each teacher participant. During analyses of the teachers' classroom lessons, pre-post questionnaires, and interviews, we determined most of the teacher participants enhanced their understandings of how scientists work and use evidence and logic to develop scientific models (Capps & Crawford, 2013). During the professional development sessions, we identified incidents when teachers experienced the messiness of science, and scientists were able to help them recognize that changing a research question, or revising a model, connects with the real work of paleontologists. The Fossil Finders professional development program aligns with best practices of supporting teachers (Capps, Crawford & Constas, 2012). In the professional development program, we aimed to model the kinds of interactions between teacher and students that offer opportunity to engage in scientific practices, including investigating and grappling with data, developing and using scientific models, analyzing and interpreting data, constructing explanations, and engaging in argument from evidence, and to understand that science is not absolute and there is no *one* scientific method (Crawford & Jordan, 2013).

10.5.5 Implications for Pre-service Teacher Educators and Professional Developers

> We must either accept that different conceptions of knowledge could develop in the context of different practices or suppose that there is some subset of practices belonging to all knowledge-productive practices. (Longino, 1990, p. 19)

Different contexts of practice lead to different types of knowledge. *Multiple models*, one of the aspects for modeling, demonstrates that models can be variable for a singular phenomenon (Upmeier zu Belzen & Krüger, 2010). These multiple models can come from a variety of understandings and viewpoints that are used to construct a more holistic understanding of the phenomena we observe. In the study reported above on prospective science teachers, the teachers viewed models as primarily fixed representations of fixed phenomena. This limited perspective of models is concerning, as a major aspect of modeling is the variety and adaptability used for knowledge-making (Gilbert 2004). It is important for teachers and students to understand the role of scientific models. Only if teachers themselves understand the epistemic aspects of models and modeling will true engagement of students in the epistemic aspects of modeling occur. The practices of science involve using models to gain understanding of the natural world. Associated with science is the recognition that there is value in holding multiple and alternative perspectives. To fully understand the uses of modeling one must first recognize science as situated in contexts that allow it to shift and change over time, and recognize that there is value in multiple and alternative perspectives. As such, models are a means to test and adapt to multiple content areas and contexts for learning. This is especially important for K-12 students as they come from diverse backgrounds, cultures, and understandings that will factor into their context of learning (Lee & Fradd, 1998).

Effective professional development should strive to support teachers in teaching the nature of science, including how engaging in the dynamic aspects of scientific models help us develop knowledge about the natural world. Teachers must have a deep understanding of all aspects of models, not just a few aspects (i.e. Justi & Gilbert, 2002, Justi & Gilbert, 2003). During teacher education and professional development, it will be important for teachers to actively reflect on their own knowledge and their teaching of models and modeling. Figure 10.2 offers a framework of how teachers can actively engage in inquiry, building and using models and reflecting on modeling and the teaching of modeling. Each cycle incorporates investigation, construction of a model and reflection on the model, and these cycles can occur iteratively. The reflection aspect includes explicit thinking about epistemology.

Identifying the centrality of scientific models and modeling and advocating their teaching in science classrooms is one step towards enhancing the teaching of school science. The next important step is ensuring teachers can effectively carry out this kind of instruction in classrooms. In reality, prospective teachers' lessons often begin, not with a question that might motivate children eliciting their mental models and/or building scientific models, but with a list of scientific terms and definitions, albeit these might be embellished with images from the Internet (Crawford, 2007). Anecdotally, prospective teachers reported they were very reluctant to prepare and

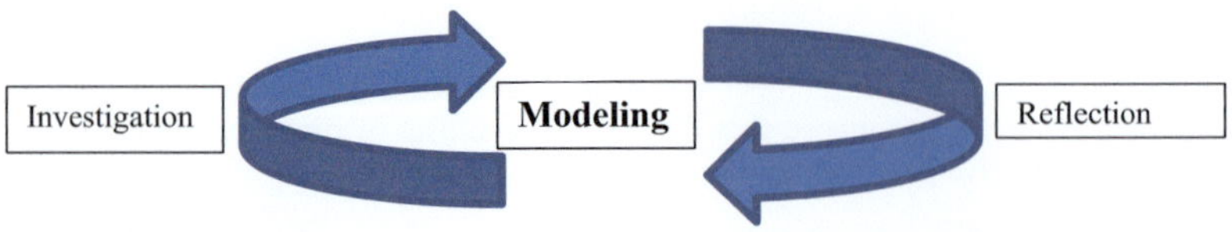

Fig. 10.2 Dual cycles of investigation leading to modeling and reflection on modeling

teach lessons rich in inquiry and use of models. They cited many roadblocks. Common responses were, "my mentor teacher told me we do not have the time to spend doing inquiry-based lessons. There is too much to cover. If it is not on the state assessment test, we cannot spend time on it" (Personal communication with prospective teachers in southeastern state, USA, October 2015).

In summary, science teacher educators and policymakers in all countries cannot afford to overlook the importance of investing in robust and carefully designed science teacher education programs and professional development opportunities, involving sustained and meaningful experiences related to developing teachers' modeling competence. Ultimately, engagement in all aspects of models and modeling by teachers will contribute to young people developing critical thinking skills and scientific literacy, useful to citizens in a world in which decisions count related to environmental crises.

References

American Association for the Advancement of Science (AAAS). (1993). *Project 2061: Benchmarks for science literacy*. New York: Oxford University Press.

Bybee, R. (2015). Scientific literacy. In R. Gunstone (Ed.), *Encyclopedia of science education*. Dordrecht, The Netherlands: Springer. https://doi.org/10.1007/978-94-007-2150-0

Capps, D. K., & Crawford, B. A. (2013). Inquiry-based professional development: What does it take to support teachers in learning about inquiry and nature of science*? International Journal of Science Education*. https://doi.org/10.1080/09500693.2012.760209

Capps, D. K., Crawford, B. A., & Constas, M. A. (2012). A review of empirical literature on inquiry professional development: Alignment with best practices and a critique of the findings. *Journal of Science Teacher Education, 23*(3), 291–318.

Crawford, B. A. (2000). Embracing the essence of inquiry: New roles for science teachers. *Journal of Research in Science Teaching, 37*(9), 916–937.

Crawford, B. A. (2007). Learning to teach science as inquiry in the rough and tumble of practice. *Journal of Research in Science Teaching, 44*(4), 613–642.

Crawford, B. A. (2012). Moving the essence of science into the classroom: Engaging teachers and students in authentic science. In K. C. D, Tan & M. Kim (Eds.), *Issues and challenges in science education: Moving forward* (VII, 350 p. 44 illus, pp. 25–42). Dordrecht, The Netherlands: Springer.

Crawford, B. A. (2014). Invited chapter. From inquiry to scientific practices in the science classroom. In N. Lederman & S. Abell (Eds.), *Handbook of research on science education* (Vol. II). New York: Rutledge.

Crawford, B. A., Capps, D.K., van Driel, J., Lederman, N., Lederman, J., Luft, J., et al. (2014). Learning to teach science as inquiry: Developing an evidence-based framework for effective teacher professional development. An invited chapter for the Edited Book: European Science Education Research Association 2011 Conference. C. Bruguière et al. (eds.), Topics and Trends in Current Science Education: 9th ESERA Conference Selected Contributions, Contributions from Science Education Research 1, https://doi.org/10.1007/978-94-007-7281-6_12, © Springer Science+Business Media Dordrecht 2014.

Crawford, B. A. & Cullin, M. J. (2004). Supporting prospective teachers' conceptions of modelling in science. *The International Journal of Science Education, 26*(11), 1379–1401.

Crawford, B. A., & Cullin, M. .J. (2005). Dynamic assessments of preservice teachers' knowledge of models and modelling. In K. Boersma, M. Goedhart, O. De Jong & H. Eijkelhof (Eds.), *Research and the quality of science education* (pp. 309–323). Dordrecht: Springer.

Crawford, B. A., & Jordan, R. (2013). Inquiry, models, and complex reasoning to transform learning in environmental education. In M. Krasney & J. Dillon (Eds.), *Trading zones in environmental education. Creating transdisciplinary dialogue*. New York/Bern, Switzerland/Berlin, Germany/Bruxelles, Belgium/Frankfurt am Main, Germany/Oxford, UK/Wien, Austria: Peter Lang Publishing Group.

Deboer, G. (2000). Scientific literacy: Another look at its historical and contemporary meanings and its relationship to science education reform. *Journal of Research in Science Teaching., 37*, 582–601.

Dewey, J. (1916). *Democracy and education*. New York: The Free Press.

Driver, R., Asoko, H., Leach, J., Mortimer, E., & Scott, P. (1994). Constructing scientific knowledge in the classroom. *Educational Researcher, 23*(7), 5–12.

Falk, A., & Brodsky, L. (2013, September). Incorporating models into science teaching to meet the Next Generation Science Standards. *Science Scope,* 61–69.

Flanagan, K. P., & Crawford, B. A. (March 2018). Prospective teachers' resistance to teach inquiry in secondary science classrooms. A paper presented at the 2018 NARST: A Worldwide Organization for Improving Science Teaching and Learning through Research International Conference in Atlanta, Georgia.

Gilbert, J. K. (2004). Models and modeling: Routes to more authentic science education. *Instructional Journal of Science and Mathematics Education., 2*(2), 115–130.

Gouvea, J., & Passmore, C. (2017). 'Model of' versus 'Models for': Toward an agent-based conception of modeling in the science classroom. *Science & Education, 26*, 49–63.

Grünkorn, J., Upmeier zu Belzen, A., & Krüger, D. (2014). Assessing students' understandings of biological models and their use in science to evaluate a theoretical framework. *International Journal of Science Education*. https://doi.org/10.1080/09500693.2013.873155

Jungck, J., & Calley, J. (1985). Strategic simulations and post-socratic pedagogy: Constructing computer software to develop long-term inference through experimental inquiry. *American Biology Teacher, 47*, 11–15.

Justi, R. S., & Gilbert, J. K. (2002). Modelling, teachers' views on the nature of modelling, and implications for the education of modellers. *International Journal of Science Education, 24*(4), 369–387.

Justi, R. S., & Gilbert, J. K. (2003). Teachers' views on the nature of models. *International Journal of Science Education, 25*, 1369–1386.

Krell, M., Upmeier zu Belzen, A., & Krüger, D. (2014). Students' levels of understanding models and modelling in biology: Global or aspect-dependent? *Research in Science Education, 44*(1), 109–132.

Lederman, N. G. (1992). Students and teachers' conceptions of the nature of science: A review of the research. *Journal of Research in Science Teaching, 29*(4), 331–359.

Lee, O., & Fradd, S. H. (1998). Science for all, including students from non-English language backgrounds. *Educational Researcher, 27*(3), 12–21.

Longino, H. E. (1990). *Science as social knowledge: Values and objectivity in scientific inquiry*. Princeton, NJ: Princeton University Press.

Mahr, B. (2011). On the epistemology of models. In G. Abel & J. Conant (Eds.), *Rethinking epistemology* (pp. 252–253). Göttingen, Germany: Hubert and Co..

National Research Council. (1996). *National science education standards*. Washington, D.C.: National Academy Press.

National Research Council. (2012). *A framework for K-12 science education: Practices, crosscutting concepts, and Core ideas*. Washington, DC: The National Academy Press.

Schwartz, R. S., Lederman, N. G., & Crawford, B. A. (2004). Developing views of nature of science in an authentic context: An explicit approach to bridging the gap between nature of science and scientific inquiry. *Science Education, 88*, 610–645.

Schwarz, C. (2009). Developing preservice elementary teachers' knowledge and practices through modeling-centered scientific inquiry. *Science Education, 93*(4), 720–744.

Taber, K. S. (2013). *Modelling learners and learning in science education: Developing representations of concepts, conceptual structure and conceptual chance to inform teaching and research*. Dordrecht, The Netherlands: Springer.
Upmeier zu Belzen, A. & Kruger, D. (2010). Modelkompetenz im Biologieunterricht [Modeling competence in biology education]. *Zeitschrift fur Didaktick der Naturwissenschaften, 16*, 41–57. Retrieved from http://www.ipn.uni-kiel.de/zfdn/jg16.html
Van Driel, J. H., & Verloop, N. (1999). Teachers' knowledge about models and modelling in science. *International Journal of Science Education, 21*, 1141–1153.
Vygotsky, L. S. (1978). *Mind in society: The development of higher psychological processes* (M. Cole, V. John-Steiner, S. Scriber & E. Souberman Eds., & Trans.). Cambridge, MA: Harvard University Press.

Chapter 11
Using Epistemic Considerations in Teaching: Fostering Students' Meaningful Engagement in Scientific Modeling

Li Ke and Christina V. Schwarz

11.1 Introduction

Despite the increasing emphasis on scientific modeling in science education, some research indicates that modeling and other reform-based practices can become proceduralized (Baek, Schwarz, Chen, Hokayem, & Zhan, 2011; Cohen & Ball, 2001). For example, modeling can become a practice that merely involves drawing pictures of pre-determined components (e.g., the parts of a cell or the planets in the solar system) or processes that specified components and relationships (e.g., include all the steps of the water cycle or an insect life cycle). Proceduralizing or essentializing any practice into *merely routines* or *processes* without asking students to consider their purpose for meeting their sense-making goals in a classroom learning community renders that practice scientifically meaningless. In other words, while proceduralized or routinized practices serve the purpose of 'doing school,' they serve little purpose for meaningfully making sense of the world. By meaningful, we refer to students' productive disciplinary engagement (Engle & Conant, 2002) that is important to the discipline of science and to the classroom community's knowledge building goals. In particular, teachers should "hold students accountable to others and to shared disciplinary norms" (Engle & Conant, 2002; p.406) and create a learning environment where students are active epistemic agents (Stroupe, 2014) who construct and evaluate knowledge collectively.

Our current learning progression of scientific practices embodies these "disciplinary norms" by highlighting epistemic considerations relevant to the classroom

L. Ke (✉)
University of North Carolina at Chapel Hill, Chapel Hill, NC, USA
e-mail: lke@unc.edu

C. V. Schwarz
Michigan State University, East Lansing, MI, USA
e-mail: cschwarz@msu.edu

A. Upmeier zu Belzen et al. (eds.), *Towards a Competence-Based View on Models and Modeling in Science Education*, Models and Modeling in Science Education 12, https://doi.org/10.1007/978-3-030-30255-9_11

community and the discipline of science that are critical for engaging students in scientific practices (Berland et al., 2016). These epistemic considerations include a focus on (a) the nature of the knowledge product (What kind of answer should our knowledge product provide?), (b) justification of the knowledge product (How do we justify the ideas in our knowledge product)?, (c) generalizability of the knowledge product (How does our knowledge product relate to other scientific phenomena and ideas?), and (d) the audience of the knowledge (Who will use our knowledge product and how?) Our prior work indicates that students use epistemic considerations as they engage in modeling and other practices. We also posit that epistemic considerations guide knowledge building and sense-making to support meaningful engagement in scientific practices.

While some research has emphasized meaningful engagement in modeling (e.g., Lehrer & Schauble, 2006; Manz, 2015; Schwarz, Passmore, & Reiser, 2017), it is critical to understand how teachers can support students' engagement in scientific modeling practices. While teachers are central to enacting practices in classroom, they are often unfamiliar with models and what modeling practices entail (e.g., Danusso, Testa, & Vicentini, 2010; Justi & Gilbert, 2003; van Driel & Verloop, 1999). Further, teachers who think of models as repositories of information (e.g., "models of" approach, Gouvea & Passmore, 2017) focus on the representational nature of models, as opposed to attending to how models are developed and used, and for what purposes. With this "models of" perspective (Gouvea & Passmore, 2017), modeling becomes a procedural exercise where students figure out the right things to include in the models without using models as tools for making sense of phenomena. Similarly, in some of our prior work, we found that teachers' explicit verbal messages about models and modeling regarding the purpose of the practice shaped students' modeling practices (Ke & Schwarz, 2016). In that work, one teacher who framed modeling as "putting what you learned into the models" and emphasized that students should finish the models in a timely manner led to students finishing the task regardless of substance and including particular information in their model, regardless of the relevance of that information.

While we know that teachers play a significant role in how practices are established and pursued in classroom communities, much prior research focuses on teachers' understanding of models and the practice of modeling (e.g., Henze & van Driel, 2011; van Driel & Verloop, 1999) rather than the role of teachers in guiding students' engagement in scientific modeling (See Kelly, McDonald, & Wickman, 2012; Lidar, Lundqvist, & Östman, 2006). Knowing that the way a teacher frames and engages students in modeling largely impacts the meaningfulness of modeling practice, we pose the question: *what are productive ways in which teachers can support students' meaningful engagement in scientific modeling?* In this chapter, we draw on Epistemologies-in-Practices framework (EIP; Berland et al., 2016) built from our prior learning progressions work (Schwarz et al., 2009) to consider how teachers might support students' engagement in modeling practices from a disciplinary perspective as well as a community knowledge-building perspective. In particular, we present examples of how a 5th grade teacher used epistemic considerations to support her students' modeling practices. We do so because this teacher worked with our modeling research group for a number of years and leveraged the epistemic considerations from our learning progression to guide her students' modeling work. In order to understand how she supported this discourse-rich engagement in modeling, we share how this teacher enacted modeling practices in her classroom. In particular, we share how she supported students at engaging in scientific modeling to generate mechanistic models that can be generalized to other phenomena. Given how likely modeling practice is to become rote in school contexts as it become more widely emphasized in standards, we argue that understanding how teachers can support meaningful student engagement in modeling practice using epistemic considerations is critical for advancing the field. The examples of dialog in this chapter illustrate how paying attention to teaching and using epistemic considerations matters for advancing modeling if we care about doing so in a meaningful way.

11.2 Conceptual Framework

In this chapter, we use EIP framework (Berland et al., 2016) to examine classroom engagement in modeling practice. The EIP framework is a revised version of our modeling learning progression (Schwarz, Reiser, Acher, Kenyon, & Fortus, 2012; Schwarz et al., 2009) aimed at supporting modeling, explanation-construction, and argumentation. Since that time, we determined that particular epistemic considerations derived from our hypothetical learning progression were helpful in supporting students to meaningfully engage in modeling, explanation and argumentation practices as they unfold over time (e.g., Baek & Schwarz, 2015; Berland et al., 2016). By epistemic considerations, we refer to the purposes and goals of the work in which students are engaged. An important aspect of studying engagement in practices is to understand what epistemic considerations are guiding the work and in what ways those align with the norms and values of science (Ford & Forman, 2006). For example, developing and revising models that address the mechanism of phenomena lies at the core of the scientific endeavor; therefore, considering the degree to which an explanation is mechanistic should guide learners who are engaged in modeling practice.

These epistemic considerations that frame and guide practices are called epistemologies-in-practice. This EIP framework is important because it can capture the nature of students' engagement in modeling practices and how teachers might be supporting students' development of modeling in the classroom community. Our research acknowledges the context specificity of considerations in practices and describes how the practices of modeling, explanation and argumentation might become more meaningful from a disciplinary and classroom perspective through epistemic considerations (Berland et al., 2016). Table 11.1 is a summary of the four epistemic considerations and some of the ways that students might address each consideration.

How are epistemic considerations related to models and modeling competence within this book? Our EIP framework overlaps with the framework for modeling competence (FMC; Chap. 1) in that they both emphasize purposes of modeling as

Table 11.1 Epistemic considerations in students' epistemologies in practice.

Epistemic Consideration	Range of Students' Considerations
What kind of answer should our model provide? (*nature*)	• Our model should *describe* what happened. • Our model should *explain how or why* something happened. In other words, it should articulate a step-by-step mechanism.
How do we justify our model? (*justification*)	• We include the information in our models that others tell us to include (so it does not need to be justified). • We construct, evaluate, and justify our models using our interpretation of the available information (e.g., data, scientific theories, personal experiences, etc).
Who will use our model and how? (*audience*)	• Our model is for the teacher to evaluate our understanding. • We collaboratively construct and use our model with our audience.
How does our model relate to other scientific phenomena and ideas? (*generality*)	• Specific scientific phenomena do not relate to one another, so our model should characterize the specific nature of each individual phenomenon. • Generalized science ideas have little relationship to specific experiences or phenomena so our model should not connect across these ways of thinking. • Our models are created from and should explain a range of phenomena, so our model should show these connections.

(Table from Berland et al., 2016)

the key to guide students' modeling processes. While we identified four epistemic considerations concerning the purposes and goals of modeling building upon our hypothetical learning progression of modeling, the FMC prioritizes the predictive nature of the models as research tools. In particular, the aspect of purposes of models in the FMC is well aligned with our *nature* epistemic consideration as they both capture the explanatory and predictive power of a model as a sense-making tool. However, there is a key difference in how the two frameworks conceptualize the term "explain." In the FMC, "explain" is at level II because it only serves a intermediate function of describing something. In our work, an explanation is part of the explanation practice which goes beyond describing because "the known relationships and correlations between variables in an initial object" are often not observable to students, and these are critical to the hidden-mechanism of how and why certain phenomena occur. Reasoning systematically through underlying factors and relationships that give rise to phenomena is a powerful thinking strategy that also allows one to make predictions about phenomena (Krist, Schwarz, & Reiser, 2019). Therefore, in our EIP framework, using models to develop a mechanistic account about phenomena is one productive epistemic consideration that is aligned to the values of both disciplinary science and the classroom knowledge building community.

Another key difference between the EIP framework and the FMC is that, while the FMC highlights students' knowledge about models such as "nature of models" and "multiple models," the EIP framework focuses not on students' ideas about models, but how students are developing and using models with respect to their goals of modeling. While students' epistemic considerations may reflect their MC in terms of the nature of models, the goals that guide students' modeling work may be less like formal dispositions that students develop over time and more about contextual considerations drawn on and supported by the teacher or others in class when doing scientifically meaningful modeling practice in classrooms.

11.3 Two Illustrating Examples

11.3.1 An Example Illustrating How Modeling Instruction Can Become Unproductive

Below we present a short classroom conversation to illustrate how modeling instruction can become proceduralized or unproductive in authentic classroom settings, even when the intent of teacher's practice may be well aligned with the disciplinary norms of scientific modeling. The conversation took place in one of our participant teacher's classroom and the teacher, Mr. L was reviewing the criteria of a good model before asking students to revise their model of evaporation.

Mr. L: What's a good model? What do good models have in common?
Student A: Evidence.
Mr. L: Good models need evidence, right? So where do we get our evidence from?
Student B: Maybe from the labels?
Mr. L: Ok, you could have labels, but I don't know if they show evidence necessarily.
Student C: From the science lab? The fancy little things we used...
Mr. L: Yes, we used the humidity detectors in the experiments we did. That's a piece of evidence we could put into our model. What else?
Student D: Molecules?
Mr. L: Are you talking about the simulations we saw on the computers?
Student D: Yeah, the one with molecules going all over the place.
Mr. L: Exactly. Remember, the experiments we did, the humidity level, the computer simulations are all evidence you want to make sure you put into your model if you haven't done so already because you guys just told me that we need evidence in our model.

In this excerpt, Mr. L seemed to be emphasizing the importance of including evidence in models, which also aligns with our Justification epistemic consideration, that students need to support their models with evidence. However, we argue that this interaction in the enactment was not particularly productive because the conversation did not help students figure out why the "humidity detector" or "computer simulations" could count as forms of evidence as well as why they needed to include them in the first place. Rather, Mr. L was the one who made the conclusion that students needed to put the "humidity detector" and "computer simulations" into their models. By doing so, students lost the opportunity to figure out what they needed to include as evidence in their particular models (and we assume each student had their own ideas about how to construct a model for a particular phenomenon) so that they could show other people that their models were correct. Later in the model revision session following the conversation, we observed many students putting experiment details including humidity level into their models. However, from our analysis of our focus student interviews, we found that some of our focus students seemed to have followed the teacher's instructions without thinking much about why they needed to do so.

11.3.2 An Example Illustrating How a Teacher Supports Modeling Through Epistemic Considerations

Context In this illustrative example, we highlight the pedagogical practice of a 5th grade science teacher, Mrs. M, from a Midwest suburban elementary school during the 2012–2013 academic year. While we worked with several different science teachers throughout our research (Ke & Schwarz, 2016), this example highlights Mrs. M's teaching and her classroom as our work indicated she was particularly effective at engaging students in epistemically-rich modeling practice in ways that aligned with disciplinary and classroom knowledge-building norms.

The transcripts of classroom dialog were derived from a 6 to 8-week model-based unit (Baek et al., 2011; Kenyon, Schwarz, & Hug, 2008) about evaporation and condensation. In the unit, students were asked to address a driving question of whether or not they would drink the liquid from a solar still. To answer this driving question, students constructed an initial diagrammatic model to explain the phenomenon, and continuously evaluated and revised their models using evidence from their empirical investigations and scientific information.

When we introduced the epistemic consideration framework (Table. 11.1) to Mrs. M about 4 years into our 9-year collaboration, she took on the framework as they provided goals for students' work during various modeling activities. She thought of the framework and the epistemic considerations as a way to evaluate whether the models were moving in a good direction. Our analysis of Mrs. M's teaching indicated that she particularly emphasized two epistemic considerations – *nature* and *generality*. The *nature* consideration focuses on what counts as a sufficient answer to scientific questions. The *generality* consideration focuses on connecting understanding about specific phenomena and more general scientific ideas. Mrs. M's approach to using the *nature* consideration was to work with students to figure out "how and why" phenomena occur. She also told us that she was going to try to support the *generality* consideration because she thought it was the most challenging epistemic consideration for students to use. In this example, we highlight what teacher practices Mrs. M employed that supported students in generating and revising mechanistic models that could be generalized to other phenomena.

"Remember to explain how and why!" One of the ways in which Mrs. M supported students' meaningful engagement in scientific modeling practice was by emphasizing the nature and purpose of the model (*nature*) throughout the unit and in different ways. This idea links to the disciplinary norms of developing and revising scientific knowledge that is explanatory in nature and address both the processes of the phenomena as well as the potential mechanisms of the phenomena. For instance, for every phenomenon of evaporation or condensation that the class generated, Mrs. M would always ask the similar "how and why" questions such as, "how and why does water appear on the cold can?", "How and why does dew form on the grass in the morning?" to reinforce the idea that it was important for students to attend to the mechanism of the phenomena with their models. Below is an excerpt that exemplifies the way in which Mrs. M emphasized explaining "how and why." The conversation occurred in the second lesson when she asked her students to construct their very first model of evaporation. In the excerpt, we bolded the utterances of Mrs. M that we coded as addressing the *nature* consideration.

Mrs. M:	You guys are going to be constructing an initial model [of evaporation]. So what changes did you observe over time? **How and why do you think those changes happen?** Those [questions] are hard to explain, aren't they? In other words, **how and why do you think the liquid seemed to disappear? How and why**, Jack?
Jack:	It's probably just natural.
Mrs. M:	OK, so **how did it happen?**
Emma:	They just evaporated.
Mrs. M:	What does that mean? **Where did it go?**
Emma:	Into the air.

At the beginning of the episode, Mrs. M asked the class "how and why do you think the liquid seemed to disappear?" Students had known the word 'evaporation' from earlier schooling, but had never unpacked what it meant or how it happened nor had they conducted any investigation to validate their hypothesis of evaporation. Therefore, Mrs. M's questions not only directed students to think about the mechanism of the phenomena, but the question was also open-ended to allow students to share their initial ideas. After Jack's answer "it's just natural," which is a typical answer at the beginning of the unit, Mrs. M problematized the phenomena by asking the class "how did it happen?" Her question sent the message to the class that "it's just natural" was not a sufficient answer and students needed to further unpack the mechanism to figure out the "how and why." As another student, Emma, responded with "it just evaporated," Mrs. M asked her to clarify what she meant by "evaporated." She also probed her about "where did the water go?" to scaffold her thinking about the process happening at the microscopic level, which is critical for understanding the mechanism. In this way, Mrs. M's question about tracing matter (e.g. water molecules in this case) across levels (macroscopic to microscopic) may have been especially helpful for students who did not think about the process going on at the microscopic level or who failed to make the connection between observable phenomena to a non-visible mechanism.

In addition, the fact that Mrs. M followed the more general question of "how and why do you think it happened?" with the more specific question of "where did the water go" also conveyed the message to the class that one way students could achieve the goal of explaining "how and why" was to figure out where the water went in the phenomena. In other words, Mrs. M not only established "explaining how and why" as the primary goal of their modeling, she also helped her students think about how to meet the goal by asking them specific questions of tracing water molecules. In fact, this was a common practice Mrs. M took throughout the unit. Whenever she was addressing the "how and why" of students' models, she followed up with questions with regard to tracing water molecules, either "where did the water go" or "where did the water come from" for phenomena of evaporation and condensation.

Not only did Mrs. M emphasize "explaining how and why" during model creation, she also prioritized "explain how and why" as one of the most important criteria for evaluating a scientific model since she understood that the goal of models is to explain phenomena. Whenever students showed their models to the class or their group members, she often asked students to show how their models addressed "how and why." The following episode exemplifies how Mrs. M supported her students to engage in model evaluation with a focus on "explaining how and why." The conversation occurred in the seventh lesson after students had constructed their initial models of condensation.

Mrs. M:	Would you please come up here and share your model with us, Tom? **I'm looking for the how and why pieces. I'm going to be asking you [class] to help Tom to tell us the how and why piece, the mechanism piece.**
Tom:	This is my model and this is the mirror and this is the water molecules evaporated from the shower. And then they'll come around to the mirror and there's a fog and there's some water droplets.
Mrs. M:	Is there anyone in here who has anything to say about this model? Whether we have any stars, any wishes, anything like that?
Tom:	Sarah?
Sarah:	I really like how you put that they are attracted to it [mirror], **but I'd like to know why.**

Starting at the beginning of this episode, Mrs. M was explicit about "explaining how and why" as the focus of the activity of model sharing and evaluation. She asked Tom to present his model with special attention to the "how and why" piece of his model. In the meantime, Mrs. M also directed class as the audience to pay attention to the mechanism piece as well. In this way, the class shared the common goal of explaining how and why as Tom began to share his model. It is important to note that Tom's presentation was no easy task, albeit brief, as it required him to understand (1) what Mrs. M meant by "how and why", and (2) what counted as the "how and why" in his own model. The fact that Tom was able to consider explaining how and why condensation occurred when creating the model and communicated how he met the goal of "explaining how and why" to the class indicated that he was engaged in modeling practices in a non-procedural way that aligned with epistemic goals of developing a mechanistic model. We suspect that Tom's mechanistic presentation of his model may be related to Mrs. M's foregrounding of the epistemic consideration of *nature*. In particular, Mrs. M's frequent questions about "how and why" in combination with specific matter-tracing questions throughout the unit may have contributed to his tracing of water molecules in his model.

After Tom's presentation, Mrs. M turned to the class for model evaluation. She asked if anyone had any "stars and wishes." (stars = praise, wishes = criticism). More importantly, the stars and wishes should be based on the epistemic considerations of modeling practices. In this conversation, since Mrs. M had made explicit that she was "looking for the how and why, the mechanism piece," the expectation she set up for the class was to provide any "stars" or "wishes" concerning how well Tom's model explained how and why condensation occurred. It is worth noting that this framing of the activity was not only meaningful from the disciplinary perspective, but more importantly, it is also accessible to students since Mrs. M had already set up the norms of evaluating models. As evident in Sarah's comment about "why water molecules are attracted to the mirror," she was familiar with the norms of giving "stars and wishes" based on "how and why." Additionally, Sarah's comment also showed that she had her own understandings of what counted as "how and why" and Tom's model did not address the "why" part. This is another indication of students' meaningful engagement in the modeling practices as Sarah was able to take up the epistemic consideration of "explaining how and why" and use it to evaluate other students' models.

"Your model should be able to explain ALL these phenomena!" Besides emphasizing the goal of "explaining how and why," Mrs. M also highlighted "explaining multiple phenomena" as one of the primary goals when engaging students in modeling practices. For example, when students were revising their models or constructing consensus models, she constantly challenged her students to use their models to explain other relevant phenomena that were familiar to students. Often times after a student presented their models in front of class, Mrs. M would follow up with questions about how that particular model was able to explain other similar phenomena. The following episode is representative of how Mrs. M set up the goal of "explaining multiple phenomena" in the modeling activities. The conversation occurred in lesson six when Mrs. M asked students to construct their consensus models of evaporation with group members.

Mrs. M:	**"Generality." A model needs to be used to explain multiple phenomena.** So what are some phenomena of evaporation? Give me some examples of evaporation that your model can explain.
Todd:	How hair dries.
Mrs. M:	So I would use my model to explain how hair dries. What else?
Cathy:	How nail polish dries.
Mrs. M:	Nail polish dries. I use my model to explain how nail polish dries.
Sam:	How the towel dries.

Mrs. M: So hanging your clothes out to dry. **So your model, needs to be able to, not just be used for one of those things, but your model needs to be able to be used to explain all of those things!** You guys are going to be given a task of creating a consensus model. **You need to decide, as a group, how you are going to create a model that represents the big idea of evaporation that can be used for all of the phenomena we were talking about earlier.** Tomorrow, I'm just gonna draw out of a hat, I'm just gonna say, [with] **your model, you're gonna tell me how you can smell dirty feet. [With] your model, you're gonna tell me, how the paint in Alicia's room dried. [With] your model, you're gonna tell me, how your bathing suit dries.** You have no idea of what I'm going to tell you, but your model is going to have to show me that.

In this episode, Mrs. M started with the overall statement of "a model needs to be used to explain multiple phenomena." Then she asked students to list examples of evaporation that they can use their model to explain. By doing this, Mrs. M provided scaffolding to help students better understand what *generality* meant in terms of students' own models and what specific phenomena their models should account for. Towards the end of the episode, Mrs. M made "explain multiple phenomena" the primary goal of consensus model construction, "You need to decide, as a group, how you are going to create a model that represents the big idea of evaporation that can be used for all of the phenomena we were talking about earlier." We argue that this framing of the activity was epistemic in nature, and fundamentally different than some other instructional goals teachers may have when engaging students in consensus model building. For instance, other teachers might be worried about student participation or group dynamics under such circumstances. The nature of students' engagement of modeling may look vastly different if those goals were set up as the primary goals of consensus model building.

Similar to how she helped her students to contextualize the goal of "explaining how and why," Mrs. M also provided scaffolding in her instruction to help her students better understand the epistemic consideration of *generality*. In particular, Mrs. M frequently asked her students to compare different phenomena so that they could recognize the connection between phenomena that share similar mechanisms and then figure out what aspects of the mechanism could be generalized to account for different phenomena. The following episode illustrates how Mrs. M scaffolded students to see the connection between two phenomena, "fog on the mirror" and "cold can." The conversation occurred in lesson seven when students were first introduced to condensation.

Mrs. M: **How are these two phenomena different?** The Coke can and the fog on the mirror. **Think about how they happen, think about where it came from.** There's enough room to explain your ideas so please do that.

(after students worked on the responses for 2 minutes)

Mrs. M:	**Now think how they are similar then.**
Mrs. M:	Jack, **tell me how they are different.**
Jack:	The Coke can is cold and the bathroom is hot and steamy.
Mrs. M:	OK. So **what is the same**, Anna?
Anna:	Well, on the mirror, you will find more than just fog, you will find droplets of water. So those droplets of water are very similar to what's on the can.

At the beginning of the episode, Mrs. M first asked students to identify the similarities and differences between the two phenomena, "cold Coke can" and "fog on the mirror in the bathroom." She further directed students to think about the "how and why" piece, "think about how they happened" and "where it came from." As demonstrated in the previous section, Mrs. M often asked general "how and why" question in combination with specific matter-tracing mechanism-oriented question such as "where did the water come from." In this case, by focusing on the similarities and differences of the two phenomena in terms of "how and why," Mrs. M intended to guide students to not only identify the analogous parts in the two phenomena, but also extract the aspect of the mechanism that could be applied to both scenarios. We argue that Mrs. M's emphasis on identifying the similarities and differences between two phenomena concerning the mechanism may potentially help students shift from not recognizing generality should/can happen, to mapping components to another context, and eventually to applying generalized mechanism to different contexts. While Anna's response did not address the mechanism of the two phenomena, she identified that the liquid water seemed to be appearing in both contexts. It is understandable that Anna did not identify the generalized process/mechanism that could explain both phenomena, since the activity occurred at the beginning of the condensation unit when students were just starting to make sense of phenomena of condensation without having learned much content knowledge.

As shown from the excerpts above, Mrs. M prioritized "explaining multiple phenomena" as one of the primary goals of modeling practices. She also scaffolded students to achieve the goal by comparing the similarities and differences between phenomena. Further, it is interesting to note that the scaffolding Mrs. M provided often involved examining the "how and why" piece of the phenomena, which may help students recognize and generalize the key components and processes that can be applied to various phenomena.

11.3.3 Teaching Using Epistemic Considerations and Its Potential Impact on Students' Modeling Practices

From the above analysis, we see how Mrs. M prioritized the epistemic considerations of *nature* and *generality* for modeling activities in class. However, how did this emphasis on "explaining how and why" and "explain multiple phenomena" impact students? Analysis of students' interviews clearly indicates that it did. Here we present one of our focus students (Jeanette)' post-interview analysis to illustrate the nature of students' modeling practices and discuss in what ways Mrs. M's instructional support might be productive at engaging students in the practice. We chose Jeanette as an illustrating example because, similar to other focus students, Jeanette made progress in both *nature* and *generality*. Also, she was reflective about her modeling experiences in Mrs. M's class.

The excerpts below highlighted what Jeanette's modeling practices looked like with respect to the epistemic considerations of *nature* and *generality*.

Interviewer: **Can you use your model to explain how condensation happens?**

Jeanette: So these are the water molecules in the air and they're flying around. But when they get closer to the can they slow down. And then they begin to clump together on the can and then more and more of them clump together and we see the fog.

Interviewer: **Why do they come to the cold can and come together?**

Jeanette: Because heat makes them go really fast. And when they get near a place that's colder, they lose the heat that makes them go fast and so they slow down.

Interviewer: **So what changes did you make to revise your initial condensation model?**

Jeanette: Well, in my first model there was no *how and why*. It didn't say *why* the molecules were doing that and *how* they were doing that.

Interviewer: **Let's try to use your model to explain some situations. For example, can you explain the phenomenon of why and how does the mirror in your bathroom fog up when you fill the bathtub with hot water?**

Jeanette: When you fill up a bathtub, the water is hot. And so that water evaporates and since the mirror is colder than the water molecules are, they will move around by it and slow down and form water droplets on it.

Interviewer: **OK, can you use your model to explain how and why rain falls? Do you think it's related or not?**

Jeanette: I don't think rainfall is related to condensation. Because when rain happens, it goes into a cloud and there's nothing solid or anything colder that they would actually go to and form on.

When asked to use her model to explain how condensation happens, Jeanette focused on the water molecules and she was able to describe the hidden mechanism at the molecular level. With her model, she identified an explanatory process, water molecules first "flying around in the air," and then "slowing down when they get closer to the can," and finally "clumping together on the can." She also made the connection between micro-level molecular movement (e.g. "clump together") to macro-level phenomena (e.g. "we see the fog"), which is challenging for students to conceptually comprehend, especially for young learners. In addition, Jeanette also made a chain of reasoning to further explain why water molecules slowed down.

Altogether, Jeanette's model-based explanation included an explanatory process at the molecular level to show how water condensed onto the cold can, a chain of reasoning to explain why the process happened, and a connection between micro-level molecular movement and the macro-level phenomena. It is also worth noting that she was not satisfied with her initial model of condensation as "there was no how and why." She elaborated on what she thought as "how and why". In contrast, in her revised model, Jeanette had both how they were doing that (water molecules slowed down and clumped together) and why they were doing that (water molecules lost heat as they come near a cold object). We argue that Jeanette was taking up the goal of "explaining how and why," which Mrs. M prioritized as the primary goal of modeling, and used the epistemic consideration to guide her revision of her initial model of condensation.

Later in the interview, Jeanette was asked to use her model to explain two other phenomena. When trying to use her model to explain how and why the bathroom mirror fogged up, Jeanette successfully generalized the key process going on in her own model, "water molecules slow down because of a colder object and form water droplets". When asked to use her model to explain rainfall, although she did not think her model was able to, it was evident that Jeanette was again trying to apply the same generalized process as she was looking for a colder solid object where water can go to. Jeanette was not only able to identify the analogous parts in both situations, she also went beyond the macro-level elements of the phenomenon. We argue that Jeanette's proficiency in *generality* may have been influenced by Mrs. M's instructional emphasis on "explaining multiple phenomena" as one of the primary goals of developing models.

11.4 Discussion

11.4.1 Epistemologies-in-Practice Framework and the Framework for Modeling Competence

We conclude this chapter by discussing how it contributes to this book and the larger modeling community in general. We also address how the EIP framework is similar or different from the FMC.

We see the frameworks are fundamentally different regarding their theoretical foundations as well as their intended use. While the EIP framework focuses on epistemic questions that inform the practice of modeling when figuring out phenomena from the situated learning perspective (Lave & Wenger, 1991; Wenger, 1998) the FMC foregrounds an individual's knowledge about models and modeling abstracted from the modeling practice and the need as provided by the context in which the modeling is taking place. In other words, while the EIP framework focuses on the purposes or goals of modeling that guide students' sense-making in particular modeling settings, the FMC points to individuals' general understanding about models (e.g., nature of models and multiple models) and the process of modeling (e.g., testing and changing models) assuming they can be applied to different modeling contexts. These are different perspectives as the EIP framework is situated in context and practice while the FMC is focused on abstracted knowledge about models and modeling in general.

Each of these frameworks aims for a different purpose. Our EIP framework, derived from our previous empirically-validated learning progression for scientific modeling (Schwarz et al., 2009; Schwarz et al., 2012), was designed to capture and characterize ways in which students engage in those key aspects of modeling (argumentation and explanation) practices. The EIP framework can also be applied to an individual or a community of learners as questions and goals that individual and communities are working towards collectively. As such, teachers can use the EIP framework as questions and goals to support students' development in modeling practices as Mrs. M's case illustrates. In contrast, the FMC captures knowledge and information about models and modeling as theoretically constructed across studies. Its purpose seems to align with diagnosing whether or not students have achieved certain levels of modeling competence through problem-solving.

Second, while the two frameworks differ with respect to their theoretical foundations and intended use, some of the sub-categories of focus relate to one another. The most obvious connection between the two frameworks is the focus on using models as sense-making tools to explain and predict phenomena: This focus is found in the *nature* epistemic consideration in the EIP framework and the *purpose of models* in the FMC. Nevertheless, as we discussed earlier, the word "explain" in the FMC is focused on describing the known relationship, as opposed to reasoning mechanistically, which is how it is often used in the U.S. Science Education Standards documents such as the Next Generation Science Standards (NGSS; NGSS Lead States, 2013). Therefore, the "how and why" message that Mrs. M was emphasizing throughout the unit goes beyond level II under the *purpose of models* in the FMC.

Another possible connection between the two frameworks lies in between *testing models* in the FMC and the *generality* consideration in our EIP framework. For example, Mrs. M's message about "explain multiple phenomena" involved the process of testing or applying the model with a context that is parallel with the initial context, which might be a level II in the FMC. However, what Mrs. M highlighted with respect to the *generality* consideration was more than just testing models with similar phenomena about evaporation. It also called for students to figure out a hidden mechanism that were general enough so that the model could be applied to those parallel situations. In that sense, the EIP framework is more fine-grained for capturing what is happening in the classroom, especially at the elementary or middle school level where prediction (level III, *purpose of models*; Chap. 1) may not be the ultimate goal of using models as a sense-making tool.

11.4.2 Teaching with Epistemic Considerations

In this chapter, we argue that EIP framework can be productive for teachers who seek to engage students in modeling practices in an epistemically-rich way. Mrs. M's case illustrates how it can be done in an authentic classroom context. In particular, the epistemological messages that Mrs. M consistently emphasized (e.g., "explain how and why" and "explain multiple phenomena") seemed to be especially productive. We hypothesized that it might be so because those messages together with Mrs. M's other pedagogical strategies (e.g. scaffolding, unpacking) may be accessible to students and open-ended enough so that students can see how the particular goals can be applied to their own models. Going back to Mr. L's case, one of the reasons why students did not seem to understand what counted as evidence for their models may be due to the fact that "good models need evidence" did not guide students in unpacking why that was important or how to apply the goal to students' own models. Imagine if Mr. L asked students instead how they could show their models are correct, that water molecules were actually going up to the air rather than seeping through the container. Students themselves may have referred back to the experiments involving using humidity detectors and decided to include the "humidity level" into their models to show the non-visible water molecules were actually going up into the air. Then the goal of the modeling activity would have become trying to figure out how to prove their models were correct as opposed to remembering what could count as evidence that should be included in the models.

While teachers sometimes struggle with how to enact scientific modeling in their classroom, Mrs. M's case shows evidence that there are disciplinarily productive ways in which teachers can engage students in modeling practices to potentially support the development of students' modeling competences. In particular, we found that Mrs. M's prioritizing the epistemic considerations of *nature* and *generality* as the primary goals of modeling practices seemed to productively influence students' modeling work. We see evidence from classroom interactions (e.g. Sarah's question about why) that Mrs. M's students had developed a strong sense of examining the "how and why" piece of phenomena and used that to navigate their own modeling work to create a more mechanistic model to better account for the phenomena. Our findings are consistent with and build on prior work around how teachers' framing and epistemological messages (Berland & Hammer, 2012; Russ & Luna, 2013; Russ, 2018) of activities in science classroom affects student participation and learning.

Moreover, our findings also suggest that simply emphasizing the epistemic considerations (e.g. "models should be able to explain and predict phenomena") is not likely to be sufficient. Our analysis of Mrs. M's teaching indicates that, in order to productively engage students in modeling practices in a meaningful way, a teacher also needs to provide scaffolding (e.g. asking students to trace water at the molecular level for *nature*; asking students to compare between two similar phenomena for *generality*) to help students contextualize the epistemic goals with their own models. In that way, students may have a better sense of how to achieve those abstract epistemic goals within the learning context that is relevant to them. We find that Mrs. M's blend of foregrounding the overarching goals and providing specific scaffolding consistently throughout the unit seem to be particularly productive in supporting students' development in modeling competences. Our result echoes with what Lidar and her colleagues (2006) found regarding the interplay between teachers' epistemological moves and students' practical epistemologies. In their study, they conjectured that some of the teacher's epistemological moves were not taken up by students probably because they were too challenging for students. Our study showed evidence that students were able to take up the "epistemological move" as Mrs. M's epistemological message was appropriately scaffolded and contextualized in a way that was approachable for students.

In addition to emphasizing epistemic considerations consistently throughout the unit, our analysis indicated that Mrs. M was successful in setting up the norms of modeling activities (e.g. "stars and wishes" for model evaluation) as well as the social interactions among students (e.g. consensus model building within groups). This seemed to be another critical aspect of Mrs. M's instruction that may potentially influence students' modeling practices. Setting up social norms around modeling practices is critical because it facilitates engagement with epistemic ideas across students. While our analysis did not specifically focus on how Mrs. M set up the norms of different modeling practices (e.g. how to give each other feedback about models, how to make a consensus model with group members), during our analysis, we often found that the norms had already been set up in the background of the activities as Mrs. M engaged students in the epistemic aspects of modeling. We point out that this is not a common practice shared by all teachers. The findings indicated that, in order to engage students in modeling in a disciplinarily meaningful way, it is necessary for students to be familiar with the norms of practice. Our finding is also consistent with what Colley and Windschitl (2016) found, that higher-rigor sense-making talks often occurred in association with conditions where the teacher was successful in engaging students in the social practice of talking to a partner or commenting on each other's ideas. More research needs to be done to further unpack the relationship between the social and epistemic aspects of engaging students in productive scientific practices.

Overall, our findings suggest that the combination of the following three elements are critical for supporting students in meaningful engagement in modeling practice, (1) prioritizing epistemic considerations as the primary goals of model development and revision, (2) providing scaffolding to contextualize epistemic considerations in concrete and approachable ways for students, and (3) setting up the norms of modeling practices within a community to support the development and revision of modeling practice/and ideas. In other words, there may not be a universal, but rather a suite of concerted instructional supports that teachers need to employ in order to productively engage student in modeling practices in a meaningful way.

11.5 Conclusions

As the research community studies modeling competence, our work contributes towards understanding how scientific modeling can productively unfold in classroom environments. In particular, the teacher's role and interactions are paramount for how modeling takes place and whether it becomes proceduralized school science or meaningful for a knowledge-building classroom community. This is particularly important given the "practice turn" and the epistemic nature of this critical practice.

What role do teachers play, and how does this matter? Our studies and those of others (e.g., Lehrer & Schauble, 2010; Vo, Forbes, Zangori, & Schwarz, 2015) indicate that teachers' enactments of modeling have real ramifications on students' modeling practices and whether modeling is taken up as a school practice or as a disciplinarily meaningful one. As such, these findings move beyond documenting teachers' knowledge of modeling and towards understanding how the practice unfolds in the classroom. Mrs. M guided students by highlighting and scaffolding epistemic messages while engaging them in the social work needed to do so.

Our work illustrates how one teacher productively engaged her students in modeling to support them in developing mechanistic models that generalized to other phenomena. Her teaching scaffolded students in developing modeling competence using epistemic considerations. There is still much more to be learned about how to meaningfully engage students in scientific modeling and how to support teachers in doing so. Understanding the interactions of teaching and learning along with supporting teachers to enact these productive interactions will be critical to advancing the field. In this way, all learners will be able to engage meaningfully in scientific modeling to make sense of the world.

Acknowledgments This work was supported by the Division of Research on Learning in Formal and Informal Settings of the National Science Foundation under Grant No. DRL-1020316 to the Scientific Practices Project at Northwestern University. The opinions expressed herein are those of the authors and not necessarily those of the National Science Foundation.

References

Baek, H., Schwarz, C., Chen, J., Hokayem, H., & Zhan, L. (2011). Engaging elementary students in scientific modeling: The MoDeLS 5th grade approach and findings. In M. S. Khine & I. M. Saleh (Eds.), *Models and modeling: Cognitive tools for scientific enquiry* (pp. 195–218). New York: Springer.

Baek, H., & Schwarz, C. V. (2015). The influence of curriculum, instruction, technology, and social interactions on two fifth-grade students' epistemologies in modeling throughout a model-based curriculum unit. *Journal of Science Education and Technology, 24*(2–3), 216–233.

Berland, L. K., & Hammer, D. (2012). Framing for scientific argumentation. *Journal of Research in Science Teaching, 49*(1), 68–94.

Berland, L. K., Schwarz, C. V., Krist, C., Kenyon, L., Lo, A. S., & Reiser, B. J. (2016). Epistemologies in practice: Making scientific practices meaningful for students. *Journal of Research in Science Teaching, 53*(7), 1082–1112.

Cohen, D. K., & Ball, D. L. (2001). Making change: Instruction and its improvement. *Phi Delta Kappan, 83*(1), 73–77.

Colley, C., & Windschitl, M. (2016). Rigor in elementary science students' discourse: The role of responsiveness and supportive conditions for talk. *Science Education, 100*(6), 1009–1038.

Danusso, L., Testa, I., & Vicentini, M. (2010). Improving prospective teachers' knowledge about scientific models and modeling: Design and evaluation of a teacher education intervention. *International Journal of Science Education, 32*(7), 871–905.

Engle, R. A., & Conant, F. R. (2002). Guiding principles for fostering productive disciplinary engagement: Explaining an emergent argument in a community of learners classroom. *Cognition and Instruction, 20*(4), 399–483.

Ford, M. J., & Forman, E. A. (2006). Chapter 1: Redefining disciplinary learning in classroom contexts. *Review of Research in Education, 30*(1), 1–32.

Gouvea, J., & Passmore, C. (2017). 'Models of' versus 'models for.'. *Science & Education, 26*(1–2), 49–63.

Henze, I., & van Driel, J. H. (2011). Science teachers' knowledge about learning and teaching models and modeling in public understanding of science. In *Models and modeling* (pp. 239–261). Amsterdam: Springer.

Justi, R., & Gilbert, J. (2003). Teachers' views on the nature of models. *International Journal of Science Education, 25*(11), 1369–1386.

Ke, L., & Schwarz, C. V. (2016). Examining the influences of teacher's framing of modeling practices on elementary students' engagement in scientific modeling. In C. Looi, J. Polman, U. Cress, & P. Reimann (Eds.), *Transforming learning, empowering learners: Proceedings of the 12th international conference of the learning sciences (2)*. Singapore, Singapore: ISLS International Society of the Learning Sciences.

Kelly, G. J., McDonald, S., & Wickman, P.-O. (2012). Science learning and epistemology. In *Second international handbook of science education* (pp. 281–291). Dordrecht, The Netherlands: Springer.

Kenyon, L., Schwarz, C., & Hug, B. (2008). The benefits of scientific modeling. *Science and Children, 46*(2), 40–44.

Krist, C., Schwarz, C. V., & Reiser, B. J. (2019). Identifying Essential Epistemic Heuristics for Guiding Mechanistic Reasoning in Science Learning. *Journal of the Learning Sciences, 28*(2), 160–205.

Lave, J., & Wenger, E. (1991). *Situated learning: Legitimate peripheral participation*. Cambridge, UK: Cambridge University Press.

Lehrer, R., & Schauble, L. (2006). Scientific thinking and science literacy: Supporting development in learning in contexts. In W. Damon, R. M. Lerner, K. A. Renninger, & I. E. Sigel (Eds.), *Handbook of child psychology* (Vol. 4, 6th ed.). Hoboken, NJ: Wiley.

Lehrer, R., & Schauble, L. (2010). What kind of explanation is a model? In *Instructional explanations in the disciplines* (pp. 9–22). Boston: Springer.

Lidar, M., Lundqvist, E., & Östman, L. (2006). Teaching and learning in the science classroom: The interplay between teachers' epistemological moves and students' practical epistemology. *Science Education, 90*(1), 148–163.

Manz, E. (2015). Resistance and the development of scientific practice: Designing the mangle into science instruction. *Cognition and Instruction, 33*(2), 89–124.

NGSS Lead States. (2013). *Next generation science standards: For states, by states*. Washington, DC: National Academies Press.

Russ, R. S. (2018). Characterizing teacher attention to student thinking: A role for epistemological messages. *Journal of Research in Science Teaching, 55*(1), 94–120.

Russ, R. S., & Luna, M. J. (2013). Inferring teacher epistemological framing from local patterns in teacher noticing. *Journal of Research in Science Teaching, 50*(3), 284–314.

Schwarz, C. V., Passmore, C., & Reiser, B. J. (2017). *Helping students make sense of the world using next generation science and engineering practices*. Arlington, VA: NSTA Press.

Schwarz, C. V., Reiser, B., Acher, A., Kenyon, L., & Fortus, D. (2012). MoDeLS: Challenges in defining a learning progression for scientific modeling. In A. Alonzo & A. Gotwals (Eds.), *Learning progressions in science (LeaPS)* (pp. 101–137). Boston: Sense Publishers.

Schwarz, C.V., Reiser, B.J., Davis, E.A., Kenyon, L., Achér, A., Fortus, D., et al. (2009). Developing a learning progression for scientific modeling: Making scientific modeling accessible and meaningful for learners. *Journal of Research in Science Teaching*, *46*(6), 632–654.

Stroupe, D. (2014). Examining classroom science practice communities: How teachers and students negotiate epistemic agency and learn science-as-practice. *Science Education, 98*(3), 487–516.

van Driel, J. H., & Verloop, N. (1999). Teachers' knowledge of models and modelling in science. *International Journal of Science Education, 21*(11), 1141–1153.

Vo, T., Forbes, C. T., Zangori, L., & Schwarz, C. V. (2015). Fostering third-grade students' use of scientific models with the water cycle: Elementary teachers' conceptions and practices. *International Journal of Science Education, 37*(15), 2411–2432.

Wenger, E. (1998). *Community of practices: Learning, meaning, and identity*. Cambridge, UK: Cambridge University Press.

Chapter 12
A Responsive Methodological Construct for Supporting Learners' Developing Modeling Competence in Modeling-Based Learning Environments

Todd Campbell, Thomas J. McKenna, Jihyun An, and Laura Rodriguez

12.1 Introduction

Science studies research has revealed how models serve as context dependent tools for organizing the day-to-day sensemaking work of scientists (Passmore, Gouvea, & Giere, 2014). When this is considered in recent calls for science learning environments to position students as legitimate participants in the social, epistemic, and material dimensions of science (Ford & Forman, 2006; Lehrer & Schauble, 2006), modeling-based learning (MBL) classroom environments (Louca & Zacharia, 2012) emerge as important for more authentically representing scientific activity and supporting students' developing modeling competence as they focus on explaining events that happen in the world. Here, the notion of modeling competence is framed in alignment with the functional-pragmatic concept of competence, especially since the main focus is on supporting students' abilities to cope with challenges (e.g., explaining phenomena) and using models as epistemic tools across a range of contexts (Klieme, Hartig, & Rauch, 2008). Therefore, the importance of MBL environments lies in how models are used by students as epistemic tools for organizing their day-to-day work across instructional units. In this, students are positioned as epistemic agents (Scardamalia, 2002; Stroupe, 2014) to work at knowing with modeling as a central knowledge development practice (Sandoval, 2015) that is stabilized through the interplay of their (i.e., students') ideas, the material

---

T. Campbell (✉) · J. An
University of Connecticut, Storrs, CT, USA
e-mail: todd.campbell@uconn.edu

T. J. McKenna
Boston University, Boston, MA, USA·

L. Rodriguez
Eastern Connecticut State University, Willimantic, CT, USA

A. Upmeier zu Belzen et al. (eds.), *Towards a Competence-Based View on Models and Modeling in Science Education*, Models and Modeling in Science Education 12, https://doi.org/10.1007/978-3-030-30255-9_12

world, evidences, and other practices in representations of scientific activity (Ford, 2005; Manz, 2015; Pickering, 1995; Rouse, 2007).

However, because the work of supporting students' engagement in modeling is complex, increased attention is needed for understanding the role of teachers in these learning environments (Griesemer, Lo, Guy, Harris, & Passmore, 2016; Kahn, 2011; Schwarz et al., 2016), especially when teachers work to position students as epistemic agents and their ideas as resources for sensemaking within these environments. Given this, a methodological construct that teachers can use in MBL environments to support sensemaking with students' ideas is the focus of this chapter. More specifically, redirection is the primary methodological construct that is explored in this chapter in the context of a high school physics classroom. It is foregrounded in the study of a MBL environment because of how it was previously found to be useful in support of sensemaking with students' ideas in classroom communities (Lineback, 2015), especially since sensemaking with students' ideas (e.g., partial understanding of scientific ideas, nonstandard ideas, and everyday experiences) is at the very heart of the epistemic practice of modeling (Gouvea & Passmore, 2017). Further, responsive teaching is prioritized as a central priority of redirection that supports teachers as an approach for taking up their students' ideas 'in the moment' and helping foreground those ideas so that the anchoring practice of modeling in the MBL environment supports the refinement of students' ideas. More specifically, Lineback described *redirection* "as instances when a teacher invites students to shift or redirect their attention to a new locus" (p. 419). Through redirection, teachers can support students by foregrounding (un)productive ideas so that they can use localized ways of knowing, especially modeling as a knowledge development practice in MBL environments, to determine the appropriateness of foregrounded focal ideas in the specific context in which they are being used (Lineback, 2015).

Because of the potential promise of redirection and the need to better understand and make explicit ways in which teachers can support student sensemaking and development of modeling competence in MBL environments, this chapter explores how teachers might be responsive to students' sensemaking with redirection in MBL environments.

12.2 Theoretical Perspectives

12.2.1 MBL Environments and Representations of Scientific Activity

MBL learning environments, for the purpose of this research, are defined in alignment with Louca and Zacharia (2012) as "an approach for teaching and learning in science whereby learning takes place via student construction of models as representations of physical phenomena" (p. 471). These environments are important in science education, since researchers like Manz (2015) point to the need for careful consideration of "what features of scientific activity systems situate the meaning of the practice in professional activity as well as whether and how we can represent those features in classroom environments" (p. 556). Given that Passmore, Gouvea and Giere (2014) note that science studies research has revealed how models serve as context dependent tools for organizing the day-to-day work of scientists, these features of scientific activity can be said to situate the meaning of practice in scientists' professional activity. Further, researchers have previously documented how models can serve as context dependent tools for organizing the day-to-day work of students across units of instruction in the service of representing scientific activity (e.g., Manz, 2015; Passmore et al., 2014; Stroupe, 2015). Here, scientific activity is understood in terms of activity theory (Vygotsky, 1978). More specifically, just as activity theory is concerned with the dialectic of subjects, tools, and objects in human pursuits (Roth & Lee, 2004), scientific activity is concerned with the dialectic of scientists, tools useful in scientific pursuits (e.g., science practices, disciplinary knowledge), and the objects of their pursuits (i.e., constructing and critiquing explanations of things that happen in the world (Ford, 2008)). Consequently, with MBL as one example of a representation of scientific activity in science classrooms, students (subjects) engage in developing and using models (tools) in concert with other science practices (e.g., engaging in argumentation) across an instructional unit to iteratively explain a unit anchoring phenomena (object) (Melville, Jones, & Campbell, 2017).

12.2.2 Modeling Competence

As alluded to earlier, the conception of modeling competence adopted in this chapter is aligned with the functional-pragmatic conception of competence (Klieme et al., 2008), whereby modeling competence can be understood as students' abilities to cope with challenges (e.g., explaining phenomena or solving problems) using modeling as an epistemic tool across a range of contexts (Chap. 1). In this conception, unlike decontextualized cognitive systems that are developed in isolated contexts and later deployed, competence is considered a context-specific ability that is sensitive to contextual demands and acquired by learners in situ (Klieme et al., 2008).

In Germany, competence models were adopted to take into account the shift in classrooms away from solely focusing on the acquisition of disciplinary science concepts to focusing more on the application of these concepts in meaningful contexts. In the U.S., this shift emphasizes the movement away from a focus in science classrooms on students 'learning about' disciplinary scientific concepts outside of meaningful contexts to a focus on students 'figuring out' how to use disciplinary science concepts and science practices to explain phenomena that happen in the world or to solve real world problems of consequence (Krajcik, 2015; NGSS Lead States, 2013). In this shift to 'figuring out' highlighted in the most recent U.S. standards documents, disciplinary scientific knowledge remains centrally important, however from the perspective of students its usefulness becomes apparent or functionally pragmatic in as much as it is helpful in explaining phenomena, generating new knowledge through predictions and investigations, and solving problems (Chap. 1). In addition, the focus on 'figuring out' in the newest standards documents has elevated the importance of science practices, like developing and using modeling competence being explored in this volume, as the tools students use to critique and refine explanations or solutions to problems as part of engaging in more authentic representations of scientific activity (Ford, 2015; Stroupe, 2015). Additionally and more specifically to the focus on modeling competence in this volume, the importance of developing students' modeling competence as a research tool is well aligned to the emphasis in the newest U.S. standards documents, especially since Chap. 1 reveals how modeling competence considers, among other things, the extent to which models "are used as tools in the acquisition of new insights" (p. 1) and reveals how "the goal . . . is to gain insightful knowledge with models" (p. 5).

In this chapter, the focus on student modeling competence was approached by examining a promising teacher-enacted methodological construct (i.e., redirection) that could potentially support the condition (i.e., an environment where ideas are foregrounded and scrutinized) under which student modeling competence might flourish. Such a condition, we postulate in alignment with others (Coffey, Hammer, Levin, & Grant, 2011; Lineback, 2015; Thompson et al., 2016), is one that takes into account the extent to which teachers are responsive to students' ideas and sensemaking practices and the relation of their ideas to disciplinary scientific ideas. Further, this positions students to coordinate their sets of ideas (Stewart, Cartier, & Passmore, 2005) to meet the cognitive demands of explaining a range of phenomena across contexts in more authentic or closer to 'real life' settings, so that student modeling competence is activated and further developed in situ (Koeppen, Hartig, Klieme, & Leutner, 2008).

12.2.3 Responsive Instructions and Redirection as a Responsive Methodological Construct

> The pedagogical task for teachers, then, is … to build upon students' initial ideas, partial understandings, and everyday experiences to support construction of on-going, evidence-based, and generalizable explanatory accounts of natural phenomena. (Thompson et al., 2016, p. 4)

This quote exemplifies the complex role of teachers in classrooms where supporting students in making progress connecting their ideas and experiences to developed and refined disciplinary science ideas over time is prioritized (Coffey et al., 2011). The growing body of research in science and mathematics education focused on understanding and supporting teachers' roles in recognizing, taking up, and assisting students in developing and critiquing theirs' and their peers' ideas over time is grounded in research such as teacher noticing (e.g., Sherin & van Es, 2005) and formative assessment (e.g., Ruiz-Primo & Furtak, 2007), and can be referred to as responsive instruction. Thompson et al. (2016) further connect responsive instruction to culturally responsive teaching (e.g., Gay, 2000), especially to the extent that it moves away from deficit framing of students' ideas, capabilities, and lived experiences. At the same time, responsive instruction presses against historical structural ways knowledge in classrooms is produced through classroom interactions in efforts to support more equitable opportunities for all students to participate. In other words, responsive instruction can be understood as providing students space and support to reason about phenomena or events that happen in the world by proposing their own ideas developed from prior experiences. Further, as part of responsive instruction, student ideas proposed as useful for reasoning about phenomena are subsequently foregrounded in discursive exchanges with peers and the teacher so that these ideas can be scrutinized and refined as they are connected with additional evidences and canonically relevant disciplinary science concepts.

However, as acknowledged by Lineback (2015), "[t]eachers are thereby placed in the rather challenging position of navigating among the various thoughts and viewpoints present, weighing the merits of pursuing one or more of those ideas and making a decision as to how to follow up those ideas in the moment during class" (p. 420). Given this, Lineback's work identified redirection as one responsive methodological construct that could be potentially useful for researchers and teachers alike for understanding how teachers navigated supporting student idea refinement in sensemaking classrooms.

Further, Lineback noted that to be considered a redirection, the teacher's attempt must communicate or provide the impression that a response from students is desired and space for that response be offered. Lineback identified two different types of redirection, an activity redirection and a focus redirection. The activity redirection can be understood as the teacher's bid to shift students from one activity to another (e.g., a discussion to designing an experiment), while the focus redirection can be understood as a teacher's attempt to shift the focus of students' attention from one scientific phenomenon or question to another. Each of these different types of redirection can be further nuanced according to Fig. 12.1 and it should be noted that Lineback reported that activity redirections occur less frequently and are easier to assess in terms of their presence in comparison to focus redirection. Finally, Lineback identified additional focus redirection codes that considered the extent to which and how the redirection was connected to students' comments (e.g., F_1-responsive directly connect to student idea(s) in previous exchange; F_2-delayed responsive disconnected from students' previous exchange, but revisited previously discussed phenomena, idea, or question).

Because of previous collaborations with the teacher in this research to explore and refine instructional approaches for supporting the development of student modeling competence within MBL environments (e.g., Campbell, Oh, & Neilson, 2013; Campbell & Neilson, 2009, 2012), as well as previous research demonstrating the teacher's developing facility engaging students in MBL environments (i.e. Campbell, Oh, & Neilson, 2012; Campbell, Zhang, & Neilson, 2011), we believed that closely examining this instruction using the redirection coding scheme could reveal informative strategies within the MBL environment that may have remained unnoticed.

Consequently, the following questions, focused this study:

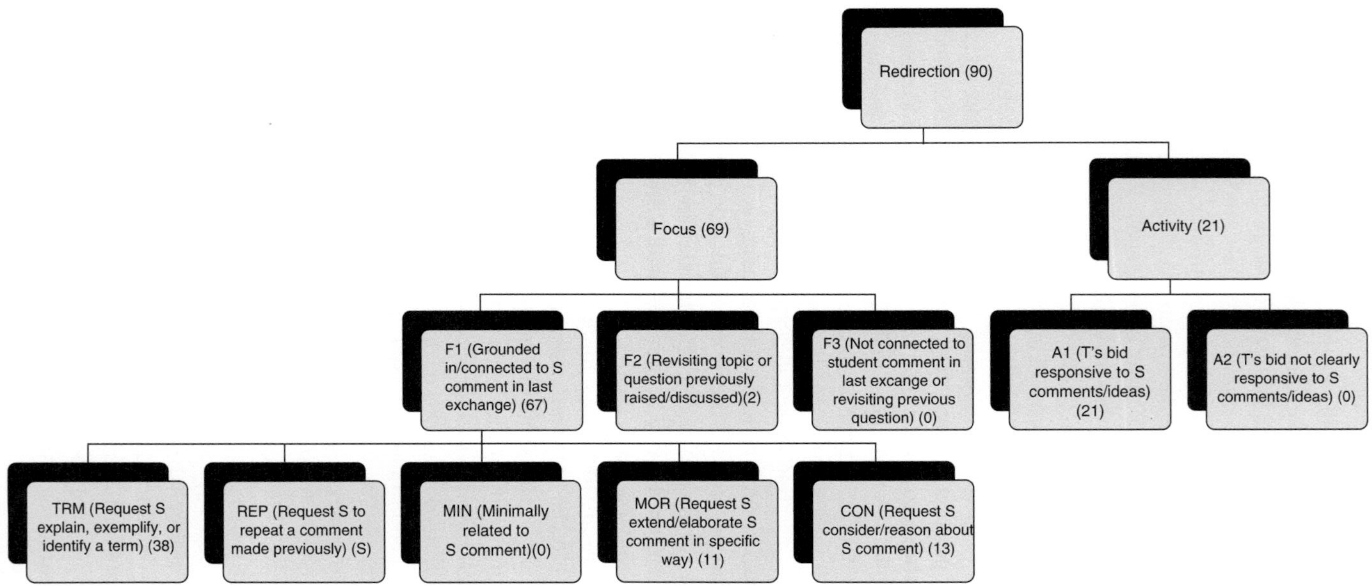

Fig. 12.1 Redirection Coding Scheme (adapted from Lineback, 2015). (Note: *T* Teacher, *S* Student, (#) Each Redirection Type Found)

1. To what extent and in what ways was redirection used as a responsive methodological construct in an MBL environment?
2. How can the use of redirection in an MBL environment be characterized?
3. What factors interact with a teacher's use of redirection within an MBL environment?

12.3 Methods

12.3.1 Context

The context of this investigation into the ways in which teachers might use a responsive methodological construct to support students' development of modeling competence in MBL environments was Mr. Bird's (pseudonym) physics classroom in the spring of 2013. Grades 10–12 students (age 15–18), with a few Grade 9 students (age 14–15), enrolled in this physics course. At the time of this data collection, Mr. Bird had taught physics for 15 years.

Mr. Bird collaborated extensively with the first author over the previous 7 years to develop MBL instructional units, refining them in response to classroom enactments, and working to understand their importance in science teaching and learning. More specifically, as the first author and Mr. Bird iterated over time how they planned units of instructions that constituted the MBL environment examined in this research they relied more generally on the Ambitious Science Teaching (AST) framework put forth by Stroupe and Windschitl (2015), whereby they (a) AST1-plan[ned] a unit around a "big science idea", (b) AST2-elicit[ed] and activat[ed] students' ideas about a puzzling phenomenon (for the purpose of adapting instruction), (c) AST3-help[ed] students make sense of science activities (with the aim of using science principles behind the selection of the activities to explain unit-anchoring phenomena), and (d) AST4-press[ed] students to construct evidence-based explanations" (p. 181).

Buoyancy was the focus of the MBL unit examined in this research, since it was believed to provide a context for applying Newton's Laws and supporting students in better understanding fluids. Figure 12.2 outlines the buoyancy unit investigated.

12.3.2 Data Collection and Analysis

Digital recordings of the five class periods served as the primary data for this research. All five class periods were transcribed so that both the recordings and transcripts could be used for analysis. An adapted version of Groenwald's (2004) phase strategy for explicating data was adopted for this research. In this, the first author completed the initial data coding and analysis, before the other three authors

Class period	Topic	Main activity	Instructional Purpose
1	Introduction to phenomena and initial explanatory models	Students discussed the phenomena, developed a 'Gotta Have List' and started development of initial explanatory models	Eliciting students initial ideas about the unit anchoring phenomena
2	Completed initial explanatory models and planned investigations to inform developing models	Students completed initial explanatory models, identified areas of their models that would benefit from additional evidence that could be gathered from investigations and planned investigations that could be carried out in the laboratory	Supporting students in identifying possible factors they thought would affect buoyancy
3 and 4	Laboratory investigations	Students completed design of investigations, carried out investigations in laboratory and prepared to share results of investigations in whole-class discussion	Investigating factors thought to affect buoyancy
5	Whole-class discussion of experimental findings	Students shared their experimental findings with the class in a whole-class setting.	Pressing students to share their laboratory findings and make sense of them in terms of how their findings might contribute to their evolving model of the unit anchoring phenomena

Fig. 12.2 Buoyancy unit

reviewed the coding in the context of transcripts. If disagreements or questions arose about a specific code or theme as the final three authors reviewed the first author's coding in the context of the transcripts, the researchers revisited the original archived videos and student artifacts and sought consensus of interpretation before finalizing the codes and themes. Analysis proceeded through a recursive process where emergent findings were continually checked and revisited as new findings emerged until it was believed that consistency of interpretation was accomplished.

12.4 Findings

The findings are organized by the research questions and include relevant transcript excerpts from Mr. Bird's classroom of evidence used for the claims presented.

1. *To what extent and in what ways was redirection used as a responsive methodological construct in an MBL environment?*

To answer the first question, a report of the general number and kinds of redirection instances found is shared. First, however the following is offered as one instance of redirection to provide a sense of what redirection looked like in Mr. Bird's classroom:

Mr. Bird: *Buoyancy, that's a term that we haven't defined yet, so it worries me. You said—*

Student 1: *Floating ... The ability of the object or something to float ...*

Mr. Bird: *Can you have different degrees of buoyancy? If you sink, you're not buoyant, is that right? ...*

Student 2: *I think there are different levels. Like, there can be something that just goes right to the bottom of whatever and stays there and will hardly move. Whereas, something else could bounce around I think, so it has the ability to move around, not just sink in water ...*

Mr. Bird: *Tell me what happens when you get in water ...*

Student 2: *Depends on who you are. Some people float, some people sink ...*

Student 3: *You are going to float. Water pushes you up ... Buoyancy would be how much the fluid pushes you up—if something's more buoyant, then the fluid will push it more. If it's less buoyant, the fluid won't push it as much*

Mr. Bird: *What if we say that's what buoyancy is there? It's the amount that water pushes you up.*

This instance of redirection occurred as the class was working to create a list of ideas that might be important when developing models to explain the three, related unit-anchoring phenomena. In this, when Mr. Bird found students frequently using the term buoyancy he redirected their attention to defining the term buoyancy, since it appeared that he did not yet feel comfortable that they were using the term in a consistent way. This example was coded as 'request[ing] the students consider/reason about student comment(s)' (i.e., F_1 CON). The other types of redirection identified can be found in Fig. 12.1.

Beyond these more general descriptors of the types of redirection reported in Fig. 12.1, it was noted that most of the ways that Mr. Bird interacted with students in the unit were characterized as redirection. Exceptions to this were minimal and included only when he introduced the demonstrations or phenomena on the first class period of the unit or outlined logistical directions to organize students' engagement in the unit either towards the beginning or end of each class period within the unit. This is perhaps most evident in the large number of instances of redirection (i.e., 90) found across the 5-class period unit.

2. *How can the use of redirection in an MBL environment be characterized?*

To answer this question, we examined more closely the types of redirection Mr. Bird used. Based on this, the following trends were noted: (a) F_1 TRM redirection was the most common type of redirection used across the unit and it was used almost exclusively on the final class period of the unit; (b) more variability in the types of redirection used occurred during Class Periods 1 and 2 compared to Class Periods 3–5; (c) activity redirection was used more during Class Periods 2–4 as students were identifying possible factors they thought affected buoyancy, carrying out investigations about these variables, and collecting data.

Notably, F_1 TRM redirection was widely used by Mr. Bird across the unit (i.e., 38/67 instances of redirection). As a reminder, this type of redirection was characterized as clearly grounded in or connected to student comments that emerged during the last exchange sequence that further requests students explain, exemplify, or identify a term. The fact that F_1 TRM redirection was so prevalent provides some insight into the responsive practices Mr. Bird used. In this, he was frequently found asking students to further explain their ideas or the basis of their claims. Additionally, as noted, Mr. Bird relied almost exclusively on this type of redirection in Class Period 5 of the unit with 22 instances of redirection identified and 18 of these being F_1 TRM redirection. Interestingly, the variability of the types of redirection Mr. Bird used across the unit decreased. This did not mean that less redirection was found later in the unit, since 22 instances of redirection were identified during Class Period 5 were comparable to the number of instances found in Class Periods 1 and 2 (i.e., 18 instances and 24 instances, respectively). Instead it merely signaled that there was more variability in the types of redirection used earlier in the unit. More specifically, Mr. Bird relied more on F_1 TRM type redirections in Class Periods 3–5, whereas earlier in the unit he relied on F_1 REP, F_1 MOR, and F_1 CON either more or equally compared to his use of F_1 TRM in the earlier class periods of the unit.

Lastly, activity redirection was used more during Class Periods 2–4 of the unit. These class periods coincided with the time students spent identifying possible factors students thought would affect buoyancy that would be tested in the laboratory as part of designing investigations and collecting data. The following is one example of an activity redirection from Class Period 4:

Mr. Bird:	*I got a test for you. We got—I think it's a great test, but what would one of our controls have to be?*
Student 6:	*The depth?*
Mr. Bird:	*The depth. You guys need to get a big tote if you can find one.*
Student 7:	*Like that?*
Mr. Bird:	*Even bigger. We'll try and keep the depth, okay?*
Student 7:	*Okay.*
Mr. Bird:	*Okay, so let's clear this out of the way. The problem that we had before was that one floated, right? We need to figure out a way to measure the buoyant force and so we're gonna use this pulley to help us. Here's what I'm thinking. We could put this at the bottom and for—what we could do is we could see how much force it took to pull it under water and hold it under water, right? Then we'll know how much water's pushing up on it, right? It's gonna try and push it up. Let's see what kind of data we get.* *Okay, let's hook up our first object.*

Prior to this episode the students and Mr. Bird recognized a flaw in how they had previously been collecting data in the laboratory to try to determine whether buoyant force changed as an object was submerged at increased depths in a liquid (i.e., water). This episode exemplifies how Mr. Bird was found making bids for students to change the activity they were doing based on issues he and the group discovered in a previous exchange. In this, it appeared as if he was using activity redirection as a means for helping students identify a different strategy other than what they may have thought of by themselves to pursue an idea they initially put on the table (i.e., that the depth an object is submerged in a liquid affects the amount of buoyant force on the object from the liquid).

3. *What factors interact with a teacher's use of redirection within an MBL environment?*

When considering what factors interacted with Mr. Bird's use of redirection, it was noted that an increased amount of redirection was used during Class Periods 1, 2, and 5 of the unit when compared to Class Periods 3 and 4. The decreased use during Class Period 3 might be attributed to a shortened class period in comparison to the other class periods, yet these trends also made sense in context with the different instructional purposes framed for each lesson explicated in Fig. 12.2. Redirection was found during Class Periods 3 and 4, but this occurred less often, mainly in small groups, and as a way for Mr. Bird to better understand how individual groups were thinking about buoyancy related to the laboratory investigations they had designed and were completing.

12.5 Discussion

At a time when increased attention has been given to supporting students' engagement in functionally pragmatic science practices including modeling competence, as is the focus of this volume, in the service of developing explanatory mechanistic accounts of real world phenomena, there is growing recognition of the need for increased attention to the role of teachers in such environments (Griesemer et al., 2016; Kahn, 2011; Manz & Renga, 2017; Schwarz et al., 2016). This research begins to address this need through revealing the extent and manner in which one teacher used redirection as a responsive methodological construct across an MBL instructional unit. In this section, we revisit our research questions to consider the extent to which our analysis of Mr. Bird's use of redirection revealed nuances of his responsive commitment to student idea refinement and how this commitment played out in terms of the transactional role he engaged in with students across the unit of instruction. We end the chapter considering potential implications of our analysis related to making more explicit the teacher's role in supporting students' development of functionally pragmatic modeling competence through redirection in MBL environments and additional research that may prove useful in building on what was learned in this research.

1. *To what extent and in what ways was redirection used as a responsive methodological construct in an MBL environment?*

Lineback (2015) proposed redirection as one methodological construct that might begin to characterize the role of teachers in responsive classrooms committed to student sensemaking. In this research within the MBL unit, redirection was found to be an important methodological construct for not only foregrounding ideas to help set the stage for agentic student pursuits during Class Periods 1 and 2 of the 5-class period unit, but also for helping them navigate investigations during Class Periods 3 and 4, and pressing them for evidence-based explanations in Class Period 5 of the unit. Evidence for this lies in how redirection was found as a mechanism that could, with the exception of logistical directions or what others have referred to as meta-talk (e.g., Campbell, Oh, & Neilson, 2013; Manz & Renga, 2017) be used early or later in each class period of the unit and be applied to characterize what Mr. Bird did throughout the unit. Researchers like, Thompson et al. (2016), point to how curriculum is necessary, but not sufficient for supporting rigor and responsiveness in science classrooms. In this, they suggested that "the interactions within the classroom are essential for sustaining the highest quality of scientific practice and sensemaking" (p. 52). In this research, the interactions that Mr. Bird engaged in with students within the classroom suggested possible mechanisms he used in his attempts to sustain the highest quality of scientific practice (i.e., students' iterative engagement in developing and using models across the unit) and sensemaking. Specifically, Mr. Bird relied on redirection, mainly in the form of focus redirection whereby he attempted to shift the focus of students' attention from one scientific phenomenon or question to another. And, as evidenced from a large majority of the types of redirection Mr. Bird used (i.e., 67 F_1 Focus Redirections, compared to 2 F_2 and no F_3 Focus Redirections), he was almost exclusively found shaping his response or the redirection he used in response to the ideas of students that emerged in the previous exchange.

When these findings are considered in context of the limited amount of other research on responsive instruction in classrooms with redirection more about Mr. Bird's use of redirection can be understood. More specifically, Lineback (2015) identified two different ways in which the same teacher was interacting with students in sample episodes she analyzed. In one episode, the teacher she followed prompted students to share their ideas, but was not found following up on her initial questions in ways that pressed students to elaborate on their thinking or to pursue a particular path of thinking, instead in this type of episode the teacher did not request that students respond to anything in particular that they or their peers may have said. Consequently, they were permitted to "take up any topic they wished", something that in the end resulted in "conversation[s] . . . meander[ing] without pushing her students to pursue any particular student's idea extensively" (p. 426). However, in other instances, Lineback found episodes where the teacher pursued "clarifications and/or elaboration from individual students on their own contributions" . . . she "actively encourage[d] her students to extend one another's comments" (p. 426). This second set of episodes characterized as the teacher asking for clarifications or elaborations, were more aligned with the responsive ways in which Mr. Bird was found helping students follow their lines of logic as a responsive form of instruction that was evidenced in the almost exclusive type of F_1 and A_1 types of redirection he used and the lack of F_3 and A_2 types of redirection.

2. *How can the use of redirection within an MBL environment be characterized?*
3. *What factors interact with a teacher's use of redirection in an MBL environment?*

Because the characterization of how redirection was believed to be intricately connected to the factors that interacted with Mr. Bird's use of redirection, the discussion of findings for these two research questions have been merged. Importantly the different activities planned for different purposes across the MBL unit contributed to the emergence of the characterizations of the use of redirection within an MBL unit, while also standing out as the most notable factor that interacted with the use of redirection in this current research. More specifically, as the first author and Mr. Bird iterated over time how they planned units of instructions that constituted the MBL environment examined in this research, they relied more generally on the Ambitious Science Teaching (AST) framework put forth by Stroupe and Windschitl (2015) described in more detail earlier. When this AST Framework was considered alongside the findings, a potential explanation for the emergent redirection trends surfaced. In the buoyancy unit examined in this research during Class Period 1 of the unit, Mr. Bird sought to elicit students' initial ideas for how they might explain the unit-anchoring phenomena. In this stage of the AST Framework (i.e., AST2, 2014a) there is a need for teachers to engage students in ways that will illuminate and help them understand the range of ideas, experiences, and language or ways of talking and thinking that students use in thinking about the anchoring phenomena (AST, 2014a). Class Periods 2–4 of the unit coincided with AST3 or the stage of the AST framework focused on helping students make sense of science activities. In this particular unit, Mr. Bird used Class Period 2 of the unit to draw on students' initial ideas. Students shared their models during Class Period 1 to identify factors within their initial models they proposed affected buoyancy as a focus of the activities that students engaged in during AST3. More specifically, in AST3 in this unit, students designed investigations that would allow them to collect data to determine whether or not their initial ideas were supported by evidence collected in the laboratory. This stage of the AST framework is intended to "help students develop new ideas to use in revising their explanatory models for the anchoring phenomena" (AST, 2014b, p. 1) and can involve activities that range from teacher demos to students designing their own study or working with second hand data. In this particular unit, as noted earlier, students designed their own investigations, carried them out, and used the emergent data as a mechanism for developing new ideas that were useful in revising their initial explanatory models. Finally, Class Period 5 of the unit aligned with AST4 of the AST framework. This stage of the AST framework, is designed to help students "rally different kinds of evidence in support of their culminating explanations" . . . during this stage they "construct and evaluat[e] claims" and "draw final ideas together in models and explanations" (AST, 2014c, p. 1). In

Class Period(s)	AST Framework	Types and Purpose of Redirection
1-2	AST2-Eliciting students' ideas	Ranging from F_1 TRM to F_1 MOR and F_1 CON to elicit the range of ideas students had about how to explain the anchoring phenomenon
3-4	AST3-Helping students make sense of science activities	F_1 TRM to get students to explain why they were doing the investigations they were doing or Activity Redirection (A_1) to make suggestions for changing the activity in which students were engaged if another activity thought more productive in helping students explain the anchoring phenomena
5	AST4- Pressing students to construct evidence-based explanations	F_1 TRM to ask students to explain something they said in a previous exchange at the end of the unit

Fig. 12.3 Unit class periods and AST framework connected to types of redirection used

the buoyancy unit, during Class Period 5 Mr. Bird invited students to share, in whole class discussion, their findings from their laboratory investigations with the aim of pressing them to articulate claims about their data that could be used in their final explanatory models.

Figure 12.3 provides an abbreviated summary of how the class periods of the unit connected to the AST framework and how this was found related to the types of redirection used that is further explicated next.

As can be seen in Fig. 12.3, as the different aims of the different stages of the AST framework were taken into account, some explanation for how Mr. Bird used redirection emerged. As an example, during Class Periods 1 and 2 of the unit, as revealed in the findings, more variability in the types of redirection used occurred. These were the class periods of the unit that were aligned with AST2, where Mr. Bird was trying to elicit the range of ideas students had about how to explain the anchoring phenomenon. During these class periods, he used far more different types of redirection ranging from F_1 TRM to F_1 MOR and F_1 CON. These were class periods that Mr. Bird was trying to elicit many ideas and support students in selecting among their ideas as they begin to design investigations to test their ideas as part of AST3. After Mr. Bird initially engaged to help shape their investigations during Class Period 2, students worked in the laboratory conducting their investigations during Class Periods 3 and 4 of the unit as part of AST3. During these class periods, he used F_1 TRM to get students to explain why they were doing the investigations they were doing or Activity Redirection (A_1) to make suggestions for changing the activity in which students were engaged if he believed another activity might be more productive in helping students explain the anchoring phenomena. An episode exemplifying this was shared earlier when Mr. Bird suggested students use a different experimental setup to examine an idea they initially put on the table (i.e., that the depth of an object submerged in a liquid affects the amount of buoyant force on the object from the liquid). Finally, during Class Period 5 of the unit, aligned with AST4, Mr. Bird relied mainly on F_1 TRM. While it is conceivable that other forms of redirection (e.g., F_1 MOR; F_1 CON) might also support extended turns in student discourse aimed at pressing students to construct evidence-based explanations,

Mr. Bird's use of F_1 TRM made sense within the purpose of AST4, whereby he was frequently found asking students to explain something they said in a previous exchange at the end of the unit. This occurred at a time when he was focused less on getting a wide range of student ideas on the table, as was his objective during Class Period 1 of the unit during AST2 when more variability in the types of redirection he used was found. At the end of the unit Mr. Bird, while committed to drawing on students' ideas, as evidenced in the instances of redirection during Class Period 5 (i.e., 15 instances of redirection), was also focused on AST4, supporting the class in converging on a consensus model of those factors that affected buoyancy with the data they collected, so that these factors could be taken into account in final revisions to students' final explanatory models.

12.6 Implications and Conclusion

New visions of teaching and learning outlined in national standards documents (e.g., NGSS Lead Stages, 2013; NRC, 2012) ask teachers to engage students in ways that are dramatically different than what has previously been done (NASEM, 2015; Reiser, 2013). Researchers, professional developers, and leaders will need to provide accounts of how this can look in classrooms and the roles teachers can take up to support student learning in contexts that more authentically represent scientific activity. The research conducted as part of this chapter provides the beginnings of such classroom accounts and the role one teacher took in supporting learners in an MBL environment, where models served as the context-dependent tools for supporting the everyday sensemaking work of students in the classroom. This is especially important as functional-pragmatic shifts toward developing students' modeling competence to explain a range of phenomena is increasingly prioritized (NRC, 2012). To this end, we acknowledge that the teacher's classroom that was the focus of this chapter cannot be used to generalize about the extent to which or how other teachers facilitate instruction that is responsive to student idea refinement. However, it is believed that our close nuanced analysis and interpretation can begin to address the possible productive roles teachers can play in these learning environments, so that these environments are more conducive to developing student modeling competence.

Beyond what the unit has begun to contribute to the framework for modeling competence (Chap. 1) that served as the focus of this volume, this research suggested, in alignment with what others have noted (e.g., Thompson et al., 2016), that the curriculum, or in the case of this research, the framework used to shape curriculum is intricately entangled in the pedagogical role of the teacher, Mr. Bird, in learning environments. More, specifically related to this research, it became evident that the different types of redirection found were connected to the purposes of the different activities that were strategically planned as part of the unit, especially related to the ways in which the unit unfolded with respect to the iterative development of students' models (i.e., eliciting students initial models, supporting the refinement of student models through investigations and whole class sensemaking). As alluded to earlier, Thompson et al. (2016) pointed to how curriculum is necessary, but not sufficient for supporting rigor and responsiveness in science classrooms. Based on this research, we might add this to point toward how responsive instruction is intricately bound up in curriculum or the intentions of the different activities within a curriculum framework. In fact, this research revealed how the AST framework, essentially a framework that supports students refinement of models across an instructional unit, appeared to serve as a compass for the teacher that led to the use of different forms of responsive instruction (i.e., different types of redirection). The specific forms of responsive instruction were likely not mapped out ahead of time, especially not at the grain size of the teacher committing to use F_1 TRM, as an example. Instead the AST framework and the subsequent unit and modeling focus throughout appeared to lead to a responsive commitment to student ideas and the emergent, instead of planned, use of the different forms of responsive instruction Lineback (2015) identified.

References

Ambitious Science Teaching. (2014a). *Teaching practice set: Eliciting students' ideas and adapting instruction.* http://uwcoeast.wpengine.com/wp-content/uploads/2014/08/Primer-Eliciting-Students-Ideas.pdf

Ambitious Science Teaching. (2014b). *Teaching practice set: Supporting on-going changes in student thinking.* http://uwcoeast.wpengine.com/wp-content/uploads/2014/08/Primer-Supporting-Changes-in-Thinking.pdf

Ambitious Science Teaching. (2014c). *Teaching practice set: Pressing for evidence based explanations.* http://uwcoeast.wpengine.com/wp-content/uploads/2014/08/Primer-Pressing-for-Explanations.pdf

Campbell, T., & Neilson, D. (2009). Student ideas and inquiries: Investigating friction in the physics classroom. *Science Activities: Classroom Projects and Curriculum Ideas, 46*(1), 13–16.

Campbell, T., & Neilson, D. (2012). Modeling electricity: Model-based inquiry with demonstrations and investigations. *The Physics Teacher, 50*, 347–350.

Campbell, T., Oh, P. S., & Neilson, D. (2012). Discursive modes and their pedagogical functions in model-based inquiry (MBI) classrooms. *International Journal of Science Education., 34*(15), 2393–2419.

Campbell, T., Oh, P. S., & Neilson, D. (2013). Reification of five types of modeling pedagogies with model-based inquiry (MBI) modules for high school science classrooms. In M. S. Khine & I. M. Saleh (Eds.), *Approaches and strategies in next generation science learning* (pp. 106–126). Hershey, Pennsylvania: IGI Global. https://doi.org/10.4018/978-1-4666-2809-0.ch006

Campbell, T., Zhang, D., & Neilson, D. (2011). Model based inquiry in the high school physics classroom: An exploratory study of implementation and outcomes. *Journal of Science Education and Technology, 20*(3), 258–269.

Coffey, J. E., Hammer, D., Levin, D. M., & Grant, T. (2011). The missing disciplinary substance of formative assessment. *Journal of Research in Science Teaching, 48*, 1109–1136.

Ford, M. (2008). 'Grasp of practice' as a reasoning resource for inquiry and nature of science understanding. *Science & Education, 17*, 147–177.

Ford, M. (2015). Educational implications of choosing "practice" to describe science in the next generation science standards. *Science Education, 99*(6), 1041–1048.

Ford, M. J. (2005). The game, the pieces, and the players: Generative resources from alternative instructional portrayals of experimentation. *Journal of the Learning Sciences, 14*(4), 449–487.

Ford, M. J., & Forman, E. A. (2006). Redefining disciplinary learning in classroom contexts. *Review of Research in Education, 30*, 1–32.

Gay, G. (2000). *Culturally responsive teaching: Theory, research, & practice*. New York: Teachers College Press.

Gouvea, J., & Passmore, C. (2017). Models of' versus 'models for'. *Science & Education, 26*(1–2), 49–63.

Griesemer, D., Lo, A. S., Guy, C. R., Harris, E., & Passmore, C. (2016). Successes and challenges in promoting student sense making in modeling classrooms. In C. Passmore (Chair) related paper set *Epistemic framing and agency modeling classrooms* conducted at the meeting of the National Association for Research in Science Teaching, Baltimore, MD.

Groenewald, T. (2004). A phenomenological research design illustrated. *International Journal of Qualitative Methods, 3*, 1), 1–1),25.

Khan, S. (2011). What's missing in model-based teaching. *Journal of Science Teacher Education, 22*, 535–560.

Klieme, Hartig, & Rauch. (2008). The concept of competence in educational contexts. In J. Hartig, E. Klieme, & D. Leutner (Eds.), *Assessment of competencies in educational contexts* (pp. 3–22). Göttingen, Germany: Hogrefe & Huber Publishers.

Koeppen, K., Hartig, J., Klieme, E., & Leutner, D. (2008). Current issues in competence modelling and assessment. *Zeitschrift für Psychologie / Journal of Psychology, 216*, 60–72.

Krajcik, J. (2015). Three-dimensional instruction: Using a new type of teaching in the science classroom. *The Science Teacher, 83*(8), 50–52.

Lehrer, R., & Schauble, L. (2006). Scientific thinking and science literacy: Supporting development in learning in contexts. In W. Damon, R. M. Lerner, K. A. Renninger, & I. E. Sigel (Eds.), *Handbook of child psychology* (Vol. 4, 6th. ed.). Hoboken, NJ: Wiley.

Lineback, J. E. (2015). The redirection: An Indicator of how teachers respond to student thinking. *The Journal of the Learning Sciences, 24*(3), 419–460. https://doi.org/10.1080/10508406.2014.930707

Louca, L. T., & Zacharia, Z. C. (2012). Modeling-based learning in science education: Cognitive, metacognitive, social, material and epistemological contributions. *Educational Review, 64*(4), 471–492.

Manz, E. (2015). Representing student argumentation as functionally emergent from scientific activity. *Review of Educational Research, 85*(4), 553–590.

Manz, E., & Renga, I. P. (2017). Understanding how teachers guide evidence construction conversations. *Science Education, 101*, 584–615.

Melville, W., Jones, D., & Campbell, T. (2017, September). *Building the science department*. Arlington, VA: National Science Teachers Association Press.

National Academy of Sciences, Engineering, and Medicine [NASEM]. (2015). *Science teachers' learning: Enhancing opportunities creating supportive contexts*. Committee on strengthening science education and teacher advisory council, division of behavioral and social science education. National science education standards. Washington, DC: The National Academy Press.

National Research Council [NRC]. (2012). *A framework for K-12 science standards: Practices, crosscutting concepts, and core ideas*. Washington, DC: National Academy of the Sciences.

NGSS Lead States. (2013). *Next generation science standards: For states, by states*. Washington, DC: National Academies Press.

Passmore, C., Gouvea, J. S., & Giere, R. (2014). Models in science and in learning science: Focusing scientific practice on sense-making. In M. R. Matthews (Ed.), *International handbook of research in history, philosophy and science teaching* (pp. 1171–1202). Dordrecht, The Netherlands: Springer.

Pickering, A. (1995). *The mangle of practice: Time, agency, and science*. Chicago: University of Chicago Press.

Reiser, B. J. (2013). *What professional development strategies are needed for successful implementation of the next generation science standards?* Paper presented at the Invitational

Research Symposium on Science Assessment, ETS, Washington, DC. Retrieved from http://www.k12center.org/rsc/pdf/reiser.pdf

Roth, W.-M., & Lee, S. (2004). Science education as/for participation in the community. *Science Education, 88*, 263–291.

Rouse, J., (2007). Practice theory. *Division I Faculty Publications* (Paper 43). Retrieved from http://wesscholar.wesleyan.edu/div1facpubs/43

Ruiz-Primo, M. A., & Furtak, E. M. (2007). Exploring teachers' informal formative assessment practices and students' understanding in the context of scientific inquiry. *Journal of Research in Science Teaching, 44*(1), 57–84.

Sandoval, W. A. (2015). Epistemic goals. In R. Gunstone (Ed.), *Encyclopedia of science education* (pp. 393–398). Dordrecht, The Netherlands: Springer.

Scardamalia, M. (2002). Collective cognitive responsibility for the advancement of knowledge. In B. Smith (Ed.), *Liberal education in a knowledge society* (pp. 67–98). Chicago: Open Court.

Schwarz, C., Calabrese Barton, A., Tzou, C. T., Colley, C., Warren, B., … Haverly, C. (2016). *Science learning contexts, characteristics, and activities*. Symposium conducted at the meeting of the National Association for Research in Science Teaching, Baltimore, MD.

Sherin, M. G., & van Es, E. A. (2005). Using video to support teachers' ability to notice classroom interactions. *Journal of Technology and Teacher Education, 13*, 475–491.

Stewart, J., Cartier, J. L., & Passmore, C. M. (2005). Developing understanding through model-based inquiry. In S. Donovan & J. Bransford (Eds.), *How students learn* (pp. 515–565). Washington, DC: National Academies Press.

Stroupe, D. (2014). Examining classroom science practice communities: How teachers and students negotiate epistemic agency and learn science-as-practice. *Science Education, 98*(3), 487–516.

Stroupe, D. (2015). Describing "science practice" in learning settings. *Science Education, 99*(6), 1033–1040.

Stroupe, D., & Windschitl, M. (2015). Supporting ambitious instruction by beginning teachers with specialized tools and practices. In J. Luft & S. Dubois (Eds.), *Newly hired teachers of science: A better beginning* (pp. 181–196). Amsterdam: Sense Publisher.

Thompson, T., Hagenah, S., Kang, H., Stroupe, D., Braaten, M., Colley, C., et al. (2016). Rigor and responsiveness in classroom activity. *Teachers College Record, 118*(5), 1–58.

Vygotsky, L. S. (1978). *Mind and society: The development of higher mental processes*. Cambridge, MA: Harvard University Press.

Part IV
Developing Students' Modeling Competence

Chapter 13
Learning to Play the Modeling Game

Richard Lehrer and Leona Schauble

13.1 Introduction

Because we work with elementary school-aged children and their teachers, we are particularly concerned with how to support the *development* of modeling (Lehrer & Schauble, 2003). That is, we seek to understand what it takes to initiate youngsters into the practices that comprise what Hestenes (1992) refers to as "the modeling game," so that over time, children come to appreciate what models are intended for, how models are constructed, that they entail decisions about what and how to represent, and that multiple models of the same natural system are possible and even desirable (Chap. 1). It is especially important for students to grasp that any modeling choice could potentially have been made in a different way and that decisions about models, therefore, involve tradeoffs in utility, efficiency, precision, and message in relation to a question. The framework for modeling competence (FMC; Chap. 1) frames these ideas as meta-modeling knowledge, but as we will describe, young children seem to display tacit understanding of these issues before they are prepared to defend them as explicit criteria.

To provide contact with productive ideas like these, in the design research that we pursue with students and teachers we aim to support youngsters as they begin to *participate in modeling practices* (Lehrer & Schauble, 2006). This entails a deliberate departure from the more common emphasis on learning canonical scientific models and applying them to solve problems. We often encourage students to struggle with inventing models before we introduce ready-made or conventional models,

This material is based on work supported by the National Science Foundation under grant no. DRL-1316312.

R. Lehrer (✉) · L. Schauble (✉)
Peabody College, Vanderbilt University, Nashville, TN, USA
e-mail: rich.lehrer@vanderbuilt.edu; leona.schauble@vanderbilt.edu

A. Upmeier zu Belzen et al. (eds.), *Towards a Competence-Based View on Models and Modeling in Science Education*, Models and Modeling in Science Education 12, https://doi.org/10.1007/978-3-030-30255-9_13

because we find that problematizing the modeling process enhances students' opportunities to understand the nature and status of models, which are not transparent to children (Grosslight, Unger, Jay, & Smith, 1991), and to learn about characteristics and functions of the natural systems represented by models (Lehrer, Schauble, Carpenter, & Penner, 2000).

13.2 Modeling as a Scientific Practice

Like all scientific practices, modeling makes sense only when one understands the goal structure within which it is embedded (Sandoval, 2014). Scientific practices rely for their vitality on their epistemic function within knowledge-making communities (Rouse, 2007). Describing their superficial structures to students is not usually an effective way of teaching them, because practices are both more variable and more context sensitive than they are often portrayed in school science (Chinn & Malhotra, 2002; Gooding, 1990). Yet it is common for educators to focus primarily on teaching those structures in the form of strategies, skills, inquiry cycles, and conventional models, but often with insufficient concern for first generating and sustaining the epistemic context within which scientific practices meaningfully function. As a result, students may learn to reproduce the structures they are taught, but fail to understand the conditions under which those strategies or procedures are useful, or how to adapt them when it is appropriate to do so. Students who are taught general rules of reasoning tend to interpret them as recipes, focus unduly on duplicating the rule or strategy transmitted by the teacher, and hence, develop a distorted picture of scientific thinking (Berland & Reiser, 2011; Manz, 2015). For example, they may worry about producing the teacher-required three pieces of evidence to support every claim but think little about how an observation or event legitimately assumes the status of evidence. They may become preoccupied with designing controlled experiments but fail to consider the more problematic elements of experiment, such as whether their proposed variables and measures are trustworthy stand-ins for their still-emerging constructs of interest. There is nothing inherently wrong with teaching skills of scientific reasoning, but we should not be surprised that if skills are taught as domain-general solutions to problems that youngsters have not yet sufficiently conceived, students respond by "doing school" rather than "doing science" (Berland & Reiser, 2011). To foster the development of scientific practices and to promote the emergence of what Ford (2015) calls a "grasp of practice," a productive first goal is to engender classroom communities whose members share the task of working collaboratively to develop, revise, and critique knowledge about the natural world (Lucas, Broderick, Lehrer, & Bohanan, 2005). Scientific reasoning, at least as experienced by young children, should be tied tightly to local forms of meaning making. In schools, local means at the classroom level; accordingly, this goal is accomplished by organizing the activity of classroom communities around extended knowledge-building experience within specific content domains.

13.3 Bringing Modeling to the Early Grades

We conduct our research as extended design studies in which our aim, working with collaborating teachers, is to create classrooms that reflect these priorities and then longitudinally study the thinking that develops as students are promoted across elementary grades (Cobb, Confrey, diSessa, Lehrer, & Schauble, 2003; Sandoval, 2004). As researchers learn about the resources and difficulties that students bring to this enterprise, we work with participating teachers to identify and test curricular and pedagogical means of adjusting instruction in response.

When teachers and students pursue this work, they inevitably confront a series of decisions that are typically "black boxed" in school science. Scientists must figure out ways to navigate the many sites of contingency that are an inherent part of scientific practice (Manz, 2015; Metz, 2004). Among others, these include how we determine the scientific relevance of questions being proposed for study (Lehrer, Schauble, & Lucas, 2008) and how to invent *conditions for seeing,* which include measures, but also material configurations, observational protocols, instruments, experimental or naturalistic comparisons and designs, and transformations and displays of data (Lehrer & Schauble, 2015). This is not to argue that students should reinvent all these things on every occasion, but taking a long-term view suggests that beginners should come to appreciate that people create these epistemic tools, which, for that reason, are subject to critique and challenge by others who may advocate for alternatives. Resolving these sites of indeterminacy raises both conceptual and material challenges – how to symbolize and talk about natural systems, but also, how to practically set up and maintain conditions in the material world that allow nature to be effectively studied (e.g., Latour, 1999; Pickering, 1995). Science rests on the understanding that nature is not always self-revealing (Shapin & Schaffer, 1985). Developing this appreciation requires experiencing the generation of questions, investigations, and data that are made, not given or found, and using these experiences to ground and critique questions and data generated by others. Unfortunately, much of school science experience asks students to think about artificial conditions in which Nature has already been silenced (Chinn & Malhotra, 2002). The indeterminacy that is central to scientists' decisions and the contingencies of the material world are frequently obliterated, perhaps to achieve efficiency or to ensure that students will get the "correct" answer.

How and in what sequence an educator should expose these sources of indeterminacy (e.g., which questions are apt to be most productive) and contingency (e.g., how particular tools facilitate some outcomes but close off others) to students, and how students can best be supported in understanding and resolving them, are pressing pedagogical questions. Our studies focus on how students, with teacher assistance, grapple with these issues, and we find not only that they are accessible to youngsters, with appropriate pedagogical support, but also that working through them at some level seems to play a critical role in helping students to appreciate and understand both their own invented solutions, strategies, and heuristics, and more conventional ones.

13.4 Focus on Representational Competence

Although we have studied several of these sources of indeterminacy, we have focused especially on supporting children's representational competence (Danish & Enyedy, 2007; diSessa, 2004; Lehrer & Schauble, 2013), which we regard as a foundational underpinning for modeling. Exploring and reflecting on the implications of a variety of ways of representing the natural world are key components of modeling and provide an accessible way to introduce youngsters to modeling practices. Symbolic representations of all kinds – physical models, drawings, diagrams, maps, mathematical descriptions, including computational simulations – rest on decisions about what the representation should include and what should be eliminated from the natural system, so that features considered theoretically central are highlighted and the potential distraction of irrelevant aspects is dampened (Latour, 1990). Repeatedly generating and interpreting symbolic representations fosters in students the gradual development of a repertoire of representational forms and design tradeoffs that can be deployed to support problem solving in new situations (Lehrer et al., 2000). Teachers encourage students to compare and evaluate alternative representations and continually provide press for revisions that enhance the precision, power, and validity of representations, and thereby increase the range of questions that students can pursue (Lehrer, Schauble, & Lucas, 2008).

Models and other representations are analogies, in which components and their relations are mapped from a source (e.g., planetary motion) to a target domain (e.g. an atom). Bridging analogies (Clement, 2009) bootstrap hybrids of source and target domains (e.g., Maxwell's paddle wheel mechanisms served as a bridge between Newtonian mechanics and electromagnetism, as described by Nersessian & Chandrasekharan, 2009). Related research on the development of analogical thinking (e.g., Gentner & Toupin, 1986) provides the insight, supported by our own observations, that youngsters find it easiest to begin by generating and interpreting representations whose similarity with the phenomena being represented is reasonably evident (Chap. 1; Lehrer & Schauble, 2000). The most basic kind of model in which similarity is especially salient is models constructed from remnants, that is, those in which actual components of the original phenomena are lifted out of their context and deliberately rearranged and displayed to call attention to key relationships. Remnant models are accessible even to the youngest students. For instance, first graders in one of our studies investigated the effects of soil composition and moisture on the growth of prairie plants (Lehrer & Schauble, 2012). They initially encountered these plants in visits to a restored prairie, and shortly thereafter participated in the construction of a rain garden near their classroom, intended as a model for studying how soils and moisture affect plant growth. The rain garden featured nine plots laid out in a grid, and within each plot students planted a standard set of prairie plants. The garden was intended as a model of conditions in the prairie, but those conditions were implemented so that variation in key variables could be more clearly identified (e.g., both soil and moisture were systematically varied across the grid, the latter by a gradient in elevation). The site was thus a simplification and standardization of the prairie and served as a context in which even young children could develop and explore conjectures about relationships between environmental factors and plant growth. Note that although the model was composed of elements of the prairie, it also deliberately omitted many species and brought into immediate contact variations in soil moisture that would occur at greater remove in the prairie. Such deliberate selectivity indicates that physical microcosms exemplify (Goodman, 1976) and do not simply copy, despite the origins of this form of model in resemblance.

In a second-level model the teacher gathered remnants from the garden—in this case, clippings from the plants—and glued them to a classroom display that highlighted the 3 × 3 design of the garden. This second-level remnant model served as an always-present mnemonic device that highlighted and reminded youngsters about the underlying structure of the garden outdoors, and thereby facilitated indoor planning for subsequent studies of plant growth and insect population in the outdoor garden.

Remnants, like the plant components in the indoor secondhand model, are literally parts of the original phenomena. But other forms of representations also preserve similarity with the modeled world. Drawings, diagrams, and maps do so as well, although not to the same degree. Creating scientific drawings introduces children to important principles of inscription in that drawings amplify some aspects of a system while simultaneously reducing others (Latour, 1999; Tytler, Prain, Hubber, & Waldrip, 2013). For example, children may use bubbles or other visual devices to magnify the structure of a leaf or to make visible a root structure that would ordinarily not be visible. More realistic drawings would not employ these devices. To emphasize important components and relationships, diagrams usually omit details that are found in drawings, and maps preserve some aspects of a space, such as relative distance and orientation, but leave out many others (e.g., lines represent highways, dots stand in for entire towns, and lesser details like telephone poles and stop signs are usually omitted altogether). For novices of all ages, and especially for youngsters, perceptual similarity seems to play a supportive role in initially establishing and then maintaining the mapping relationships between the source and the potential target. Reasoning with representations can be cognitively demanding, especially for children, and, as is the case in analogical reasoning, similarity seems to support the mapping process in modeling.

13.5 Provoking Change: What Develops?

To support developing representational competence, teachers position children in lines of inquiry where representations are re-used, extended, and expanded in ways that help make visible new aspects of a natural system. These extensions and innovations may rely less on similarity and more on relationships that are not immediately available to perception. The first graders in the prairie study were surprised to see a large, to-scale drawing of prairie plants, illustrating that much of the mass of the plants is underground, in their root systems. During a routine maintenance prairie burn conducted by a teacher, the children learned that the root mass prevents the native plants from being destroyed by fire by supporting the plant's later regrowth. To pursue questions about root growth, the students subsequently grew a variety of seeds in root chambers filled with soils of different composition. Students debated over ways to record changes in the length of the roots as they grew. Their initial displays were made with string, which resembles roots and, moreover, can be hung below a baseline representing ground level. Because these displays looked like roots, children readily accepted them as representations. But as the roots grew and students tried to compare the lengths of string, a disadvantage became evident—the string tended to curl and was difficult to see, even from a moderate distance. As a result, it did not adequately support the comparison of root lengths across days of growth. To resolve this problem, students agreed to substitute strips of paper, which were easier to see and compare, but less directly resembled roots. Moreover, instead of dangling the strips below a baseline representing ground, students now arranged them above the line, consistent with an intent to communicate *increase* in length and to describe growth by the differences in length over time.

In related studies of plant growth in other classes (Lehrer et al., 2000), students coordinated drawings and silhouettes created by pressing plants at different points of growth to interpret change of plant height as a Cartesian graph. The changing ratios of height to elapsed time made visible an "S-shape" of growth that seemed to be common across all the plants. As students noted, the S-shape was consistent with drawings and silhouettes, indicating that plants grew "slowly at first, then much faster, and slowed down again at the end." This shape was identified as a common pattern and in later grades was generalized to other contexts of growth, including growth of individual organisms (such as hornworms) and populations (e.g., bacteria). As this narrative illustrates, a representation that initially relies on similarity can be gradually transformed, via analysis, use, and feedback, into a more conventional display that no longer relies on direct resemblance to its referent. Students gradually augment their preference for similarity as they develop a perceived need for increased explanatory power and increased ability to look through other symbolic descriptions to see new aspects of a system. For example, conceiving, measuring, and representing intervals on a graph that represents plant growth was provoked by the question, "What is the same and what is different in the way all our plants are growing?" Note, too, that the phenomena in question, plant growth, circulates among the representations, so that children's appreciation of the characteristics of growth originates in the network of relations described by these representational re-descriptions of it (Latour, 1999).

Perhaps because they have a history with representations of all kinds, adults sometimes underestimate the cognitive work required to grasp why a convention like a coordinate graph is a reasonable way of representing a concept like plant growth. Yet representations, even those explicitly taught in school, are not automatically connected in novices' minds to a history of useful applications. Even when a representation is familiar, considerable domain-specific knowledge may be required to support its interpretation in novel contexts. Moreover, youngsters' initial grasp of the overall purposes for representations may be vague. If one's goal is to learn about the world, why attend to a representation as a source of information or as a site for investigation—that is, why rely on a data display, a diagram, or a model—rather than the actual phenomena? Until they have had repeated opportunities to observe the conceptual advantages that representations provide, it is little wonder that youngsters tend to confuse representations with copies or depictions (for purposes of illustration or artistic expression), as documented by Grosslight et al. (1991).

Youngsters seem especially resistant to omitting information in representations, even though simplification is required to achieve amplification of the features that are considered important (Latour, 1990). But this does not necessarily mean that youngsters cannot distinguish models from copies; what may seem to be a general preference for copy may actually be a signal that children have not yet grasped what the model or representation is intended to accomplish.

For example, we observed a different class of first graders using a tub of hardware equipment to construct models that "work like your elbow" (Penner, Giles, Lehrer, & Schauble, 1997). Although there was considerable variability in the constructions that pairs of children produced, none of the initial models actually functioned like an elbow. Instead, all of them were depictive, rather than functional. The models featured Styrofoam™ balls to depict "the bump where your elbow goes" (as one child said), along with elaborately constructed "hands" and fingers represented with Popsicle™ sticks. These initial models seemed to express a concern for reducing representational ambiguity by making a persuasive case to peers for the representational validity of the constructions. When the teacher reminded children that the goal was to produce models that "worked like" their elbows, a girl pointed out that elbows, unlike the models in the classroom, bend. This observation instigated a flurry of revision. The next round of models all included bending "elbows," but the models were now constructed with springs or pipe cleaners and rotated through a full 360-degree range. When challenged, children insisted that their arms also could move freely. To problematize this solution, a co-teaching researcher duct-taped children's upper arms to their bodies and invited them to try to move their forearms through a 360-degree trajectory, resulting in the discovery that "our real elbows get stuck right here!" (e.g., they do not move backward). The final models constructed by the children all bent, but also included ways to constrain the range of motion. In closing interviews, a researcher attempted to learn the extent to which children were aware of this distinction between duplicating and modeling. During the interview, as one boy proudly showed off his construction, the researcher provoked this distinction by objecting, "But it doesn't *look* like an elbow to me." The student explained gently to the researcher (who he clearly considered in need of enlightenment), "Well, it's only a model."

This example illustrates two points. First, the goal children initially adopted in their constructions was to communicate persuasively what their display was intended to represent. We feel that it would be a mistake to prematurely override that concern, because it addresses a critical aspect of representational competence, namely, sense of audience—that is, how others are likely to "read" what one's representation communicates. This is an important objective for early modeling instruction, one that requires time and pedagogical attention. It is to be expected that youngsters might focus on the insight that the interpreter needs to understand what the representation stands for. Only subsequently does children's concern shift to representing less immediately visible aspects, such as function. The second point is the critical role of model test and revision. Especially when young students' sense of modeling is fragile, modeling instruction needs to be pursued within contexts that support the generation of clear feedback about model fit (and misfit). In this case, children could easily see whether their constructions adequately reproduced the motion of their forearms. Young students may not be accustomed to holding their explanations to account. The ability to test one's model provides clear feedback about whether and the extent to which the model works. Opportunities and encouragement for students to compare solutions and to revise models in response to feedback are arguably as important as having opportunities to engage in the initial stage of model development.

In these examples, youngsters required sufficient time to apprehend the affordances of representations, to explore the tradeoffs of using different conventions for representing, and to begin to develop a sense of audience, that is, to appreciate the range of ways that peers may make meaning from one's attempts to represent. These accomplishments can be challenging for any novice who has weak domain knowledge, not just for young children. A novice of any age is often unsure what to emphasize in a representation and what can be omitted; struggling with these questions is an integral part of modeling practice. Therefore, we feel that rather than hurrying students past these uncertainties, teachers should engage students in addressing them directly. Designing representations, using representations to support an argument to an audience, and revising representations in response to feedback from peers are processes that help students see what *a particular representation* is intended to accomplish with *a particular audience*, and also eventually cumulate to *a more general sense of representational competence*.

Developing representational competence is a lifelong task; although it may be an especially pertinent agenda for children, adult professionals continue to invent and interpret representational re-descriptions (e.g., Vertesi, 2014). As students develop their grasp of representational practice, they become increasingly prepared to confront more complex challenges that go beyond simply representing events and objects and center more explicitly on questions about model fit and misfit—especially, whether and to what extent a model legitimately stands for the phenomenon it is intended to represent.

13.6 Reflecting on the Nature and Status of Models

As suggested earlier, models represent natural systems that are materially modified in some way. Thus, they inherently entail indeterminacy in the sense that representations and material arrangements have an open texture (Hesse, 1962). Other representations and configurations are always possible, so both students and scientists sometimes struggle to understand whether and how representations and material arrangements can be taken to support inferences about the modeled phenomenon. In the following we review two occasions on which upper elementary grade students encountered uncertainty about the nature and status of models. Considered together, these events provoke reflection about the characteristics of modeling tasks that may make modeling more and less challenging.

The first example comes from third graders' visits to a stream to tabulate its aquatic organisms. The students observed that crayfish were present in some parts of the stream but entirely absent in others. They conjectured that perhaps the crayfish were clustering in places where they could hide in the substrate from predators. However, they were unsure how to test this assumption, because the stream was deep and most of the streambed was inaccessible to observation.

To support investigation, the teacher brought a plastic wading pond to the classroom, and students used duct tape to mark off four equal-sized quadrants. They then installed different types of substrate within the quadrants to explore their hypotheses about crayfish preference. These included white rocks, mixed rocks, mixed rocks with plants, and no substrate at all. The wading pool was filled with stream water and ten small crayfish were installed. Over the next weeks, students took turns visiting the pool, once in the morning and once in the afternoon, to count and record the number of crayfish observed in each quadrant. They initially displayed the results of the counts by affixing stickers to a circular display sectioned into quadrants, like the pool. This initial data display was deliberately designed by the teacher to capitalize on its similarity to the wading pool model. Once students became familiar with the data and what it represented, the stickers were replaced by a data table that more effectively supported cumulating and displaying counts of crayfish by quadrant over time.

The counts confirmed children's initial expectations; more crayfish were indeed observed among the mixed rocks with plants than in any of the other quadrants. Students took this finding as confirmation of their expectation that the crayfish were actively hiding in that quadrant because the plants and rocks provided protection. The teacher, however, asked children whether they could be sure that the crayfish were not just randomly wandering around, meaning that there was no choosing going on, but rather, that the results were simply due to chance. To pursue this alternative interpretation, the teacher (supported by researchers) introduced a model of chance, with the intention of persuading children to apply it to the crayfish context.

The teacher began by asking class members to conduct repeated trials of ten spins (representing the ten crayfish) with equally partitioned, two-color spinners. This exercise helped children work through many of the naïve conceptions that frequently surface in investigations with chance devices (Lehrer, Horvath, & Schauble, 1994; Metz, 1998). For example, many students believed at first that qualities of the spin—to the left vs. to the right or fast vs. slow—would allow them to systematically control the outcome. By comparing trials run under different conditions, the students eventually concluded that it was impossible to predict the outcome of any particular spin. They also noted that with increasing numbers of trials, the outcomes came close to 50% in each half.

Once these conceptions were addressed, the teacher shifted to evenly partitioned, four-color spinners to reflect the four quadrants of the wading pool. Each student predicted the results of spinning ten trials, ran the spinner, and recorded the outcomes. The class accumulated the data, displayed it in a pie chart, and observed that about a fourth of the outcomes occurred in each of the quadrants. Next, the teacher asked each student to design a unique spinner, using anywhere from 2 to 4 colors, and to spin 20 times, record the results, and display them in a graph. A follow-up discussion focused on how the outcomes reflected, but did not precisely copy, the design of the spinners.

Finally, the teacher felt the class was ready to connect these ideas about chance back to the original question about the crayfish. To contextualize these ideas, the teacher sought to highlight the difference between choice and chance by placing a different snack in each corner of the classroom. She produced a spinner with four quadrants, each labeled by a small picture: of grapes, crackers, broccoli, and cookies. As each student spun the indicator on the spinner, he or she went to stand in the corner representing the outcome. The class observed that approximately equal numbers of people ended up in each corner. Then the teacher asked students to move to the corner where she had placed the snack they *preferred*. As she had planned, almost all of the children went to the corner representing cookies. These results were also graphed and displayed, and the two displays, considered together, supported a lively discussion about the differences between choice and chance.

Having navigated this extended development of ideas about chance and preference, the teacher urged students to apply this thinking to the original question about crayfish. As students inspected their crayfish data once again, the teacher asked them to reconsider their original conclusion: Did the crayfish end up in certain quadrants by chance, or were they really actively choosing to go to the mixed rocks with plants? To her surprise, most students refused to reconsider their initial interpretation. They argued that it was evident that the crayfish preferred mixed rocks with plants, and the plausibility of their conjectured explanations (e.g., to hide from predators or to avoid predation by "camouflage" within similarly colored rocks) seemed to cement those opinions in place and render them impervious to further review.

In retrospect, there are several features of the chance model that, we believe, made it difficult for children to accept as a satisfactory model of crayfish behavior. The first is the well-established difficulty that people of all ages experience when thinking about chance, including the counterintuitive notion that although no one outcome can be predicted, structure will nonetheless emerge across a large enough sample of outcomes. It is not surprising that students initially entertained many of the naïve conceptions about chance that have been documented in previous research (e.g., Konold & Kazak, 2008; Lehrer et al., 1994; Metz, 1998, 2004). However, their clear advancement in thinking about chance devices suggests that this is not the entire explanation. We suspect that another barrier to assuming a modeling perspective was that children had previously formulated and convinced themselves of a plausible reason to support their initial interpretation. There is considerable evidence that plausibility affects reasoning (e.g., Klayman & Ha, 1987; Schauble, 1990). Third, our follow-up interviews with children revealed that they conceived of animals as agentive and intentional, making it especially counterintuitive to apply a model of blind chance as an explanation of behavior. Doing so would require a form of counterfactual reasoning that is critical to experimentation, but challenging. Finally, the chance model, unlike the representations described earlier, bore an especially low level of similarity with its target. A spinner simply bears no resemblance to a crayfish. Hence, preserving the mapping between the model and the target might have been difficult even if it had not violated students' conceptions about animal behavior. However, during interviews conducted with a sample (n = 10) of children later in the year, participants were asked to consider a hypothetical bird feeder loaded with both walnuts and peanuts. Children were told that a biologist wanted to learn which food the birds preferred, so she carefully counted the number of times crows picked walnuts and the number of times they picked peanuts. After watching 20 crows, she found that 12 picked walnuts and 8 picked peanuts. We asked children what they thought of the results, in light of what the biologist wanted to learn. When asked what they thought of the results, four of the ten students spontaneously expressed doubt about what could be inferred about preference on the basis of the data we described. As one said, "But that's still not a very big difference. If it was just by chance, then, well, that was still pretty close." Another remarked, "You still wouldn't be sure, because it was a close tie. And 8 is just 4 less than 12. It would have to be a few more walnuts to make me sure that it was walnuts and not peanuts." Another said: "It (the difference) could be real; it could be just by chance." Hence, a minority of the children began to entertain difference in light of chance variability, perhaps because they did not have firm preconceptions about the food preferences of crows.

A second occasion when students questioned the status of models occurred as sixth graders were investigating seasonal change in a local retention pond and its surrounding shoreline (Lehrer & Schauble, 2017). On both fall and spring visits, students collected samples of plants and animals in both aquatic and terrestrial contexts and recorded their findings to support conclusions about the number and diversity of species living in these locations. Back in their classroom, the sixth graders developed ecosystem column models to support more controlled testing of factors that affect the interrelationships between the pond and the plants along the shore. The ecosystem models were constructed in two connected liter bottles, one modeling an aquatic system and the other, a terrestrial system. Pairs of students made decisions about the substrate, plants, and animals to include in each part of the system, connected the ecosystem columns, and collected data on outcome variables of their own selection over time.

Thirteen different models were produced across the classroom. We were interested in whether students regarded the ecosystem columns as models or instead, thought of them as simple attempts to duplicate the original pond, albeit in a limited way. For example, we were unsure whether they expected that a model would vary if they constructed and ran it again under precisely the same conditions. To find out, we asked students: "If you build your eco-column again exactly the way you made it the first time, do you think it would come out exactly the same?" We were surprised but pleased to learn that the sixth graders expected that the model instantiations would vary; indeed, they appeared untroubled by that possibility. For example, Carlen replied, "Probably not." He went on to explain that in the next repetition, one of the organisms might eat more of the plants, which would produce changes in the eco-column that were not seen in the previous version. Another student noted that one could not expect an exact copy because "…it's nature…you never know what seed will plant (germinate) faster…'cause you can't control that. You can only control, like, the number of plants when you first planted it." Yet a third student summarized, "You have to do everything a couple of times to actually see what would really happen." Only one of the seven students interviewed entertained the possibility that replicating an identical design could produce exactly the same outcomes, but his belief was expressed only as a hypothetical: "It could."

When directly asked, "Why did your class make eco-columns?", most of the students explained that the models were intended to represent ecological processes that were operating in the pond, such as, "The elodea was a producer for the fish," and, "We, like, questioned interdependent and dependent relationships." These students acknowledged that although the ecosystem model and the pond differed in significant ways, what bound the two systems were process and relation, not literal ingredients. Students also proposed that a virtue of the model was to make some aspect of these ecological processes more visible. "Like, so see how they (fishes) live on the plants without having to, like, go under the water." Nonetheless, a minority (two of the seven) of students seemed to think of the models as copies of the pond. "We wanna have, like, a little section of the pond that we can't bring into the school. So we made one." These two students expressed concern that the differences between pond and model might be problematic for the model-status of the columns: "They kinda work differently, 'cause it (the model) didn't have a lot of animals." Given their expectation that the models could not be expected to function like the pond, it is not surprising that they did not seem to grasp that this was the intention for constructing them. In fact, when asked about the purpose of constructing the models, these students seemed to find the question confusing. One of them conjectured that the goal might have been to compare the eco-column to the pond to "… see which one's better (cleaner)."

Unlike the third graders, the sixth-grade students were more willing to accept models that violated expectations about perceptual similarity and focused instead on invisible processes and relationships that they considered more important. However, even after a year's work in the actual pond and several weeks in a follow-up modeling exercise, a couple of students remained unsure about the viability of a model that did not include all the components found in the original source. These models made variability of outcomes especially visible, in that several columns were initially designed with similar components but looked very different by the end of the semester. This kind of natural variation is often difficult for students to understand (Lehrer & Schauble, 2004), and it seemed to influence a minority of students to question the model-status of the system.

It is possible that the more sophisticated thinking about models that we observed in the sixth graders is primarily a result of their greater age and reasoning capability. However, it is also possible that the aquatic models explored by the sixth graders were better suited to provoke thoughtful consideration of the epistemic status of the model. The third graders readily dismissed the model of chance as applicable to crayfish behavior. In spite of their rich experience within both the model world and the natural world, they rejected the proposed mapping between these two worlds. Students did not regard as plausible the alternative to their initial conjecture about crayfishes' intentional choice of substrate. Doing so would have required them to entertain as a model a spinner device that did not perceptually resemble its referent, to engage in counterfactual thinking, and to tolerate a violation of a core assumption that intention underlies behavior. In contrast, although the sixth graders began by regarding their aquatic models as small-scale copies of the pond, only a minority of students retained that perspective at the end of their investigations. This may be because the variation among the models designed (e.g., students selected different substrates, animals, plants) invited generalization across the particular "ingredients" and turned students' attention to more general functional relationships.

13.7 Conclusion

Careful readers will have noticed a number of correspondences between the approach described here and the FMC introduced in Chap. 1. Interestingly, these correspondences focus on key features that signal development. For example, both approaches note the importance of the transition from an initial emphasis on similarity as a criterion for assessing model fit to a more nuanced concern with theoretically motivated relations and functions. Second, both programs acknowledge that as novices become more practiced in modeling, they begin to become aware of and eventually to apply more principled and general criteria for what counts as an adequate model, criteria that the FMC refers to as meta-modeling knowledge. Third, both approaches emphasize the importance of model test and model revision as important mechanisms for supporting learning.

At the same time, there are also some differences. For example, we do not draw a very clear distinction between representations (or model objects) and models. We use the term "representational competence" to describe our goals for young children because we are hesitant to claim that young children grasp the full panoply of understandings that a professional brings to modeling. On the other hand, we think of the relationship between representations and models as a fluid continuum, and there is no clear point on the continuum that delineates the border from one to the other. Indeed, whether a novice applies meta-modeling knowledge in a given situation may vary with domain, task, and support. Our collaborating teachers continually "push the envelope" so that representational competence is always expanding toward more elaborated forms of modeling that acknowledge multiple models of the phenomena under investigation. We have not found it productive to encourage them to worry about whether "we are there yet."

Moreover, we have not found it helpful to worry about the distinction between models "in the head" and model objects "in the world." We acknowledge, of course, that models are only models *to someone*. However, from a pedagogical perspective we emphasize that models derive their legitimacy not only from the conceptions of individuals, but from the collective engagement of a knowledge making and critiquing community. Moreover, although people's cognitive representations certainly influence the external representations they produce, the influence goes both ways: representations often influence—sometimes in fundamental ways—the conceptions of the person who originally generated them. Representations are not merely explications; they are also sometimes the source of invention, especially when they are hybrids that combine elements from a number of sources (Gooding, 2006).

As we have been arguing, one of the virtues of modeling is that it serves as a centerpiece for inducting children into productive approximations of the practice of science. Developing understanding of modeling threads an ensemble of practices highlighted by the Next Generation Science Standards (NGSS) in the United States. These standards delineate eight core practices of science, including argumentation, explanation, and modeling (NGSS Lead States, 2013). Among them, we regard modeling as occupying a central role, because it involves a dual relation between representation and material, and between individual and collective—that is, models are evaluated in light of collective understandings, which hold students accountable to the quality of evidence (measures, data, nature of investigation) and to anchoring their individual pursuits to collective endeavor and meaning-making. One aspect of high quality research questions is that they spur related questions, often raised by others to whom they are communicated, and high quality evidence is not simply an overwhelming amount of data, but rather, data constructed in light of the model proposed. We believe that these understandings are important epistemic and epistemological outcomes—epistemic in the sense that students employ models to know, and epistemological in the sense that students learn about the mechanics of model invention and revision—the signature practice of science (Nersessian, 2008).

References

Berland, L. K., & Reiser, B. J. (2011). Classroom communities' adaptations of the practice of scientific argumentation. *Science Education, 95*(2), 191–216.

Chinn, C. A., & Malhotra, B. A. (2002). Epistemologically authentic inquiry in schools: A theoretical framework for evaluating inquiry tasks. *Science Education, 86*(2), 175–218.

Clement, J. (2009). The role of imagistic simulation in scientific thought experiments. *Topics in Cognitive Science, 1*, 686–710.

Cobb, P., Confrey, J., diSessa, A., Lehrer, R., & Schauble, L. (2003). Design experiments in education. *Educational Researcher, 32*(1), 9–13.

Danish, J. A., & Enyedy, N. (2007). Negotiated representational mediators: How young children decide what to include in their science representations. *Science Education, 91*(1), 1–35.

diSessa, A. (2004). Metarepresentation: Naïve competence and targets for instruction. *Cognition and Instruction, 22*(3), 293–331.

Ford, M. (2015). Educational implications of choosing "practice" to describe science in the next generation science standards. *Science Education, 99*(8), 1041–1048.

Gentner, D., & Toupin, C. (1986). Systematicity and surface similarity in the development of analogy. *Cognitive Science, 10*(3), 277–300.

Gooding, D. (1990). *Experiment and the making of meaning: Human agency in scientific observation and experiment*. Dordrecht, The Netherlands: Kluwer Academic Publishers.

Gooding, D. C. (2006). From phenomenology to field theory: Faraday's visual reasoning. *Perspectives on Science, 14*(1), 40–85.

Goodman, N. (1976). *Languages of art*. Indianapolis, IN: Hackett Publishing Company, Inc.

Grosslight, L., Unger, C., Jay, E., & Smith, C. L. (1991). Understanding models and their use in science: Conceptions of middle and high school students and experts. *Journal of Research in Science Teaching, 28*(9), 799–822.

Hesse, M. B. (1962). *Forces and fields. The concept of action at a distance in the history of physics*. New York: Dover.

Hestenes, D. (1992). Modeling games in the Newtonian world. *American Journal of Physics, 60*(8), 732–748.

Klayman, J., & Ha, Y.-W. (1987). Confirmation, disconfirmation, and information in hypothesis testing. *Psychological Review, 94*(2), 211–228.

Konold, C., & Kazak, S. (2008). Reconnecting data and chance. *Technology Innovations in Statistics Education, 2*(1), Center for the Teaching of Statistics, Department of Statistics, UCLA, UC Los Angeles.

Latour, B. (1990). Drawing things together. In M. E. Lynch & S. Woolgar (Eds.), *Representation in scientific practice* (pp. 19–68). Cambridge, MA: The MIT Press.

Latour, B. (1999). *Pandora's hope: Essays on the reality of science studies*. Cambridge, MA: Harvard University Press.

Lehrer, R., Horvath, J., & Schauble, L. (1994). Developing model-based reasoning. *Interactive Learning Environments, 4*(3), 218–232.

Lehrer, R., & Schauble, L. (2000). The development of model-based reasoning. *Journal of Applied Developmental Psychology, 21*(1), 39–48.

Lehrer, R., & Schauble, L. (2003). Origins and evolution of model-based reasoning in mathematics and science. In H. Doerr & R. Lesh (Eds.), *Beyond constructivism: A models and modeling perspective* (pp. 59–70). Mahwah, NJ: Erlbaum.

Lehrer, R., & Schauble, L. (2004). Modeling natural variation through distribution. *American Educational Research Journal, 41*(3), 635–679.

Lehrer, R., & Schauble, L. (2006). Scientific thinking and science literacy. In W. Damon, R. Lerner, K. A. Renninger, & I. E. Sigel (Eds.), *Handbook of child psychology* (Child psychology in practice) (Vol. 4, 6th ed., pp. 153–196). Hoboken, NJ: Wiley.

Lehrer, R., & Schauble, L. (2012). Seeding evolutionary thinking by engaging children in modeling its foundations. *Science Education, 96*(4), 701–724.

Lehrer, R., & Schauble, L. (2013). Representational re-description as a catalyst of conceptual change. In B. M. Brizuela & B. E. Gravel (Eds.), *"Show me what you know": Exploring student representations across STEM disciplines* (pp. 244–249). New York: Teachers College Press.

Lehrer, R., & Schauble, L. (2015). Developing scientific reasoning: The role of epistemic practices. In R. Lerner, L. S. Liben, & U. Mueller (Eds.), *Child psychology and developmental science* (Cognitive processes) (Vol. 2, 7th ed., pp. 671–714). Hoboken, NJ: Wiley.

Lehrer, R., & Schauble, L. (2017). Children's conception of sampling in local ecosystem investigations. *Science Education, 101*(6), 968–985.

Lehrer, R., Schauble, L., Carpenter, S., & Penner, D. (2000). The inter-related development of inscriptions and conceptual understanding. In P. Cobb, E. Yackel, & K. McClain (Eds.), *Symbolizing and communicating in mathematics classrooms: Perspectives on discourse, tools, and instructional design* (pp. 325–360). Mahwah, NJ: Erlbaum.

Lehrer, R., Schauble, L., & Lucas, D. (2008). Supporting development of the epistemology of inquiry. *Cognitive Development, 23*(4), 512–529.

Lucas, D., Broderick, N., Lehrer, R., & Bohanan, R. (2005). Making the grounds of scientific inquiry visible in the classroom. *Science Scope, 29*(3), 39–42.

Manz, E. (2015). Resistance and the development of scientific practice: Designing 'the Mangle' into science instruction. *Cognition and Instruction, 33*(2), 89–124.

Metz, K. (1998). Emergent understanding and attribution of randomness: Comparative analysis of the reasoning of primary grade children and undergraduates. *Cognition and Instruction, 16*(3), 285–365.

Metz, K. (2004). Children's understanding of scientific inquiry: Their conceptualization of uncertainty in investigations of their own design. *Cognition and Instruction, 22*(2), 219–291.

Nersessian, N. J. (2008). Model-based reasoning in scientific practice. In R. A. Duschl & R. E. Grandy (Eds.), *Teaching scientific inquiry: Recommendations for research and implementation* (pp. 57–79). Rotterdam, The Netherlands: Sense Publishers.

Nersessian, N. J., & Chandrasekharan, S. (2009). Hybrid analogies in conceptual innovation in science. *Cognitive Systems Research Journal, Special Issue: Integrating Cognitive Abilities, 10*, 178–188.

NGSS Lead States. (2013). *Next generation science standards: For states, by states*. Washington, DC: The National Academies Press.

Penner, D., Giles, N., Lehrer, R., & Schauble, L. (1997). Building functional models: Designing an elbow. *Journal of Research in Science Teaching, 34*(2), 125–143.

Pickering, A. (1995). *The mangle of practice: Time, agency, and science*. Chicago: The University of Chicago Press.

Rouse, J. (2007). Practice theory. *Division I Faculty Publications* (Paper 43). http://wesscholar.wesleyan.edu/div1facpubs/43

Sandoval, W. (2004). Developing learning theories by refining conjectures embodied in educational designs. *Educational Psychologist, 39*(4), 213–223.

Sandoval, W. (2014). Science education's need for a theory of epistemological development. *Science Education, 98*(3), 383–387.

Schauble, L. (1990). Belief revision in children: The role of prior knowledge and strategies for generating evidence. *Journal of Experimental Child Psychology, 49*(1), 31–57.

Shapin, S., & Schaffer, S. (1985). *Leviathan and the air pump: Hobbes, Boyle, and the experimental life*. Princeton, NJ: Princeton University Press.

Tytler, R., Prain, V., Hubber, R., & Waldrip, B. (2013). *Constructing representations to learn in science*. Rotterdam, The Netherlands: Sense Publishers.

Vertesi, J. (2014). Drawing as: Distinctions and disambiguation in digital images of Mars. In C. Coopmans, J. Vertesi, M. Lynch, & S. Woolgar (Eds.), *Representation in scientific practice revisited* (pp. 15–35). Cambridge, MA: MIT Press.

Chapter 14
Toward an Epistemology of Modeling-Based Learning in Early Science Education

Loucas T. Louca and Zacharias C. Zacharia

14.1 Introduction

Models and modeling are considered to be integral parts of science learning, primarily because they provide a context in which students can explain and predict natural phenomena. Externalized through various means, a concrete model could provide students with a tool that enables them to understand a phenomenon, as well as to make new predictions concerning this phenomenon (Schwarz et al., 2009; Sect. A). NRC (2012) categorizes modeling as among the scientific practices that K-12 students should learn to apply. NRC argues that "models make it possible to go beyond observables and imagine a world not yet seen" (NRC 2012, p. 50). In other words, external models and modeling enable learners not only to *see* but also to *re-see* the natural world, becoming tools for scientific reasoning and for envisioning (otherwise theoretical) ideas in science.

Recognizing models and the process of modeling as core components of science education (NGSS Lead States, 2013) suggests two important issues. The first is the construction and use of external models as ways to represent the function/mechanism underlying natural phenomena at the core of learning in science (Chap. 3). Second, learning in science entails learning with and about the process of scientific modeling (Linn, 2003; Chap. 1). Science proceeds through the construction and refinement of external models of natural phenomena (NRC, 2012), and therefore, learning science includes developing an understanding of natural phenomena by constructing models as well as learning the processes of developing and refining those models (White & Frederiksen, 1998). For the purposes of this chapter, we refer to the processes of learning through and about modeling as *modeling-based*

L. T. Louca (✉) · Z. C. Zacharia
Science Education, European University Cyprus, Nicosia, Cyprus
e-mail: L.Louca@euc.ac.cy

A. Upmeier zu Belzen et al. (eds.), *Towards a Competence-Based View on Models and Modeling in Science Education*, Models and Modeling in Science Education 12, https://doi.org/10.1007/978-3-030-30255-9_14

learning (MBL; e.g., Louca, Zacharia, & Constantinou, 2011; Louca & Zacharia, 2015; Gilbert & Justi, 2016; Chap. 12), emphasizing the process of constructing models and, through this construction, learning about phenomena and the process of modeling itself. We use the term MBL to differentiate this process from *model-based learning*, which has been given a totally different meaning by other scholars [e.g., Gobert & Buckley (2000) defined model-based learning as the construction of mental models of phenomena]. In this chapter, MBL denotes that it is the process of modeling itself that we are interested in. Using the term *model-based learning* could also be misleading in other ways. For instance, it could denote learning that occurs only when a model is used as an end product.

MBL has been widely advocated as an approach that can be applied to meaningfully engage students in authentic practices of learning of and about science (Louca et al., 2011; Louca & Zacharia, 2015), and over the years, a number of studies have verified its effectiveness. However, these studies have focused mostly on high-school and university students. We know much less about MBL among K-6 students, specifically in reference to detailed descriptions of how young students work with MBL in science. Such information is vital for designing curricula and learning materials that enable younger students to develop their modeling competence. By modeling competence, we mean students' understanding of modeling, models, and the use of models in science (Upmeier zu Belzen & Krüger, 2010; Chap. 1). Finally, by modeling epistemologies, we are referring to students' views and knowledge of the models they construct and use through MBL. Their understanding of the nature of the process of model development through MBL is similar to what Louca, Elby, Hammer, and Kagey (2004) described in their view of students' personal epistemologies and their effect on students' learning of science: Students with sophisticated epistemological views use the process of learning more actively, which leads to a better conceptual understanding of the various physical phenomena.

In this chapter, we describe an investigation of whether an already established framework for modeling competence (FMC; Chap. 1) captures/describes the modeling competence of young novice modelers. In previous research, young novice modelers (i.e., K-6 students) engaged in modeling in different ways than older individuals (Louca & Zacharia, 2015). For this chapter, we analyzed data from young learners who specifically followed MBL as described by Louca and Zacharia (2015). The analyses used the FMC as a reference for investigating whether it could capture the modeling competence of young novice modelers, or if it could not, whether it could enrich the existing framework to include/accommodate young students' understandings of modeling competence.

14.2 Theoretical Background

MBL has been recognized as a learning approach that could support formal science learning as early as the pre-school years. MBL has also been highlighted in the most recent NRC Framework for K-12 Science Education (NRC, 2012). It is noted as one of the basic scientific practices, and its added value is being argued for across K-12.

In fact, the authors of this framework highlight the fact that MBL should be applied as early as possible:

> Modeling can begin in the earliest grades, with students' models progressing from concrete "pictures" and/or physical scale models (e.g., a toy car) to more abstract representations of relevant relationships in later grades, such as a diagram representing forces on a particular object in a system. Young students should be encouraged to devise pictorial and simple graphical representations of the findings of their investigations and to use these models in developing their explanations of what occurred (p. 58).

In an MBL context, science learning is accommodated through a recursive modeling process in which students are involved in several steps: constructing, using, evaluating, and revising/reconstructing models that represent physical phenomena (Lesh & Doerr, 2003; Schwarz et al., 2009). Learning is facilitated by observing the natural phenomenon or system at work and representing it through a model, which consists of "elements, relations, operations, and rules governing interactions that are expressed using external notation systems" (Lesh & Doerr, 2003, p. 10). It should be noted that learning occurs through both successes and failures. For instance, after constructing a model to represent a phenomenon, it is often the case that the students notice unforeseen effects and implications from the presence or absence of a particular representational choice, and they proceed with changes, which result in revising and reconstructing their model. The latter illustrates the recursive nature of modeling and indicates that MBL is a gradual learning process.

In addition to representing a phenomenon through a series of steps, MBL involves the development of *meta-modeling knowledge* (i.e., knowledge about how and why models are used and what their strengths and limitations are; for more details, see Schwarz & White, 2005). MBL is a construct that blends the steps of practicing modeling with meta-knowledge (Nicolaou & Constantinou, 2014; Gilbert & Justi, 2016; Chap. 3). This means that on top of following a series of steps for enacting the modeling process, learners need to understand the purpose of each of these steps as well as the characteristics (i.e., understanding the purpose of the elements, relations, operations, and interactions) of the model.

Looking across the MBL literature, various frameworks exist concerning the steps that need to be followed to enact modeling (e.g., Louca & Zacharia, 2015; Hestenes, 1997; Lesh, Hoover, Hole, Metcalf, Krajcik, & Soloway, 2000; Windschitl, Thompson, & Braaten, 2008). However, there is a significant overlap in several basic principles across all these frameworks. More or less, all frameworks involve a construction step and an evaluation step. The differences emerge when the details of each framework are unearthed. For instance, there are frameworks that support the idea that modeling is a cyclical process in which the learners go through the same steps in each modeling cycle (e.g., Constantinou, 1999). On the other hand, other frameworks depict modeling as a process that spirals; here, learners do not necessarily go through all the steps of the modeling process in each cycle (Louca & Zacharia, 2015). In addition to the significant overlap in the steps of the modeling process across the frameworks of the domain, Samarapungavan, Tippins, and Bryan (2015) argued that all these frameworks also overlap in terms of the way modeling (a) impacts children's epistemic learning goals, such that they learn to inquire on their own, (b) transforms students into producers of knowledge (i.e., inventors of models) rather than into consumers of knowledge as traditional approaches usually do, and (c) blends learning about "content" and "process" together, which enables students to view and perceive science learning in its correct dimensions.

Finally, the MBL approach falls under the inquiry-based learning approaches. Through the inquiry prism, MBL could be seen as a fine blend of cognitive (science concepts and scientific inference processes), epistemic (knowledge validation and evaluation), and social (understanding the sociocultural norms and practices of science) dimensions (for details, see Duschl & Grandy, 2008). According to Windschitl et al. (2008), model-based inquiry may be able to provide a more epistemically congruent representation of how science works nowadays. For example, it could "embody the five epistemic features of scientific knowledge: that it is testable, revisable, explanatory, conjectural, and generative" (p. 964).

14.2.1 The MBL "Cycle" *of Young Modelers*

When engaging in modeling, K-6 modelers follow a different route (Louca et al., 2011; Louca & Zacharia, 2012, 2015) than the ones described in the general literature for older modelers (e.g., Krell & Krüger 2016).[1] More specifically, the contents of the various modeling practices/steps differ as well as the sequence in which these modeling practices occur. Young students' modeling work initially follows a four-step MBL cycle (Fig. 14.1). The same steps have been found in other modeling cycles (e.g., Hestenes, 1997; Krell & Krüger 2016; Lesh et al., 2000; Windschitl et al., 2008); however, for K-6 modelers, the level of sophistication is different. For instance, these students usually start modeling at a superficial level, in which they represent only parts of the phenomenon without including any aspects of the underlying mechanism. Additionally, K-6 students' modeling begins as a cyclical process, but it usually evolves into a spiraling one. The latter implies that not all modeling steps/practices are followed in consecutive modeling rounds in which the K-6 modelers aim to improve their models (by modeling rounds, we mean a complete enactment of the four steps of the MBL "cycle").

The MBL *"cycle"* begins with young students observing and investigating a physical phenomenon or part(s) of it. In doing so, the young students start building a *story* for the phenomenon. This *story* is based on observations, prior knowledge, ideas, and experiences. It begins at a superficial level, but as the modeling progresses, its level of sophistication gradually increases (e.g., aspects of the phenomenon's underlying mechanism are added). Prior student knowledge, ideas, and experiences appear to be a significant repository of data and information for students engaging in modeling, with K-6 students heavily referring to their experience during the MBL *"cycle"*. After K-6 modelers develop a *story* that describes the

[1] The quotation marks denote that it is not always a cyclical process; it could also turn into a spiraling process as we discuss later in the chapter.

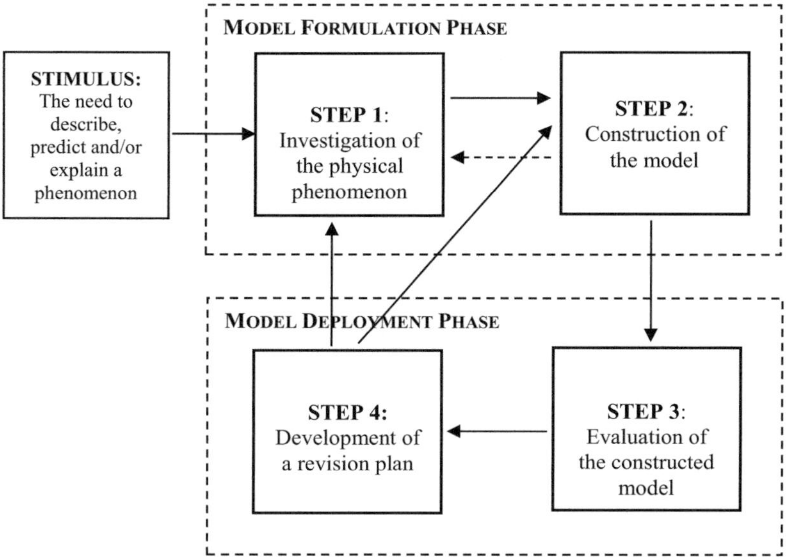

Fig. 14.1 The modeling-based learning "cycle" (adopted from Louca & Zacharia, 2015)

physical phenomenon under study, they proceed directly to the construction of an external, concrete model, while skipping the construction of an internal, mental model. In the case of older learners/modelers, the start of the modeling cycle is more sophisticated. It involves observation, understanding the purpose of the model, prior knowledge and experience, and the construction of a mental model (Nersessian, 2008) that can later be translated into an external model (Mahr, 2015).

In the construction step of the MBL "cycle," K-6 modelers follow two different practices: planning and development. Planning mostly consists of breaking down the phenomenon under study into small pieces that can be incorporated into an external model. In this sense, the process of planning for K-6 modelers is a process of identifying parts of the phenomenon and treating them separately, rather than envisioning and treating all of, or at least a number of, the phenomenon's parts together. The latter explains why these modelers skip or fail to build mental models before proceeding with the construction of an external one. Simply, when the K-6 modelers start modeling, they fail to collect the minimum amount of information needed to put a mental model of the phenomenon together. Previous research (Louca et al., 2011) has suggested that novice modelers need to start developing an external model in order to realize that they need to look for the missing components of the model.

The development of the model looks like the "writing and debugging" process of formal programming. The young modelers write their *story* (see above) and identify the components of the model described in this *story*, and then they proceed to construct a model. They talk about their model, revise their *story* and proceed with small changes, talk about the model some more, and make additional small changes until they feel that this representation matches their *story*. This back-and-forth process is not a formal evaluation of the constructed model but rather a process of reaching the representation/model they agreed to construct in their *story* through a process of trial-and-error (Chap. 13). During this step, students actually "invent" the physical objects (e.g., ball), the physical processes (e.g., moving the ball), and the physical entities (e.g., velocity, acceleration) comprising the phenomenon (e.g., free fall of objects). Invent means finding ways to represent these aspects in the models that are under construction; for example, finding a way to represent velocity. Older modelers follow a more sophisticated modeling approach right from the beginning in which the necessary model objects, processes, and entities are usually present, due to prior knowledge or observations (Krell & Krüger 2016). Furthermore, the internal and external models they construct have some sort of an underlying mechanism right from the beginning of the modeling process (e.g., older students usually know and use mathematical formulas that are related to the phenomenon). When older students feel that they have constructed a satisfactory model, they move toward a process of formal evaluation (e.g., Hestenes, 1997). For K-6 modelers, this process does not begin automatically; rather, the teacher needs to initiate it (Loucas et al., 2011). After a formal evaluation is in place, the process of model evaluation usually takes two major forms. First, learners use their model to see whether it can explain the data they collected or the experiences they recalled or used as a starting point for the model's construction. Second, they evaluate their model in terms of logic; that is, whether the model represents a plausible mechanism that can account for what is observed. For example, K-6 students start by comparing their model to the actual phenomenon and by examining whether their model represents and simulates the phenomenon under study (usually on the surface; e.g., for free fall, K-6 modelers will be happy to see their object fall to the ground. No issues of velocity or acceleration will be considered at this point, unless the teacher points them out). Finally, over the years, we have found limited data where novice modelers deploy or decontextualize their model into a new situation or phenomenon in an effort to evaluate its explanatory power, as suggested by Constantinou (1999).

Another major difference between the modeling "cycle" and other modeling cycles (e.g., Krell & Krüger 2016) is that any revisions made to the constructed model by K-6 modelers in any subsequent modeling cycle occur within the investigation (i.e., during the formulation of the story) and construction steps. In this sense, revision becomes an epistemological or a meta-modeling process (Papaevripidou et al., 2009; Schwarz et al., 2009) in which students decide which route is more appropriate to follow to revise their model (i.e., students stop following a sequential modeling procedure and pick the modeling step that needs to be revisited, that is, the investigation or the construction step, in order to revise their model).

To sum up, the MBL "cycle" begins as a cycle and gradually evolves into a spiral. In this context, K-6 students skip certain steps of the modeling process as the previous modeling rounds are enacted. (Louca et al. 2011; Louca & Zacharia, 2012, 2015) suggested that during the first round of modeling, novice modelers usually identify the physical objects and focus on obtaining a model that looks like the phenomenon they have observed in real life. Only after this, during the second modeling round, can they identify and represent the physical behaviors of the identified objects, thus moving from how their model looks to how it functions, which in fact represents a shift from an ontological to an epistemological perspective of modeling. In the subsequent consecutive modeling rounds, novice modelers can progress to identifying, characterizing, and representing physical entities, which usually consist of concepts represented as variables.

Finally, data from previous studies (Louca et al., 2011; Louca & Zacharia, 2012, 2015) have not shown that novice modelers engage in solid thought experiments during modeling (Krell & Krüger 2016). In addition, novice modelers have not been found to use the resulting models to formulate hypotheses that they later test through experimentation in the real world. The latter prevents K-6 students from understanding how the model and experiential world connect and how the model can be applied.

14.2.2 MBL in K-6 Science Education

Research focusing on the K-6 age range has shown that students can engage in the process of modeling (e.g., Acher, Arca, & Sanmartı, 2007; Forbes, Zangori, & Schwarz, 2015; Manz, 2012; Schwarz et al., 2009). For example, Schwarz et al. (2009) showed that K-6 modelers are able to enact the steps involved in the modeling process, namely, constructing, using, evaluating, and revising/reconstructing models to represent physical phenomena. The modeling-based cycle's hands-on and minds-on nature is a good fit for science learning at such young ages (Louca & Zacharia, 2015; Samarapungavan, Tippins, & Bryan, 2015; Zangori & Forbes, 2016).

According to the literature in this domain, models serve as sense-making tools that provide bridges for these students between their conceptual understanding, their observations, and the underlying scientific theory (Coll & Lajium, 2011; Gilbert, 2004). Given this, K-6 students could develop models that represent their understanding of a phenomenon or system and then use this model to engage in scientific reasoning and to form explanations for how and why the phenomenon or system works (Forbes et al., 2015; Schwarz et al., 2009; Verhoeff, Waarlo, & Boersma, 2008).

In a study of classroom discourse during MBL (Louca, Zacharia, et al., 2011), we described three distinct types of discourse (modeling frames) that learners engaged in: (a) (an initial) phenomenological description, (b) operationalization of the physical system's story, and (c) construction of algorithms. By modeling frames, we mean the different ways in which students understand the learning process that is taking place and how they participate in the process. All these findings suggest that the students who engage in MBL by following the same modeling practices may be understood as being engaged in different modeling frames. In other words, they engage in modeling with different purposes, different end goals, and different combinations of modeling practices.

In a different MBL study (Louca & Zacharia, 2015), a number of modeling practices that young students tend to follow were identified, suggesting that novice modelers appear to enact modeling in a different manner than more advanced modelers. For instance, we have argued that the revision phase of MBL is an epistemological procedure, and any revisions of the constructed model arising during a particular iteration of modeling occur within the *investigation and construction phases*. Additionally, the decomposition of the phenomenon under study into smaller parts happens within the *constructing the model phase* of the modeling process and not during the *investigating phase* (Louca & Zacharia, 2015).

These findings suggest that when students are engaged in MBL, even when they follow the same modeling practices, they might engage in modeling with different purposes, different end goals, and different combinations of modeling practices. In earlier work (Louca & Zacharia, 2008, 2012), it was suggested that student modeling may take several different forms, depending on how students frame their work: the process may become technical (with respect to the code underlying their programming decisions) or conceptual (with respect to the way causal agents such as velocity are represented through code). It can also become procedural (by describing how something happens through time) or causal (by describing how an agent affects a physical process).

A different line of research has investigated students' understanding of models and modeling in science (Chap. 1). The most important contribution of this framework is that it has identified a number of model-related issues that can be used to describe students' understanding of models and modeling in science. Equally as important, the framework's differentiation of the three different levels of student understanding proposes a differentiation between the descriptive, explanatory, and predictive natures of students' understanding of the use and function of models in science. There may be other types of differentiation that can be applied and could be valuable (e.g., the functional nature of models, that is, models that represent how a phenomenon is caused or functions, instead of simply describing the phenomenon). However, this distinction and the ability to differentiate between students' understanding of models and their use is valuable.

Given the particular ways in which K-6 students engage in modeling, the goal of the study was to investigate their understanding of models and their use in application. Thus, we examined how fifth graders' MBL experience influenced their modeling competence concerning the five described aspects (Fig. 14.2). In line with work on student epistemologies in science (e.g., Louca et al., 2004), and in an effort to account for and describe the ways students see and use models during science learning, we analyzed data that supported the ways in which students work in authentic classroom contexts as described by MBL (Louca & Zacharia, 2012, 2015; Gilbert & Justi, 2016). The idea was to enrich an existing FMC to include/accommodate young students' understanding of models and modeling.

Aspects and modeling competence	Levels of student understanding		
	Level 1	Level 2	Level 3
1.Nature of models 2.Multiple models 3.Purpose of models 4.Testing models 5.Changing models	Limited view on only the model-object	Perspective on the relation between the model and the original	Perceiving a model as a scientific idea

Fig. 14.2 The framework for modeling competence (cf. Fig. 1.3)

Physical phenomena (units) studied	Student group 1	Student group 2	Student group 3
1. Accelerated motion down an inclined plane	X		
2. Free fall	X	X	X
3. Water cycle	X		X
4. Diffusion of solid substances in water		X	
5. Projectile motion		X	X

Fig. 14.3 Physical phenomena studied

14.3 Methods, Data Sources, and Analyses

14.3.1 Study Context

This study involved three groups of fifth-grade students in two public metropolitan elementary schools in Cyprus (a total of 48 students working in groups of 2 or 3 students). Students in both classes met with the same teacher once a week for 80 min for a total of 7 months. Following a case study approach (Yin, 1994), different physical phenomena with each class of students were treated as a different case. For this study, data from nine cases (three student classes x three topics) were used in order to describe in detail the process of developing models for physical phenomena (Fig. 14.3). All students had access to a variety of modeling media (computer-based programming environments, paper-and-pencil, three-dimensional materials) to construct models for three similar and two different physical phenomena (a total of five phenomena for the entire study; Fig. 14.3). The duration of the study for each group ranged from 3 to 5 weeks.

14.3.2 Data Sources

Whereas studies that have investigated students' understanding of models and modeling have collected data from student interviews and questionnaires, in our study, we investigated students' understanding using discourse and artifact data, in an effort to inform research about findings from alternative data sources. Our effort focused on identifying elements related to the five aspects of models in FMC (Fig. 14.2).

Transcripts from videotaped conversations from all the case studies served as the primary source of data. A total of 1151 min of student conversations were analyzed. To triangulate the findings, student-constructed models collected at the end of each student meeting session were analyzed.

14.3.3 Data Analysis

Following previous research, the analyses of student discourse and student-constructed models focused on the prime constituents/players of models in physical phenomena (Louca et al., 2011), which include: physical objects, physical entities, and physical processes. As the first step of the analysis, Louca and Zacharia, (2012) coding scheme for discourse and artifact data was used. As presented in Fig. 14.4, the discourse coding scheme differentiates between the discussion of physical objects, physical entities, and physical processes amongst students, while providing the different ways that these can be characterized. The discussion of physical objects is usually about two different things: (a) the description of the story of the physical objects or physical system under study and (b) the description students' experiences in support of these stories. Then, descriptions of physical processes and physical entities included three different ways students talked about them (conceptually, quantitatively, and operationally).

Student-constructed models were analyzed using an artifact analysis adopted from another study (Louca, Zacharia, Michael, & Constantinou, 2011). Codes from this analysis included the ways in which students represented different elements in their models: physical objects (characters), physical entities (variables), physical processes (procedures), and physical interactions. Figure 14.5 presents the codes used for the artifact analysis. The findings from this analysis were added to timeline graphs, aligning the analysis and timing of the construction of each model with the graphs so that these could support the initial data.

After all discourse and artifact data were coded, the discourse data were used to develop nine separate timeline graphs, one per case study, to present the sequence of student conversation as characterized by our analysis. Coded utterances were displayed in timeline graphs to reveal the temporal interrelationships of the coded statements. For each case, one graph was developed. Then, based on previous work (Louca & Zacharia, 2015), the timeline graphs were structured on the basis of the

Coding Categories	Code Description
Description of the story of a physical object or a physical system	Students talked about the overall story of a physical system or a physical object. This usually involved descriptions about what would happen to the overall physical system, without any reference to the mechanism that was actually causing the overall phenomenon or the behavior of the object.
Description of experiences/data in support of the story of the physical system	Students used experiences from the physical world to support their answers or ideas in the conversation.
Description of physical processes …	Students described a physical process (e.g., change in position, change in velocity) …
… conceptually	… qualitatively, without any reference to the mechanism that was actually responsible for causing the changes in the physical process.
… quantitatively	… by using numerical examples. No reference was made about the mechanism that was actually responsible for causing the changes in the physical process.
… operationally defined	… by describing a series of actions that would result in the physical process.
Description of physical entities …	Students described a physical entity (e.g., velocity, acceleration) …
… conceptually	… qualitatively, without any reference to the mechanism that was actually responsible for causing changes in the physical entity.
… quantitatively	… by using numerical examples. No reference was made to the mechanism that was actually responsible for causing changes in the physical entity.
… operationally defined	… by describing a series of actions that would result in the physical entity (or the changes in the physical entity).

Fig. 14.4 Codes used to analyze modeling practices, adopted from Louca et al. (2011)

Category	Codes
1. Representation of physical objects	1.1. Physical objects internal to the physical system
	1.2. Physical objects external to the physical system
2. Representation of object characteristics (physical entities)	2.1. No representation of physical entities
	2.2. Represented with a non-variable numerical value
	2.3. Represented with both a variable & a non-variable numerical value
	2.4. Represented with a variable
3. Representation of object behaviors (physical processes)	3.1. Non-causal
	3.2. Semi-causal
	3.3. Causal

Fig. 14.5 Codes used for the analysis of student-constructed models adopted from Louca et al. (2011)

MBL "cycle" iterations that students engaged in during the study. Each time students went through a modeling "cycle" and they were about to begin a new one, we viewed this as an MBL "cycle" iteration.

Based on these data, nine case studies from the MBL "cycle" were developed, including transcript excerpts and examples of student-constructed models. During this step, descriptions of the context of each round of modeling in each case study were added, with the goal of having a detailed account for each case study to the largest possible extent. The description of context included a description of students' overall goals during the MBL "cycle." The idea of modeling frames from Louca and Zacharia, (2012) was used to describe this context: Modeling frame I: (Initial) Phenomenological description; Modeling frame II: Operationalization of the story of the physical system; and Modeling frame III: Construction of algorithms.

In the last step of the analysis, each of the nine case studies were revised, trying to apply the three levels from the FMC for each round, focusing on: (1) the nature of the models, (2) the existence of multiple models, (3) the purpose of the models, (4) the process of testing models, and (5) the process of changing models. Students' discourse and the models they developed in each round were described separately for each aspect of the framework. For this, the transcript was not coded line-by-line, but rather, a description was given for the timeline section (round of modeling) of each of the case studies.

14.4 Findings

14.4.1 Nature of Models

Both analyses (discourse and artifact) focused on elements of models that students included either in the models they constructed or in the models they discussed during their modeling group work. Physical objects were addressed in student conversations (and presented in student-constructed models) from modeling round 1 of the students' work (for all nine sub-cases with 3 rounds of modeling), appearing to suggest that the development of models first requires students to address the need to represent the "players" involved in the phenomenon (physical objects) before moving on to the rest of the model's properties. This finding was verified by the artifact analysis of the models of all student groups in all sub-cases (Fig. 14.6).

However, physical processes (including interactions) and physical entities were found to be context dependent. A discourse analysis revealed that physical processes and physical entities were discussed in most of the student groups during round 1 of modeling only at the conceptual and quantitative levels, whereas discussions about operationalizing them appeared only in modeling rounds 2 and 3. This is also supported by the artifact data analysis, which indicated that the physical processes and the physical entities appeared to be non-causal in the models in most cases and were derived as early as round 1; physical processes, however, appeared in the models as a mixture of semi-causal and causal representations in both rounds

	Modeling-based learning Round 1	Modeling-based learning Round 2	Modeling-based learning Round 3
Student work overall …	Students conceptualize models as phenomenological descriptions of the story of the physical object(s)	Students focus on operationalizing the descriptions of physical entities	Students focus on representing relationships between physical entities in a physical system (i.e., defining physical processes)
Discourse Analysis			
Description of the story of a physical object or a physical system	✔	✔	✔
Description of experiences/data in support of the story of the physical system	✔	✔	✔
Description of physical processes …	Conceptual and quantitative	Operationally defined	Mostly operationally defined
Description of physical entities …	Conceptual and quantitative	Mostly operationally defined	Mostly operationally defined
Artifact Analysis			
Representation of physical objects	✔	✔	✔
Representation of object characteristics (physical entities)	Non-causal	Semi-causal	Causal
Representation of object behaviors (physical processes)	Non-causal	Semi-causal & causal	Semi-causal & causal
Framework for Modeling Competence			
Nature of models (limited level III)	Level I	Level II	Level II & limited level III
Multiple models (no level III)	Level I & level II (only during model evaluation)	Level II	Level II
Purpose of models (limited level III)	Level I	Level II	Level II & limited level III
Testing models (no level III)	Level I & level II (only during model evaluation)	Level II	Level II
Changing models (no level III)	Level I & level II (only during model evaluation)	Level I & level II	Level I & level II

Fig. 14.6 Summary of findings

2 and 3. On the other hand, physical entities appeared in student models in semi-causal forms only in round 2 and in causal forms only in round 3. Interestingly, all of the abovementioned findings were confirmed in all student groups and in all different phenomena studied, independent of the number of previously modeled phenomena.

These findings suggest that usually modeling round 1 of student work is characterized by a process of developing models as phenomenological descriptions of the phenomena under study. Previously, this discourse was characterized as modeling frame I (Louca & Zacharia, 2012). Working within modeling frame I, students described the story of the overall physical system and/or the story of the individual physical objects involved in the phenomenon under study. These stories were described as a temporal sequence of "scenes" that captured the phenomenon, without dealing with the individual objects' behaviors that resulted in the overall phenomenon. Similarly, students did not talk about any of the necessary components of the scientific model or how the phenomenon took place. Everyday experiences were used as reality checks to support students' ideas about what would happen in the phenomenon under study. All these fit with level I of the aspect of the nature of the model in the FMC (Fig. 14.2), thus suggesting that students develop models that are, *to the greatest possible extent, copies of the reality* (phenomenon) they study.

By contrast, the data suggest that modeling rounds 2 and 3 reflect students' understanding of the nature of models at level II (Fig. 14.2), with students' focus at the end of modeling round 1 on improving the extent to which their model is good for *developing idealized representations of the phenomena* under study. Further, in subsequent rounds of modeling, there were some limited indications of level III, where students' efforts were focused on developing a representation of the phenomenon that would cause (through the relationships between physical processes and physical entities) the phenomenon instead of simply depicting the phenomenon.

14.4.2 Purpose of Models

In different rounds of modeling, the discourse data suggest that students seemed to view, use, and/or visualize the models they constructed differently. Adopting the terminology from earlier work in modeling (Louca & Zacharia, 2012), during modeling round 1, students conceptualized their models so the models would act as phenomenological descriptions of the phenomenon under study, simply describing the story of the overall physical system and/or the story of the individual physical objects involved in the phenomenon under study. This was also the case when, in subsequent rounds of modeling, students had discussions about a new phenomenon or the new features that they wanted to add to their models. An artifact analysis suggested that these discussions led to the development of descriptive models of physical phenomena that simply provided scenes from the phenomenon in a temporal sequence without any reference to the causal mechanism underlying the phenomenon.

In modeling round 2, the students' purpose seemed to focus on the description of the story of the physical entities and included the objects' characteristics (i.e., velocity and acceleration) and the objects' behaviors (i.e., accelerated motion) in an effort to operationalize the story of the physical system. This discussion led to the construction of models of physical phenomena that would have both descriptive and causal features. This view of the purpose of the models and the modeling process seemed to occur in the process of translating the story of the physical system into programmable code so that models of the phenomenon could be developed.

In modeling round 3, students identified and investigated the relationships between physical entities. The fact that their models needed to have this property motivated students to develop a construction-of-algorithms view of the model construction process, helping them to operationally define both the physical entities and the physical processes. This was done in a process of translating descriptive ideas about the phenomenon into operationally defined causal representations of relationships between different components of the phenomenon.

In terms of the FMC, the data showed that students in the study appeared to progress across different levels of understanding with respect to the purposes of the models they constructed. Students in all nine cases started at level I and progressed to level II, despite the fact that they had prior experience with the MBL "cycle." Moreover, the study revealed that data showing students progressing to level III, in which students used the models they constructed to predict something about the phenomenon under study, were not consistent. Conversations about level III occurred only in cases 3 and 4 (water cycle and diffusion). During the evaluation stages, students brought into the conversation similar phenomena in order to evaluate the accuracy of the model under evaluation. For instance, in the case of diffusion (the phenomenon studied was the diffusion of a drop of red food coloring in a beaker of water), students used their experience with the dissolution of sugar in water to test whether their model was accurate enough to predict this phenomenon. In this sense, similarly to the identification of level I and level II, the data suggested that the activation of more advanced levels seems to be context dependent. However, this dependency seems to differ: To move from level I to level II, the dependency seems to be the modeling round the students are in and, thus, the context or content of their actual modeling work. In all the cases, in the beginning of their work with a new model or phenomenon, students started with level I, and in subsequent rounds, they had some conversations that fell into level II. For level III, there seemed to be an additional layer of context dependency, seemingly related to the type of phenomenon under study. Kinematics phenomena (cases 1, 2, and 5) did not activate or "spark" level III conversations related to the purpose of models.

14.4.3 Multiple Models

In terms of the notion of the existence and usefulness of multiple different models representing the same phenomenon, the analysis suggested that prior to modeling round 2, students did not discuss this. As noted above, during modeling round 1, students focused on obtaining a model that works (Louca & Zacharia, 2008), acting as a phenomenological description of the phenomenon under study (Louca & Zacharia, 2012). The fact that modeling round 1 ended with an evaluation of the first model they constructed (Louca, Zacharia, Michael et al., 2011) seemed to work as a context in which students adopted and discussed the idea that it is possible to have multiple different models that represent the phenomenon in different ways, along with limitations and advantages. The discourse data from student modeling conversations suggested that this idea remained active for the rest of the modeling unit (for all remaining rounds) but disappeared again during the first round of the next phenomenon modeled, possibly suggesting a contextual relationship with the type of student work—particularly with the modeling rounds—along with students' notions of the purpose of the model.

During modeling round I, students seemed to view and talk about differences between multiple models of the same phenomenon as differences between the models themselves and not as differences between alternative representations of different phenomena. This depicts level I. However, the process or step of evaluating the models they constructed seemed to help students see differences in their models as alternative ways of representing different parts of the phenomenon under study, which is depicted in level II. Nevertheless, none of the data showed that students viewed models as tools for making predictions about the phenomenon under study, although this might be related to the role of the teachers and how they approached modeling in their science teaching.

14.4.4 Testing and Changing Models

Only after modeling round 1 did students start talking about testing, editing, and making changes to their models. In round 1, students focused on obtaining a model that showed reality, and their main concern was to make their models look like the real phenomenon. Once activated, the idea of revising and testing their models remained active for the rest of the modeling unit but disappeared again during the first round of the next phenomenon they were modeling.

Students did not have any conversations that reflected level III, which includes the view of models as theoretical tools that can be used to make predictions about aspects of the phenomenon under study. Given that the rest of the findings were related to other aspects of the FMC, it is still unclear whether this finding was due to the students' lack of knowledge, modeling experiences, or abilities. Rather, it might be related to the role of the teacher and the data collection period, where the emphasis was placed primarily on students' development of models for the phenomenon and not the use of models as tools for investigating and learning about phenomena. Therefore, the FMC might serve as an instructional guide for teachers and researchers in preparing or designing lesson plans for MBL in science.

Further, most student work in modeling round 1 reflected level I because students emphasized the testing or changing of a model itself. However, level II appeared in modeling rounds 1 and 2, with no apparent pattern regarding when students tested and changed models after they compared their models with the phenomenon under study.

14.5 Discussion and Conclusions

In this chapter, findings from applying the FMC to K-6 modelers were presented. Specifically, the purpose was to investigate fifth-graders' understanding of models and modeling during MBL. Adopting a discourse-based perspective instead of directly asking students for their understanding in interviews and questionnaires, students' modeling work was analyzed to identify the levels on which these students' understanding of models and modeling could be located within the various aspects of the framework.

One of the main themes that runs across all the findings is that, overall, there was not substantial evidence to show that the students could reach level III with respect to any of the five elements that were investigated. This could suggest several things. It is possible that the FMC accounts for an understanding of models and modeling in older students or in students across a wide spectrum of ages, while suggesting that for K-6 students, it might be reasonable to expect that they might not reach level III. In this sense, some understanding or abilities related to modeling processes such as using models as tools to predict natural phenomena do not develop until later ages. Of course, this needs to be investigated in more detail.

Nevertheless, throughout the various aspects, students sometimes worked at level I and sometimes at level II. At first glance, there did not seem to be a developmental pattern to this in the sense that the same students in one modeling unit started working at level I, then moved to level II, and then, in the next modeling unit (which took place 2–3 weeks after the end of the first unit), once again started working at level I. This last part is in line with the theoretical idea of resources by Hammer and colleagues (Hammer, 2000; Hammer & Elby, 2002, 2003; Louca et al., 2004) by which students may simultaneously hold an understanding of a particular idea at different levels, only activating one level at a time on the basis of the context.

Primarily derived from physics education research, the idea of modeling resources is used to identify student knowledge, abilities, or reasoning skills in relation to various modeling tasks. Instead of seeing the absence of a particular level as a need to help students develop the modeling abilities they lack, it might be more productive to view this as a need to help students develop more reliable access to the *modeling resources* they might already have and might be context-dependent (Gilbert & Justi, 2016; Krell, Upmeier zu Belzen, & Krüger, 2013; Chap. 13).

For example, as presented in the findings, the model evaluation tasks administered at the end of round 1 of MBL seemed to activate the notion that it is possible to have multiple different models for the same phenomenon, each one with its own advantages and disadvantages. This idea remained alive for the rest of the modeling unit, but it disappeared when a new modeling unit began until students again reached the evaluation point of their first models.

For instance, students' views of the nature of models as theoretical reconstructions of the phenomenon (level III) seemed to be in sync with students identifying and investigating the relationships between physical entities. For instance, a causal model in which physical entities and physical processes are operationally defined suggests that students likely viewed the process as developing a theoretical reconstruction of the phenomenon that included the mechanism that underlies/causes the phenomenon. In this study, students were not explicitly asked to reflect on those issues and articulate their understanding of the nature of the models they had developed. Overall, while the data from this study did not indicate that students had an understanding at level III within a predictive frame, it was possible to see some level III understanding (as in the case of the nature of models) in some of the aspects of the FMC (Grünkorn et al., 2014).

The context-dependencies of modeling resources (Krell et al., 2013) also deserve to be highlighted. The context of creating computer-based programs that could create general models of the phenomenon under study was vital for leading students to invent and define physical entities in the form of program variables. They would then use these in the program rules, which would include interactions between physical objects, their behaviors, and characteristics.

Given all this, we contend that instead of seeing the absence of particular levels as an indicator that there is a need to help students develop the modeling abilities they lack, it might be more productive to view this as a need to help students develop more reliable access to *modeling resources* they already have but might be context-dependent. This approach has different implications for MBL and instruction and may shape research on modeling competences in different ways. Further investigation of this issue is of course needed, particularly focusing on how novice modelers can be supported to access these resources in a better, more reliable manner.

If we start sketching a framework for modeling resources, there is at least another important relevant implication. The role of the teacher as a possible activator of different modeling resources needs to be considered (Samarapungavan, Tippins, & Bryan, 2015; Zangori & Forbes, 2016). As previously identified (Louca & Zacharia, 2012), there are instances in MBL with novice modelers where the teacher needs to push student thinking in a particular direction (i.e., toward a specific modeling step), especially when students' prior experience with the modeling process is limited. This is relevant to the findings here because it is possible that the absence of level III is related to the *way* the teacher enacted the MBL "cycle" or to his goals for each modeling unit with students.

A second but related implication is that the FMC might be used productively as a guide for teachers throughout MBL in science (Fleige, Seegers, Upmeier zu Belzen, & Krüger, 2012). In this sense, in addition to the designing of modeling units or learning sequences, this framework could help teachers identify and respond to students' modeling difficulties during teaching and learning, thus providing teachers with a productive tool for helping students reach level III with respect to various aspects of the framework.

Finally, the data from this study included only student work and conversations through MBL in science. Of course this is a limitation of the study, creating the need for a more thorough examination of MBL across other disciplines. However, as we have argued elsewhere (Hammer & Louca, 2008), different ways of investigating the same phenomenon may reveal different aspects or pictures of reality, suggesting that a detailed investigation might need to consider a number of different research methods.

References

Acher, A., Arca, M., & Sanmartı, N. (2007). Modeling as a teaching learning process for understanding materials: A case study in primary education. *Science Education, 91*, 398–418.

Coll, R. K., & Lajium, D. (2011). Modeling and the future of science learning. In M. S. Khine & I. M. Saleh (Eds.), *Models and modeling cognitive tools for scientific enquiry* (pp. 3–21). New York: Springer.

Constantinou, C. P. (1999). The cocoa microworld as an environment for modeling physical phenomena. International journal of continuing education and life-long. *Learning, 8*(2), 65–83.

Duschl, R. A., & Grandy, R. E. (2008). *Teaching scientific inquiry: Recommendations for research and implementation*. Rotterdam, The Netherlands: Sense Publishers.

Fleige, J., Seegers, A., Upmeier zu Belzen, A., & Krüger, D. (2012). Förderung von Modellkompetenz im Biologieunterricht. [Fostering Modeling Competence in Biology Classes]. *Der mathematische und naturwissenschaftliche Unterricht, 65*(01), S. 19–S. 28.

Forbes, C. T., Zangori, L., & Schwarz, C. V. (2015). Empirical validation of integrated learning performances for hydrologic phenomena: Third-grade students' model-driven explanation-construction. *Journal of Research in Science Teaching, 52*(7), 895–921.

Gilbert, J. K. (2004). Models and modelling: Routes to more authentic science education. *International Journal of Science and Mathematics Education, 2*(2), 115–130.

Gilbert, J., & Rosária, J. (2016). *Modelling-based teaching in science education*. Switzerland: Springer.

Gobert, J. D., & Buckley, B. C. (2000). Introduction to model-based teaching and learning in science education. *International Journal of Science Education, 22*(9), 891–894.

Grünkorn, J., Belzen, A. U. z., & Krüger, D. (2014). Assessing students' understandings of biological models and their use in science to evaluate a theoretical framework. *International Journal of Science Education, 36*(10), 1651–1684.

Hammer, D. & Louca, L. (2008). *Challenging accepted practice of coding*. Paper presented at the Symposium "How to study learning processes? Reflection on methods for fine-grain data analysis" at the Conference of the Learning Science Society (ICLS), The Netherlands, 24–28 June.

Hammer, D. M. (2000). Student resources for learning introductory physics. *American Journal of Physics, Physics Education Research Supplement, 68*(S1), S52–S59.

Hammer, D. M., & Elby, A. (2002). On the form of a personal epistemology. In B. K. Hofer & P. R. Pintrich (Eds.), *Personal epistemology: The psychology of beliefs about knowledge and knowing* (pp. 169–190). Mahwah, NJ: Lawrence Erlbaum.

Hammer, D. M., & Elby, A. (2003). Tapping epistemological resources for learning physics. *Journal of the Learning Sciences, 12*(1), 53–90.

Hestenes, D. (1997). Modeling methodology for physics teachers. In E. F. Redish & J. S. Rigden (Eds.), *The changing role of physics departments in modern universities: Proceedings of international conference on undergraduate physics education* (p. 935957). New York, NY: The American Institute of Physics.

Justi, R. S., & Gilbert, J. K. (2002). Science teachers' knowledge about and attitudes towards the use of models and modelling in learning science. *International Journal of Science Education, 24*(12), 1273–1292.

Krell, M., & Krüger, D. (2016). Testing models: A key aspect to promote teaching-activities related to models and modelling in biology lessons? *Journal of Biological Education, 50*, 160–173.

Krell, M., Upmeier zu Belzen, A., & Krüger, D. (2013). Students' levels of understanding models and modelling in biology: Global or aspect-dependent? *Research in Science Education, 44*(1), 109–132.

Lesh, R., & Doerr, H. M. (2003). *Beyond constructivism: Models and modeling perspectives on mathematics problem solving, learning and teaching*. Mahwah, NJ: Lawrence Erlbaum.

Lesh, R., Hoover, M., Hole, B., Kelly, A., & Post, T. (2000). Principles for developing thought revealing activities for students and teachers. In A. Kelly & R. Lesh (Eds.), *The handbook of research design in mathematics and science education* (pp. 591–646). Mahwah, NJ: Lawrence Erlbaum Associates.

Linn, M. C. (2003). Technology and science education: Starting points, research programs, and trends. *International Journal of Science Education, 25*, 727–758.
Louca, L., Elby, A., Hammer, D. M., & Kagey, T. (2004). Epistemological resources: Applying a new epistemological framework to science instruction. *Educational Psychologist, 39*(1), 57–68.
Louca, T. L., & Zacharia, C. Z. (2008). The use of computer-based programming environments as computer modeling tools in early science education: The cases of textual and graphical program languages. *International Journal of Science Education, 30*(3), 137.
Louca, T. L., & Zacharia, C. Z. (2012). Modeling-based learning in science education: A review. *Educational Review, 64*(1), 471–492.
Louca, T. L., & Zacharia, C. Z. (2015). Learning through Modeling in K-6 science education: Re-visiting the Modeling-based learning cycle. *Journal of Science Education and Technology, 24*(2), 192–215.
Louca, T. L., Zacharia, C. Z., & Constantinou, P. C. (2011). In quest of productive modeling-based learning discourse in elementary school science. *Journal of Research in Science Teaching, 48*(8), 919–951.
Louca, T. L., Zacharia, Z., Michael, M., & Constantinou, P. C. (2011). Objects, entities, behaviors and interactions: A typology of student-constructed computer-based models of physical phenomena. *Journal of Educational Computing Research, 44*(2), 173–201.
Manz, E. (2012). Understanding the codevelopment of modeling practice and ecological knowledge. *Science Education, 96*(6), 1071–1105.
National Research Council [NRC]. (2012). *A framework for K-12 science education: Practices, crosscutting concepts, and core ideas*. Washington, DC: National Academies Press.
NGSS Lead States (Hrsg.). (2013). *Next generation science standards: For states, by states*. Washington, DC: National Academies Press.
Nicolaou, C. T., & Constantinou, C. P. (2014). Assessment of the modeling competence: A systematic review and synthesis of empirical research. *Educational Research Review, 13*, 52–73.
Papaevripidou, M. Constsantinou, C. P., & Zacharia, Z. C. (2009). *Unpacking the Modeling ability: A framework for developing and assessing the Modeling ability of learners*. Paper presented at the American Educational Research Association (AERA) annual conference, April 13–18 2009, San Diego, CA
Samarapungavan, A., Tippins, D., & Bryan, L. (2015). A modeling-based inquiry framework for early childhood science learning. In *Research in early childhood science education* (pp. 259–277). Amsterdam, The Netherlands: Springer.
Schwarz, C. V., Reiser, B. J., Davis, E. A., Kenyon, L., Achér, A., Fortus, D., et al. (2009). Developing a learning progression for scientific modeling: Making scientific modeling accessible and meaningful for learners. *Journal of Research in Science Teaching, 46*(6), 632–654.
Upmeier zu Belzen, A., & Kruger, D. (2010). Modellkompetenz im Biologieunterricht. [Modeling Competence in Biology Classes]. Struktur und Entwicklung. In: Zeitschrift fur Didaktik der Naturwissenschaften *16*, 41–57. http://www.ipn.uni-kiel.de/zfdn/pdf/16_Upmeier.pdf
Verhoeff, R. P., Waarlo, A. J., & Boersma, K. T. (2008). Systems modelling and the development of coherent understanding of cell biology. *International Journal of Science Education, 30*(4), 543–568. https://doi.org/10.1080/09500690701237780
White, B. Y., & Frederiksen, J. R. (1998). Inquiry, modeling and metacognition: Making science accessible to all students. *Cognition & Instruction, 16*(11), 3–118.
Windschitl, M., Thompson, J., & Braaten, M. (2008). Beyond the scientific method: Model-based inquiry as a new paradigm of preference for school science investigations. *Science Education, 92*, 941–967.
Yin, K. R. (1994). *Case study research: Design and methods*. Thousands Oaks, CA: Sage Publications, Inc.
Zangori, L., & Forbes, C. T. (2016). Exploring third-grade student model-based explanations about plant relationships within an ecosystem. *International Journal of Science Education, 37*(18), 2942–2964.

Chapter 15
Supporting Primary Students' Developing Modeling Competence for Water Systems

Cory T. Forbes, Kim Lange-Schubert, Florian Böschl, and Tina Vo

15.1 Introduction

The last half-century has seen significant and increasing attention paid to the role of processes and practices that mimic those of science in the teaching and learning of science in K-12 classroom settings. From an emphasis on 'science as inquiry' in the United States and Germany (Duit, Gropengießer, & Stäudel, 2007; KMK, 2004; Mayer, 2007; National Research Council [NRC], 2000) as knowledge construction to the current focus on 'scientific practices' in the *Next Generation Science Standards* (NGSS Lead States, 2013) or 'scientific processes' in the *Perspectives Framework for General Studies* in Germany (GDSU, 2013), the importance of 'doing science' has been and remains at the core of science education reform efforts. While perspectives vary on the exact nature of these practice-oriented dimensions of science learning, they generally encompass a variety of scientific practices in which students should engage to learn about natural phenomena. These include conducting investigations, generating and organizing data, and formulating, communicating, and engaging in argumentation around evidence-based explanations. A core practice that spans all of these scientific meaning-making processes is scientific modeling, which involves developing and using abstracted representations or

C. T. Forbes (✉)
University of Nebraska-Lincoln, Lincoln, NE, USA
e-mail: cory.forbes@unl.edu

K. Lange-Schubert · F. Böschl
University of Leipzig, Leipzig, Germany
e-mail: kim.lange-schubert@uni-leipzig.de; florian.boeschl@uni-leipzig.de

T. Vo
University of Nevada-Las Vegas, Las Vegas, NV, USA
e-mail: tina.vo@unlv.edu

A. Upmeier zu Belzen et al. (eds.), *Towards a Competence-Based View on Models and Modeling in Science Education*, Models and Modeling in Science Education 12, https://doi.org/10.1007/978-3-030-30255-9_15

theoretical reconstructions of natural phenomena to engage in sense-making about underlying mechanisms for observable phenomena. While arguably underemphasized in science learning environments historically, scientific modeling has more recently been identified as a core practice in many scientific disciplines. As a result, it is also increasingly foregrounded as a core component of science teaching and learning, particularly in disciplinary contexts focused on complex, often large-scale system processes.

A significant body of work on scientific modeling has emerged over the years (Gilbert & Justi, 2016; Krell, Reinisch, & Krüger, 2015; Passmore, Stewart, & Cartier, 2009; Schwarz et al., 2009), including work on primary science teaching and learning (e.g., Acher, Arcá, & Sanmarti, 2007; Lehrer & Schauble, 2012; Louca & Zacharia, 2015; Manz, 2012). However, while scientific modeling is becoming an increasingly crucial curricular and pedagogical dimension of science learning environments, there is a continuing need for both theoretical development and empirical research on scientific modeling. This is particularly important as a means of providing teachers with theoretically-grounded and research-based curricula, instructional strategies, and assessments that both help foster and account for students' model-based reasoning about natural systems. Moreover, in considering the entire K-12 continuum, perhaps nowhere greater is the need for robust insights into and approaches to scientific modeling than in the elementary, or primary grades. Despite an emphasis on scientific modeling in NGSS and other international science standards, reviews of literature (Gelman & Kalish, 2006; Metz, 1995) observe that developmental constraints are commonly cited as a barrier to engaging early learners in robust model-based science teaching and learning.

In addressing this need, we believe the concept of *competence* has great potential in providing a holistic, robust conceptual and analytical framework for both fostering and studying scientific modeling in science learning environments, including those for early learners. A competence-based perspective on scientific modeling leverages insights gained from prior and current theoretical and empirical work on scientific modeling in science classrooms to provide an overarching framework that defines both cognitive and practice-based dimensions of modeling practice, as well as disciplinary content. The goal of this chapter, then, is to put forward an integrated framework for scientific modeling based upon research and development within a specific disciplinary domain – water systems – that is grounded in our ongoing collaborative research efforts with primary students (Forbes, Zangori, & Schwarz, 2015; Lange-Schubert, Schubert, Böschl, & Forbes, 2016; Vo, Forbes, Zangori, & Schwarz, 2015; Zangori, Vo, Forbes, & Schwarz, 2017). This three-dimensional framework emphasizes three elements – modeling practices, knowledge about models and modeling, and disciplinary concepts – and integrates these into target 'learning performances' that define modeling competence. Of particular importance, here are the epistemic dimensions of modeling competence, or what students should know about scientific models and modeling and how these dimensions play out in their modeling practices when learning about domain-specific natural phenomena.

To illustrate specific competences in this framework, we begin to draw upon data recently collected from primary students in Germany through purposefully designed, model-based cognitive tasks. While preliminary, these findings begin to provide a foundation upon which to begin identifying 'levels' of competences or learning performances, help clarify underlying theoretical constructs, and provide points of discussion for theorizing and measuring modeling competence across the K-12 continuum. As such, this work has the potential to inform broader efforts to support model-based teaching and learning in science classrooms, as well as aid in science education research and the evaluation of model-based curricular and instructional interventions.

15.2 Theoretical Background

15.2.1 What Is Scientific Modeling?

The chapters in this book provide a robust description of scientific modeling that reflects a wide variety of perspectives and emphases across grade levels and disciplinary domains. In synthesizing these robust descriptions, we identify a key set of features that are essential to the practice of using models scientifically. First, such models can include a host of representations, many of which are already very familiar to students and teachers, including diagrams, physical structures, computer-based simulations and visualizations, and even whole-group role plays. Each of these is meant to represent some aspect of the natural world and, as such, are by default abstracted representations or theoretical reconstructions of the world. However, second, representation or reconstruction alone does not make a model scientific. Rather, it is what students do with models that is critical. To engage in scientific modeling, students must use models to pose questions and hypothesize, design investigations, interpret observations and data, explain phenomena, and communicate about phenomena. Third, models are pliable entities, meaning they should evolve over time as students evaluate their utility and modify them iteratively as part of their use to engage with the natural world (Chap. 1). Harré's (1970) notion of 'projective convention' emphasizes that the nature of an entity as a model is entirely dependent on its use in relation to the world. This is a reciprocal relationship, one in which the model reflects some version of lived experience, but also in which the world is observed and interpreted through the lens of the model. As such, models can and should function as both representations and sense-making tools by an individual. To that end, we define scientific modeling as the use of abstracted (Chap. 17), multi-modal representations or theoretical reconstructions of systems, not exact recreations, used within communities to illustrate, explain, and predict system-specific phenomena.

15.2.2 Why Focus on Water Systems?

Water is a topic that spans disciplinary domains and is connected to virtually everything in the K-12 science curriculum. For early learners, beginning to develop 'water literacy' is critical not only to form a foundation for lifelong science learning (NRC, 2007), but also to begin developing the capacity for informed decision-making about water-related global issues and challenges (ESLI, 2009). Earth's water cycle is a large, complex system with many component parts, but one that learners experience in various forms on a daily basis. As in many other countries, students, including primary students, in the United States of America and Germany, should build knowledge about the hydrosphere and its interactions with other Earth systems (GDSU, 2013; NGSS Lead States, 2013). By the time students leave primary school, they should understand "that a system is a group of related parts that make up a whole and can carry out functions its individual parts cannot" and be able to describe both components and processes of Earth systems (NGSS Lead States, 2013, pg. 85). However, early learners often articulate alternative ideas about water systems (Dickerson, Penick, Dawkins, & Van Sickle, 2007; Forbes et al., 2015; Gunckel, Covitt, Salina, & Anderson, 2012; Zangori et al., 2017), partly because some elements of the water cycle are challenging to observe (i.e., water vapor, groundwater, etc.). Research has shown they tend to emphasize the parts of the water cycle rather than its processes and virtually ignore key elements of the water cycle, such as subsurface groundwater, human dimensions of water systems and explanatory parameters like gravity or energy. For primary students to learn about the water cycle – a model in itself, since water moves on a global scale that cannot be observed directly – and the importance of water systems in everyday life, they require opportunities to visualize and make sense of water systems and their often invisible fundamental components and processes. They do so through multiple and varied representations that foreground particular dimensions of these systems as target phenomena for observation and investigation.

15.2.3 A Framework for Modeling Competence

To be competent with scientific modeling, students should develop abilities to use models in ways that mirror science to reason about natural phenomena. This general view of modeling competence is supported by standards in the United States and Germany (GDSU, 2013; NGSS Lead States, 2013), both of which emphasize students' conceptual understanding of disciplinary concepts and practices of science. But what does this competence look like? According to Weinert (2001), who has significantly influenced the scientific discussion on competence, competence is a domain-specific disposition for problem solving in variable situations. Adapted to the field of science, the GDSU (2013) frames science competence as discipline-specific content knowledge and scientific practices coming together to solve science based problems in the real world. As Klieme, Hartig, and Rauch (2008) observes, competences can only be "conceived as resulting from learning processes where the individual interacts with his or her environment" (pg. 7). Consistent with Harré's (1970) notion of 'projective convention', modeling competence must be defined by and developed through relational interactions between the learner, the model, and the real-world phenomenon.

While there is certainly a cognitive dimension of competence (Chap. 1), we draw upon both Klieme et al. (2008) and Harré (1970), as well as situated and activity-based perspectives on learning and expertise (e.g., Engeström & Sannino, 2010), to foreground the practice-based dimensions of scientific modeling as integrated elements of a holistic perspective on modeling competence. Competence cannot be defined solely by 'knowledge of' scientific models and modeling, though this is a critical aspect. Students' knowledge is intertwined with what they do, or the ways in which they use models to engage with and make sense of natural phenomena. Knowledge is embodied in practices in which knowing is meaningful. This perspective on knowledge and practice is consistent with different models of modeling competence (e.g., Gilbert & Justi, 2016; Nicolaou & Constantinou, 2014; Schwarz et al., 2009; Upmeier zu Belzen & Krüger, 2010). What we highlight is that 'content' plays another crucial dimension in modeling competence. As Krell and colleagues (2015) argue, context dependencies or discipline-specificities of students' and teachers' understanding of science, scientific knowledge, and scientific inquiry are frequently discussed in science education. According to their summary of the existing studies, it has been shown that students and teachers do not possess a stable understanding about science, scientific knowledge, and scientific inquiry, which adds evidence to our theoretical framework that proposes context as a unique dimension in modeling competence which should be assessed accordingly.

Over the past 5 years, our team has collaborated to design model-based interventions and investigate model-based teaching and learning in primary science learning environments with a specific focus on water systems (Forbes et al., 2015; Lange-Schubert et al., 2016; Vo et al., 2015; Zangori et al., 2017). Our work has been heavily influenced by international colleagues, many of whom have contributed chapters to this book. This work has been grounded in an evolving conceptual framework informed by theory and developed through empirical research. In particular, the framework relies on similar theoretical assumptions about scientific models and modeling as that of Upmeier zu Belzen and Krüger (2010), with its focus on the nature and purpose of models. The integrated nature of modeling competence has been a core assumption of our work, which has, from its beginning, revolved around a three-dimensional framework involving crucial dimensions: disciplinary concepts, epistemic features, and modeling practices (Fig. 15.1). At the intersection of these core elements lie *learning performances*, consistent with earlier work by Krajcik and colleagues (Krajcik, McNeill, & Reiser, 2007) and now with contemporary science standards (GDSU, 2013; NGSS Lead States, 2013). These learning performances define observable outcomes for students in terms of performances through which their conceptual knowledge is made evident. Learning performances for students' model-based reasoning about water systems, as well as the broader,

		Modeling Practices		
		Construct/Revise	Use	Evaluate
Epistemic Considerations	**Nature of Models (A model is…)**			
	Evidence-based	Learner constructs or revises a model that incorporates evidence about a phenomenon	Learner uses a model to incorporate new evidence about a phenomenon	Learner evaluates a model based on the evidence provided about a phenomenon
	Appropriately detailed/ complex	Learner constructs or revises a model that is appropriately detailed/complex to represent a phenomenon	Learner uses a model that is appropriately detailed/complex to describe a phenomenon	Learner evaluates the appropriateness of the complexity of a model pertaining to a phenomenon
	Generalizable	Learner constructs or revises a model that is generalizable to/from a phenomenon	Learner uses a model to make a generalization about a specific phenomenon	Learner evaluates the generalizability of a model of a phenomenon
	Purpose of Models (A model is for…)			
	Predict/ hypothesize	Learner constructs or revises a model that aids in making predictions or hypothesizing about a phenomenon	Learner uses a model to predict and hypothesize about a phenomenon	Learner evaluates a models ability to predict and hypothesize about a phenomenon
	Explain (whole/ part)	Learner constructs or revises a model that aids in explaining some or all of a phenomenon	Learner uses a model to explain some or all of a phenomenon	Learner evaluates a models explanation of a phenomenon
	Organize	Learner constructs or revises a model to organize their ideas about a phenomenon	Learner uses a model to organize their ideas about a phenomenon	Learner evaluates a models organization of a phenomenon
	Generate	Learner constructs or revises a model to generate new information/ideas about a phenomenon	Learner uses a model to generate new information/ideas about a phenomenon	Learner evaluates a model to generate new information/ideas about a phenomenon

Fig. 15.1 Learning performance framework for students' model-based reasoning about water systems

holistic framework in which they are situated, have provided the structure through which we have designed and cultivated curriculum, pedagogy, and student learning opportunities in both the United States and Germany. They provide an important tool through which to design curriculum, instruction, and classroom learning experiences for students, as well as constructs upon which to study and evaluate questions about model-based teaching and learning. As such, they are a foundation for and infused throughout all aspects of our research and development efforts.

The learning performances in Fig. 15.1 are defined by three dimensions. First, in the framework, disciplinary concepts are represented by references to 'a phenomenon' in each of the individual learning performances. 'A phenomenon' is a placeholder for a variety of scientific concepts to which this framework could theoretically apply. In our work, the big idea foregrounded is *water is matter that, when heated and cooled, changes form and circulates through the Earth's geosphere, biosphere, and atmosphere.* This big idea is derived from a core conception in Earth Science that all geosystems are the result of energy flow and mass cycling (AAAS, 2007; ESLI, 2009; GDSU, 2013). To understand energy flow and mass cycling in hydrologic systems, students must develop robust understanding of the hydrologic cycle, thermodynamics, and properties of substrates (air, soil, biomaterial), as well as geospatial abilities (AAAS, 2007). The three target concepts underlying this big idea are (1) *water exists in different forms below, at, and above the Earth's surface* (Concept 1); (2) *water on Earth is in motion and cycles at a global scale* (Concept 2); and (3) *the cyclical movement of water on Earth shapes and impacts the geosphere* (Concept 3).

The second dimension focuses on the practices of modeling in which students engage, or what students 'do' with and to models. A variety of instructional models have been proposed for scientific modeling in classroom settings (e.g., Kenyon, Schwarz, & Hug, 2008; Passmore et al., 2009; Schwarz et al., 2009). These practices include opportunities to *develop* models and *use* them to make predictions, formulate questions, design and conduct investigations, explain phenomena, and communicate and justify ideas. Over time, students should also *evaluate* their models to understand how their ideas fit into a bigger picture and *revise* them accordingly to match their developing understanding. Over long-term learning sequences, students should engage in these practices iteratively to study and investigate natural phenomena of interest.

Finally, a third dimension focuses on the epistemic dimensions of models and modeling. To reason productively with models, students must attend to epistemic features that guide and give meaning to their work. We identify these epistemic features as core elements of "what counts as valued and warranted scientific knowledge" (Sandoval & Reiser, 2004, p. 348) generated through the use of models to engage with and explore particular disciplinary concepts (represented by the two other dimensions). Drawing from theory and research in the field, including our prior work (Forbes et al., 2015; Lange-Schubert et al., 2016; Vo et al., 2015; Zangori et al., 2017), we identify epistemic features that underlie students' scientific claims (Fig. 15.1). While each epistemic dimension represents a component of an effective model-based explanation, students' explanations should exhibit strong evidence of each epistemic dimension. Together, these seven features (Fig. 15.1) comprise epistemic components of mechanism-based explanations for water-related phenomena.

However, as with all theoretical constructs, fundamental ideas underlying each of these three dimensions continue to evolve through empirical study. While modeling practices and disciplinary concepts have remained fairly constant in our work over the past 5 years, many questions remain about core epistemic dimensions of these learning performances. What are key aspects of models and modeling that make them scientific? What levels of conceptual understanding do primary students show and what level should students have about them? How do they manifest themselves in certain modeling practices in which students engage to investigate natural phenomena, in this case, water-related concepts? A primary objective of this work is to help clarify epistemic dimensions of primary students' model-based reasoning about water systems and to use the empirical work reported in this chapter to shed first light on the question of the 'status quo' of primary students' modeling competence.

15.3 Methods: Student Assessment Task Development

To generate evidence of primary students' modeling competence, we used Evidence Centered Design (ECD; Mislevy, Steinberg, & Almond, 2003) to guide and support our development of model-centered cognitive tasks for students. ECD allows for the design of assessment tasks which link observations in what students do or say to suggestions of what they understand or know. Guided by our theoretical framework (Fig. 15.1) as a series of claims, we developed tasks for students that would provide opportunities for them to express their knowledge or general understanding on the topic. Each cell in Fig. 15.1 represents a discrete claim about students' model-based learning about water systems. Cognitive tasks were designed to elicit evidence of these claims. For example, students were assigned the task of evaluating three different water cycle models, comparing and contrasting features they valued. This task provided an opportunity for students to acknowledge the generalizability of a model and note how well a model explains hydrological phenomena. Prompts were also written into the task to ensure students were provided space to engage with each claim. Additionally, interview questions also probed students thinking. Researchers developed multiple cognitive tasks and iteratively refined those tasks to encompass every facet of the framework. Once a series of representative tasks were developed, other researchers on the project separately reviewed the tasks to ensure that cohesion and representation of the learning performance framework (Fig. 15.1) were appropriately represented. The tasks were embedded within a semi-structured interview protocol (Merriam, 2009) focused on eliciting students' ideas. Once the tasks and interview protocol were developed, piloting focused on refining and clarify the task-based assessment and interview questions. The pilot involved students between the ages of 6 and 10 (n = 8). These students were purposefully sampled (Patton, 2001) to represent a range above and below third grade students, our intended audience for the modeling tasks. By using this group to pilot the tasks, we were also able to ensure the tasks were aligned with our framework, adapt the language to be more accessible to students of that age range, and develop additional interview questions and prompts that allowed students better opportunities to display evidence of the learning performances.

The final set of cognitive tasks included students evaluating multiple water systems models, data, and evidence before revising a model to be more in line with their conceptualizations. It also provided opportunities for students to use and associate physical representations with diagrammatic models while discussing similarities and differences. Semi-structured interview questions were parsed through the tasks, asking students to reflect on their actions and decisions. The content focus ranged across the water cycle, highlighting key components in-line with national and international reform standards and documents (e.g. condensation, evaporation, precipitation, water movement, water storage, groundwater, runoff).

15.3.1 Context and Participants

The study was situated within three third-grade classrooms in a single urban, international school in Germany, working with students ages 7–9 to explore how they use models and modeling practices to engage in reasoning and thinking about water-related phenomena (n = 48). These students were purposefully chosen (Patton,

2001) due to their school's focus on teaching about water in the second and third grades allowing students to have familiarity and some foundational knowledge of the science content being investigated.

15.3.2 Data Collection and Analysis

Using the finalized cognitive tasks and interview protocol, we collected student-developed artifacts (e.g., students' evaluations and ranking of model types, where and how they would include data and evidence into a model) along with students' explanatory statements about their ideas and conceptualizations of models and modeling. Combined, the average completion time for the student tasks and interviews was approximately 20 min in total.

For analysis, students' task explanations and semi-structured interviews were audio recorded and transcribed with assigned pseudonyms (indicated by $*$) for the purpose of this research. These transcriptions were uploaded to a qualitative analysis tool (MAXQDA) and coded using *a priori* codes focused on modeling practices and epistemological considerations as outlined in the framework. Twenty percent of students' transcriptions were double-coded, with the rest evenly divided among two authors and coded to completion. Once students' ideas were appropriately allocated within different sections of the framework, we conducted code queries of the intersections between modeling practices and epistemological considerations; openly coding to begin deriving various levels of students' modeling competence. This was done to provide boundaries to the data being analyzed. Qualitative analysis was an iterative process focused on data reduction and verification bounded by the previously mentioned framework.

15.4 Results

Our data analysis resulted in the identification of learning performances reflecting primary students' competence levels around models and modeling of the water cycle. Here, we present empirical data to illustrate examples of students' model-based reasoning about water for two different learning performances presented in Fig. 15.1. Additionally, we unpack this data into potential levels of students' performance for each. These results serve as illustrative examples, with a range of performances observed, of primary students' modeling competence for water systems using the framework in Fig. 15.1.

Example 1 – The intersection of "appropriately detailed/complex models (epistemic considerations)" and "evaluation (modeling practices)" (Fig. 15.1). Inherently, scientific models possess different levels of complexity, typically determined by the models' intent or intended audience. Within our framework, we acknowledge the importance of modeling complexity and attribute this feature to the ‘nature of models,’ an aspect of modeling that allows students to understand how models are generally used, across various science concepts. Evaluation, grouped with modeling practices, is another important skill primary students are expected to learn. Being able to evaluate the accuracy or appropriateness of a model is important towards accounting for differing ideas and negotiating towards consensus. When looking at the intersection of “appropriately detailed/complex models (epistemic considerations)” and “model evaluation (modeling practices)” within the context of the water cycle, we found varying levels of sophistication with which students evaluated models for their complexity. Primary students used differing types of heuristics to evaluate the complex nature of water cycle models (Fig. 15.2). As highlighted in this example, and recurrent in other learning performances, a level 3 understanding reflects the upper anchor of students’ knowledge,

Example 1
Learning performance levels for students‘ consideration for *“appropriately detailed/complex models (epistemic considerations)”, “evaluation (modeling practices)”*, and *water-related concepts*

Level	Description	Student examples
1	Students look specifically at the number of concrete elements represented - labels/words/numbers	“This one has lots of words.” “It has all the names of things.”
2	Students look at both concrete and abstract elements and sometimes interpret meaning	“It has more labels and it has more like this one has, like those signs[symbols].”
3	Students look at the abstract elements of the model including how they are related to the concrete elements, discussing how those elements are connected.	“We are seeing what is happening, how [water] moves.” “These arrows mean the stuff is moving into the clouds…”

Example 2
Learning performance levels for students‘ consideration for *“explain (whole/part)(epistemic considerations)”, “revision (modeling practices)”*, and *water-related concepts*

Level	Description	Student examples
1	Students’ revisions focused on specific parts of a phenomenon, adding (isolated) concrete elements (e.g. drawings, labels, words, numbers) to expand explanations	“Maybe underneath some water, so you see it can go through the ground.” “[adds label] Because you don’t know what this is and you think it could be the water or sunlight.”
2	Students’ revisions loosely connected concrete and abstract components expanding and/or connecting specific explanations in isolation	“[That arrow] could be where the rain goes and then could make a line, so then would be like a river that went down.”
3	Students’ revisions considered both abstract and concrete components; expanding and/or connecting to provide or strengthen a more mechanistic or holistic explanation of a phenomenon	“The black lines mean where the gas mostly goes up. And the yellow means, just, that it’s a specific gas.” “Because if you don’t know, then you don’t know the whole theme of this model, the whole point.”

Fig. 15.2 Examples of levels for student learning performances from framework in Fig. 15.1

derived from student interview and task data. Levels 1 and 2, signify a less sophisticated idea, where critical aspects of the upper anchor are missing. While not overtly included, a level 0 exists where there is no evidence of students' knowledge or understanding, typically represented by students indicating: "I don't know."

When investigating how students were evaluating the complexity of water cycle models, initial criteria were topic relevance (e.g., focused on water) and acknowledging model components. Within the models we used, words or labels often represented more concrete components of the water cycle (e.g., precipitation, condensation, runoff) while arrows and lines depicted more abstract components. Students with a level 1 understanding often focused on the number of labels present. Variance within level 1 included if students did or did not describe the meaning of the vocabulary. Josephine∗ commented that a water cycle model was best "because there are many labels," however when prompted to elaborate further about the 'many labels' she was not able. Some students who also fell into this category could define some but not all of the vocabulary on which they based their choice. A level 2 understanding was different, in that students noticed and commented on the more abstract ideas within the model. While these students also struggled with defining some of the terms used within a scientific model of the water cycle, they recognized that the complexity of a model is also reliant on the different symbols used (e.g., arrows, dotted lines). Level 2 students acknowledge some abstractions within the model, but often vaguely, and were undecided about how they were connected. Ra∗ pointed out "you can see that water is coming in the air, but you don't really know how it happens or something." Ra∗ is acknowledging that water movement in the form of evaporation is occurring in the model but when prompted did not connect evaporation to an abstract mechanism (e.g., the sun), hidden construct (e.g., water vapor) or a concrete component (e.g., clouds). Therefore, the ability for students to recognize abstractions and/or consider a model more holistically became the upper anchor when evaluating the complexity of a model. A level 3 understanding occurred when students ascribed meaning to the abstract elements of the model. Students did so in two ways, either by recognizing the purpose of the abstract elements (e.g., water storage, heat transfer) or acknowledging how the abstract elements connected the concrete components (e.g., water movement underground, water transformation from liquid to gas). Luke∗ explained "I can tell because of the arrows going up and down, up and down. I can tell it's rainy here. Then [the water] goes down on the river, and then it gets sucked back into the clouds." Luke∗ was able to trace the path of the water using the arrows in the model, connecting multiple areas of water storage. It should be noted the distribution of students who articulated a level 3 understanding when evaluating different water cycle models' complexity was limited.

Example 2 – The intersection of "explain (whole/part) (epistemic considerations)" and "model construction/revision (modeling practices)" (Fig. 15.1). A salient perspective about scientific models is their usage as visual media and representations to adequately explain complex systems or natural phenomena. Within our framework, we attribute the explanatory power of models to the epistemic consideration category of 'purpose of models.' This category encompasses different aspects of why models are used in authentic scientific practices. This epistemic idea becomes particularly powerful for primary students when paired with the modeling practice of construct/revise. The ability to construct/revise a model is necessary to accommodate or communicate a new findings or ideas. For this particular example within our learning performance framework, we focused on how students can revise previously established models to form better explanations about water.

Similar to example 1, the analysis of the student data also revealed diverging levels of primary student articulation, when focusing on the specific intersection of models' explanatory power and model revision of the water cycle (Fig. 15.2). Example 2 parallels other performances, having level 3 represented the high anchor, and level 1 and 2 characterize differing levels of complexity and missing significant elements of the final rank. Similarly to example 1, most students were able to speak to varying degrees about how revising a model would create a better explanation, and the levels developed here highlight how students were directing their attention and the sophistication of their ideas.

Students who expressed level 1 understandings of models discussed simply adding more labels to the water cycle model. Many expressed that having more clearly labeled elements would improve the explanatory power by expanding the vocabulary base of the model and highlighting more features. Asked for a reason what they would change in the model to 'make the model better,' Jon∗ answered in a way that mirrored many of the other level 1 responses, telling that "I would put […] lot of labels just like there. Just that I would put, for example, a sentence or words here or here or here that explains what it's showing in there." Level 2 performances focused not only expanding parts of the model through revision, but these students also looked to connecting pieces of the water cycle model as well. This indicates a high level of complexity; students recognize that by connecting different components of the water cycle they can show a process, extending the explanatory power through connecting ideas. Interestingly, this level of connection occurred in isolation, focusing on a single sequence of events, Jen∗ who was adamant in drawing a line between a mountain and a lake in one model, claiming that if she "wouldn't have drawn this line, they [other people] wouldn't know what the mountain has to do and why there is a lake". This student clearly recognized that the run-off water from the mountains feeds into the lake, connecting those components, yet did not continue this train of thought, further connecting this to other processes of the water cycle, like lake water infiltrating the ground, although she talked about underground water movement later on. The final and most sophisticated level of primary students' understanding of revision to impact the explanatory and illustrative purpose of a model focuses more holistically. The few students with a level 3 understanding made larger, broader changes to the model. Bri∗ used color or dotted lines to indicate movement of water across the entire model so that it "would show like water […] steaming up [forming yellow gas he added] and the black lines mean where the gas mostly goes up. And the yellow [coloring] means, just, that it's a specific gas. It […] has rain after [refers to condensation label], runs down the mountain, and goes into the lake […]. Another student highlighted the sun's impact on the water system in her model by drawing more rays, and one added a key to explain the more abstract features of the model. Students exhibiting a level 3 understanding were better able to expand

and connect different features of their model to create stronger explanations for underlying processes of water systems.

15.4.1 Summary

The examples presented in the previous sections illustrate the complex ways in which students reasoned about water systems using models and engaging in modeling practices within the bounds of the framework presented in Fig. 15.1. Within individual learning performances, we were able to identify discernable levels of reasoning that, considered separately or collectively, help define students' overall modeling competence for water systems. For these two learning performances, as shown in Fig. 15.2, students interacted with models in ways that ranged from less sophisticated (referring to concrete – often isolated – palpable elements, such as words and labels) to more sophisticated, abstracted reasoning with models in ways that reflected their practice-based and epistemic nature and purpose.

15.5 Discussion, Implications, and Conclusion

Scientific modeling is a powerful sense-making practice that can support students' learning across disciplinary domains (Gilbert & Justi, 2016; Schwarz et al., 2009; Upmeier zu Belzen & Krüger, 2010). Existing as both a representation of students' ideas and a tool for sensemaking/reasoning about new data, a model can serve as a crucial link between the student and the natural world. Calls from international science education reform support the use of more authentic scientific practices in primary classrooms, including scientific modeling (AAAS, 2007; ESLI, 2009; GDSU, 2013; NGSS Lead States, 2013). This study provides evidence of primary students' modeling competence within a disciplinary focus on water systems. To that end, we presented the design and empirical grounding of a learning performance framework focused on primary students' competences about models and modeling. The focus on water systems provides a familiar disciplinary context through which to help support students to engage in scientific modeling. This study informs a larger body of literature focused on scientific modeling in K-12 science classrooms (Krell et al., 2015; Nicolaou & Constantinou, 2014; Passmore et al., 2009), including scientific modeling in primary classroom contexts (Acher et al., 2007; Lehrer & Schauble, 2012; Louca & Zacharia, 2015; Manz, 2012). It also builds upon our own line of prior and current collaborative research and development efforts to better support and study primary students' model-based reasoning about water systems (Forbes et al., 2015; Lange-Schubert et al., 2016; Vo et al., 2015; Zangori et al., 2017). We specifically focus on describing how primary students are engaging in scientific modeling as defined by our three-dimensional learning performances framework in Fig. 15.1.

Results from this work yield two important insights into primary students' modeling competence about water systems. First, while each dimension of the learning performances framework was represented (modeling practices, epistemic considerations, and disciplinary concepts), water-related concepts were the only dimension of each learning performance that was consistently reflected in student data. Although students consistently acknowledged water as important, they often had incomplete or incorrect ideas about it, which was consistent with literature on students' learning about water and water systems (Dickerson et al., 2007; Forbes et al., 2015; Gunckel et al., 2012; Zangori et al., 2017). Some evidence exists to indicate primary students seemed to require a familiar context as a foundation on which an advanced conceptualization of modeling could be framed. However, further investigations into that claim must be conducted focusing on how accurate students' science ideas are and if that accuracy impacts students' modeling competence. Additionally, primary students struggled to discuss scientific modeling outside the scope of the disciplinary context of water systems, even when probed during interviews. This is consistent with literature which suggests that while primary students are capable of modeling within particular disciplinary domains, they struggle and need support extending modeling past familiar ideas and across contexts (Acher et al., 2007; Lehrer & Schauble, 2012; Louca & Zacharia, 2015; Manz, 2012). Future research could investigate the impact of students modeling competence on their reasoning about familiar and unfamiliar scientific concepts.

Second, despite using ECD (Mislevy et al., 2003) to guide the development of our task-based assessments for all 21 learning performances in Fig. 15.1, only 15 of the learning performances were meaningfully represented with all three dimensions in the student data. As noted previously, although primary students in this study generally attended to the range of targeted disciplinary concepts associated with water systems, they did not always exhibit knowledge about all modeling practices or epistemic considerations. While 15 features represent a majority of the facets that comprise our conceptual framework, it also demonstrates a lack of evidence for 6 specific areas – the intersections (1) *use* and *evidence-based*, (2) *use* and *appropriately detailed/complex*, (3) *construction/revision* and *predict/hypothesize*, (4) *evaluation* and *predict/hypothesize*, (5) *construction/revision* and *organize*, (6) *evaluation* and *generate*. It should be noted that all practices and considerations were represented, simply not at the same time. For example, students' data accounted for multiple instances of '*using*' a model to '*predict/hypothesize*.' Students were asked to use a water cycle model to make a prediction. Answers included "[the model] helps predict the evaporation from the lakes and rivers," "maybe [the model] shows the temperature helping get [water] up into the sky? Like, predicts something about how the water moves underground." These student quotes provide insight into how students conceptualize the use of models to predict water movement. However, when students were asked to *revise* a water cycle model with new information to make it more *predictive* of given water phenomena, all students in the study opted out of the question or discussed a different facet of the framework. Thus, no meaningful information focused on students' ability to *revise* the *predictive* purpose of a model was observed. Interestingly, no single practice or epistemic feature was represented disproportionately in the group of learning performances for which no evidence was observed. These 'missing' dimensions merit further study, not only with primary students but across the K-16 continuum, to fully explore a learning performance-based learning progression for model-based reasoning about water systems.

This research has multiple implications for the design of primary science learning environments and assessment of students' modeling competence. First, when thinking about supporting and providing opportunities to develop primary students' modeling competence, this research identifies potential areas of engagement and topics in need of additional assistance. Our framework and associated tasks identified that some primary students can think more holistically across the water cycle using scientific models engaging in multiple dimensions and sophisticated ways. Educational supports that provide opportunities for all students to recognize and engage with multi-dimensional perspectives on modeling are necessary to help further students' scientific sense-making and reasoning (NRC, 2007; NGSS Lead States, 2013). Second, when thinking about assessing students' modeling competence, we see direct implications for primary classrooms. Creating and providing teachers' tools to quickly and efficiently assess students' modeling competence and epistemologies and would allow for more tailored appropriate student support (Krajcik et al., 2007; Sandoval & Reiser, 2004). However, in this vein, it is important to acknowledge the limits of our modeling tasks, as they failed to elicit evidence of 6 out of the 21 learning performances. This dearth of information could indicate a flaw in the task-based assessment, a lack of saturation within our population, or potentially indicate areas of particular difficulty for primary age students engaging in multiple dimensions of scientific modeling.

The work presented here lays the foundation for our continuing collaborative research efforts focused on investigating primary students' modeling competence. More research and tools need to be developed to help support teachers' and students' engagement in scientific modeling at the primary level. Ongoing testing and refinement of our task-based assessment will provide better and more clear measures of primary students' understandings and conceptualizations around both modeling and disciplinary concepts. Driving this research and refinement are multiple questions about primary students' use of scientific models across different content domains, types of models, and within different modeling practices and epistemic considerations. For example, while the focus of our work is on model-based teaching and learning about water-related concepts, how does the learning performances framework in Fig. 15.1 apply to other scientific concepts? If primary students are capable of engaging in all of the learning performances, are there some that prove to be more foundational to children's understanding and conceptualization around the dimensions of scientific modeling and domain-specific content than others? Moreover, how will the transfer and adoption of the interview protocol and embedded task-based assessment to other primary school science content areas impact the learning performances levels? There also remain questions about teachers' knowledge and pedagogical reasoning within the learning performances framework who, if they are to create supporting learning environments to foster students' modeling competence, must be prepared and supported themselves to do so. The need for continued work in this area is evident and therefore remains a major aim of this ongoing collaboration.

References

Acher, A., Arcá, M., & Sanmarti, N. (2007). Modeling as a teaching learning process for understanding materials: A case study in primary education. *Science Education, 91*, 398–418.

American Association for the Advancement of Science (AAAS). (2007). *Atlas for scientific literacy*. Washington, DC: National Science Teachers Association Press.

Dickerson, D., Penick, J. E., Dawkins, K., & Van Sickle, M. (2007). Groundwater in science education. *Journal of Science Teacher Education, 18*(1), 45–62.

Duit, R., Gropengießer, H., & Stäudel, L. (Hrsg.). (2007). *Naturwissenschaftliches Arbeiten. Unterricht und Material 5–10.* 2. Auflage. Seelze-Velber: Ernst Friedrich Verlag.

Earth Science Literacy Initiative (ESLI). (2009). *Earth science literacy principles*. Retrieved from http://www.earthscienceliteracy.org/document.html

Engeström, Y., & Sannino, A. (2010). Studies of expansive learning: Foundations, findings and future challenges. *Educational Research Review, 5*, 1–24.

Forbes, C. T., Zangori, L., & Schwarz, C. V. (2015). Empirical validation of integrated learning performances for hydrologic phenomena: 3rd-grade students' model-driven explanation-construction. *Journal of Research in Science Teaching, 52*(7), 895–921.

Gelman, S. A., & Kalish, C. W. (2006). Conceptual development. In W. Damon, R. M. Lerner, D. Kuhn, & R. S. Siegler (Eds.), *Handbook of child psychology: Vol. 2 cognition, perception, and language* (6th ed., pp. 687–733). Hoboken, NJ: Wiley.

Gesellschaft für Didaktik des Sachunterrichts [GDSU – Society for Teaching of Science Education]. (2013). *Perspektivrahmen Sachunterricht* [Perspectives framework for general studies]. Bad Heilbrunn, Germany: Klinkhardt.

Gilbert, J. K., & Justi, R. (2016). *Modelling-based teaching in science education*. Bern, Switzerland: Springer International Publishing.

Gunckel, K. L., Covitt, B. A., Salinas, I., & Anderson, C. W. (2012). A learning progression for water in socio-ecological systems. *Journal of Research in Science Teaching, 49*(7), 843–868.

Harré, R. (1970). Constraints and restraints. *Metaphilosophy, 1*(4), 279–299.

Kenyon, L., Schwarz, C. V., & Hug, B. (2008). The benefits of scientific models. *Science and Children, 46*(2), 40–44.

Klieme, E., Hartig, J., & Rauch, D. (2008). The concept of competence in educational contexts. In J. Hartig, E. Klieme, & D. Leutner (Eds.), *Assessment of competencies in educational contexts* (pp. 3–22). Cambridge, MA: Hogrefe & Huber.zangori.

Krajcik, J., McNeill, K. L., & Reiser, B. (2007). Learning-goals-driven design model: Developing curriculum materials that align with national standards and incorporate project-based pedagogy. *Science Education, 92*(1), 1–32.

Krell, M., Reinisch, B., & Krüger, D. (2015). Analyzing students' understanding of models and modeling referring to the disciplines biology, chemistry, and physics. *Research in Science Education, 45*(3), 367–393.

Kultusministerkonferenz (KMK). (2004). *Bildungsstandards der kultusministerkonferenz. Erläuterungen zur konzeption und entwicklung*. Berlin, Germany: Luchterhand.

Lange-Schubert, K., Schubert, J., Böschl, F., & Forbes, C. T. (2016). Wasser – Boden – Interaktionen: Durch wissenschaftliches Beobachten, Untersuchen und Modellieren über den Wasserkreislauf lernen [Water – soil interactions: Learning through scientific monitoring, investigating, and modeling on the water cycle]. In M. Adamina, M. Hemmer, & J. C. Schubert (Eds.), *Die*

Geographische Perspektive Konkret – Begleitband 3 zum Perspektivrahmen Sachunterricht (pp. 29–42). Bad Heilbrunn, Germany: Klinkhardt.
Lehrer, R., & Schauble, L. (2012). Seeding evolutionary thinking by engaging children in modeling its foundations. *Science Education, 96*(4), 701–724.
Louca, L. T., & Zacharia, Z. C. (2015). Examining learning through modeling in K-6 science education. *Journal of Science Education and Technology, 24*(2–3), 192–215.
Manz, E. (2012). Understanding the codevelopment of modeling practice and ecological knowledge. *Science Education, 96*(6), 1071–1105.
Mayer, J. (2007). Erkenntnisgewinnung als wissenschaftliches Problemlösen. In D. Krüger & H. Vogt (Eds.), *Handbuch der Theorien in der biologiedidaktischen Forschung* (pp. 179–188). Berlin, Germany: Springer.
Merriam, S. B. (2009). *Qualitative research. A guide to design and implementation* (3rd ed.). San Francisco: Wiley.
Metz, K. E. (1995). Reassessment of developmental constraints on children's science instruction. *Review of Educational Research, 65*(2), 93–127.
Mislevy, R. J., Steinberg, L. S., & Almond, R. G. (2003). Focus article: On the structure of educational assessments. *Measurement: Interdisciplinary Research and Perspectives, 1*(1), 3–62.
National Research Council (NRC). (2000). *Inquiry and the national science education standards: A guide for teaching and learning*. Washington, DC: The National Academies Press.
National Research Council (NRC). (2007). *Taking science to school: Learning and teaching science in grades K-8*. Washington, DC: National Academies Press.
NGSS Lead States. (2013). *Next generation science standards: For states, by states*. Washington, DC: National Academies Press.
Nicolaou, C. T., & Constantinou, C. P. (2014). Assessment of the modeling competence. A systematic review and synthesis of empirical research. *Educational Research Review, 13*, 52–73.
Passmore, C., Stewart, J., & Cartier, J. (2009). Model-based inquiry and school science. Creating connections. *School Science and Mathematics, 109*(7), 394–402.
Patton, M. Q. (2001). *Qualitative research and evaluation methods*. Thousand Oaks, CA: Sage.
Sandoval, W. A., & Reiser, B. J. (2004). Explanation-driven inquiry: Integrating conceptual and epistemic scaffolds for scientific inquiry. *Science Education, 88*(3), 345–372.
Schwarz, C., Reiser, B., Davis, E. A., Kenyon, L., Acher, A., Fortus, D., et al. (2009). Developing a learning progression for scientific modeling: Making scientific modeling accessible and meaningful for learners. *Journal of Research in Science Teaching, 46*(6), 632–654.
Upmeier zu Belzen, A., & Krüger, D. (2010). Modellkompetenz im Biologieunterricht. [Modeling competence in biology classes]. *Zeitschrift für Didaktik der Naturwissenschaften, 16*, 41–57.
Vo, T., Forbes, C. T., Zangori, L., & Schwarz, C. (2015). Fostering 3rd-grade students' use of scientific models with the water cycle: Elementary teachers' conceptions and practices. *International Journal of Science Education, 37*(15), 2411–2432.
Weinert, F. E. (2001). Leistungsmessungen in Schule. Vergleichende Leistungsmessung in Schulen: eine umstrittene Selbstverständlichkeit. In F. E. Weinert (Ed.), *Leistungsmessungen in Schulen* (pp. 17–31). Weinheim, Germany: Beltz.
Zangori, L., Vo, T., Forbes, C. T., & Schwarz, C. V. (2017). Supporting 3rd-grade students' model-based explanations about groundwater. A quasi-experimental study of a curricular intervention. *International Journal of Science Education, 39*, 1–22.

Chapter 16
Designing Technology Environments to Support System Modeling Competence

Tom Bielik, Lynn Stephens, Dan Damelin, and Joseph S. Krajcik

16.1 Introduction

The modeling practice, which includes the elements of constructing, using, evaluating and revising models, has always been a central practice used by scientists and has recently gained more prominence in science classrooms. In parallel, the prominence of systems thinking and system modeling has grown in science education as students are expected to investigate complex systems to make sense of phenomena (National Research Council [NRC], 2012). Since technologically advanced modeling tools have become more widely used in STEM education, the question arises as to how these tools influence learning and how they can be used to probe theories about learning through modeling. However, the reverse may also be asked: how do theories of modeling influence the development of digital modeling tools? Our prior experiences and understandings of scientific modeling, its importance in the development of many areas of science, and our belief in the educational value of system modeling in particular inspired us to develop a modeling tool that could provide a better onramp to system thinking. In turn, classroom use of the modeling tool provided us with a new and detailed view into student modeling practices and challenges.

We begin with a brief overview of the theoretical framework related to the development of system modeling tools, then describe the tool itself and several aspects of system modeling competence that it is designed to support. The chapter is concluded

T. Bielik (✉) · J. S. Krajcik
Michigan State University, East Lansing, MI, USA
e-mail: tbielik@msu.edu

L. Stephens
University of Massachusetts, Amherst, MA, USA

D. Damelin
The Concord Consortium, Concord, MA, USA

A. Upmeier zu Belzen et al. (eds.), *Towards a Competence-Based View on Models and Modeling in Science Education*, Models and Modeling in Science Education 12, https://doi.org/10.1007/978-3-030-30255-9_16

by discussing how the aspects of system modeling competence correspond with the ideas of modeling competence presented in this book.

16.2 A Framework for System Modeling Competence

The Next Generation Science Standards (NGSS Lead States, 2013), a set of internationally benchmarked standards for science education widely used in the United States, defines the modeling practice at the secondary school level as developing, revising, using, and evaluating models to predict and explain phenomena (Appendix F, p. 6). Lehrer and Schauble (2015) point out that many philosophers and researchers have identified modeling as the signature practice of science, and that the other seven science and engineering practices of the NGSS are "deployed in the goal of constructing, revising, critiquing, and contesting models of the natural world" (p. 1241). Furthermore, the modeling practice is viewed as critical for advancements in science knowledge and critical for students to make sense of phenomena and share their ideas (Harrison & Treagust, 2000; Passmore, Gouvea, & Giere, 2014; Schwarz et al., 2009). Engaging learners in constructing, using, evaluating and revising models is key in helping them build useable knowledge (Pellegrino & Hilton, 2012) that can explain and predict phenomena and solve problems.

An important part of modeling stressed by *A Framework for K-12 Science Education* (NRC, 2012) is the ability to *generate* models. Schwarz et al. (2009) maintain that it is crucial to involve learners in the construction of models rather than primarily working with models provided by teachers or scientific authorities, and that by doing so students can articulate their own understanding. Clement (2000, 2008) argues for the importance of having students construct explanatory models, but found that most students did not do this without appropriate support. In the Schwarz et al. (2009) study, with the right supports, elementary and middle school students were able to develop a more sophisticated view of modeling, construct models that included explanatory mechanisms, and use these models to make predictions about closely related phenomena. However, students face several challenges when constructing and using models and have few opportunities to engage in this practice (Harrison & Treagust, 2000; Schwarz et al., 2009). Engaging in the modeling practice often means thinking about phenomena from a system perspective, commonly referred to as 'systems thinking.' Systems thinking encompasses the cognitive processes involved in understanding and working with complex systems. It includes consideration of the system boundaries, components of the system, interactions between components in the system and between different systems, and that systems have emergent properties based on the behavior of the system (Passmore et al., 2014). Models are tools that represent the investigated system and can support students in figuring out how complex systems behave and predict the outcome of changes in complex systems (Yoon et al., 2015). As described in *A Framework for K-12 Science Education* (NRC, 2012), "*An explicit model of a system under study can be a useful tool not only for gaining understanding of the system but also for conveying it to others. Models of a system can range in complexity from lists and simple sketches to detailed computer simulations or functioning prototypes. Models can be valuable in predicting a system's behaviors or in diagnosing problems or failures in its functioning, regardless of what type of system is being examined*" (pp. 91–92). The basic structure of a system model is essentially a network of causal links. Systems thinking is challenging for students. The dynamic nature of these models and difficulties people have with causal reasoning interfere with the ability to design and predict the outcomes of complex system models (Zimmerman, 2007; Chinn & Brewer, 2001). We may be able to conceptualize each component of the model, but 'running' the entire model in our heads is nearly impossible. The outcomes of complex system models can be counterintuitive and it is difficult to know a priori which components of the system will have significant impact. The difficulties of understanding complex systems are well documented. Engaging in the modeling practice through system modeling can provide a support to help students develop a systems thinking perspective (Booth Sweeney & Sterman, 2000, 2007; Dörner, 1980, 1996; Fretz et al., 2002; Hmelo-Silver & Pfeffer, 2004; Jacobson & Wilensky, 2006).

Students can engage in modeling in a number of different ways. Commonly this occurs through the construction of some physical representation, illustration, or model diagram (Krajcik & Merritt, 2012). It also occurs through the exploration of simulations designed and produced by others (Rutten, Van Joolingen, & Van Der Veen, 2012). However, it is uncommon for students to create their own testable models from scratch. These are usually manifested as computational models, and require significant expertise in either computer programming, writing mathematical equations, or both. Our challenge was to make the construction, testing, sharing, and revising of computational models accessible to many more students by overcoming these barriers. By opening up the full range of engagement in the modeling practice, growth in student modeling practices and the possibility of achieving competence in modeling that would not be otherwise possible is within reach.

A framework for system modeling competence should encompass key features of how students build, evaluate, use, and revise models. We have identified four aspects of system modeling competence that appear to be necessary in order for students to construct system models useful for understanding natural phenomena:

1. Defining the boundaries of the system by including components in the model that are relevant to the phenomenon under investigation.
2. Determining appropriate relationships between components in the model.
3. Using evidence and reasoning to build, evaluate, use, and revise models.
4. Interpreting the behavior of a model to determine its usefulness in explaining and making predictions about phenomena.

The first two aspects of the framework for system modeling competence encompass the most common challenges we have observed regarding students building models (Damelin, 2017). These relate primarily to model structure and provide insight into students' system thinking and causal reasoning. The third aspect stems from the question that always arises after a student has defined a relationship between components in a model: 'How do I know this relationship is right?,' To answer that question students can compare the output of a model to validating data sources such as publicly available datasets, results from their own experiments, teacher demonstrations, and readings. The fourth aspect comes from how models are used in the real world to explain and make predictions about phenomena or to solve a problem. All four aspects of the framework for system modeling competence are important for students to engage in when designing, testing, and revising models, and when building their understanding of the purpose and nature of models. In this chapter we provide a description of the four aspects of system modeling competence, and illustrate them with several examples from students' models developed in a high school curricular unit.

16.3 Development of the Modeling Tool

As part of a U.S. National Science Foundation funded project,[1] we developed a modeling tool called SageModeler and embedded it in an environment that would allow the model output to be compared with external validating data sources. Our hypothesis was that an iterative approach to model construction that uses real-world data and experiences as evidence for the relationships between components of the model would result in students creating models they could use to explain and make predictions about the phenomenon under study.

SageModeler,[2] a free, web-based tool, is designed to support students, beginning in middle school, to engage in systems thinking. SageModeler was inspired by a previously designed modeling tool, Model-it (Metcalf-Jackson, Krajcik, & Soloway, 2000). It facilitates the diagramming of a system and makes it possible to calculate and visualize model output without requiring students to write equations or code. Several scaffolds were built into the software to achieve these goals (Damelin, Krajcik, McIntyre, & Bielik, 2017).

Students begin by dragging images that represent model components to the canvas and linking them together to represent a relationship between those components. This initial system diagram provides an opportunity for students to make their first choices about what should be included within the boundaries of the system and to indicate how the causal chains will direct model behavior. At this point the model diagram is a visual representation of a student's systems thinking. This feature of the software supports students in engaging in aspect 1 above, defining the boundaries of the system.

[1] This material is based upon work supported by the National Science Foundation under Grant Nos. 1417900 and 1417809. Any opinions, findings, and conclusions or recommendations expressed in this material are those of the author(s) and do not necessarily reflect the views of the National Science Foundation.

[2] SageModeler can be freely accessed at https://learn.concord.org/building-models

In order for the system diagram to become a runnable model, each component is treated as a variable that can be calculated by the modeling engine. The next step is to define each relationship link in the model such that the impact of one variable on each of the other variables to which it is linked can be calculated. In order to do this without requiring coding or writing equations, students construct a verbal description of how one variable affects another. For example, in a model of gas properties, students could use the relationship inspector to construct a sentence such as the following: An increase in *Volume* causes *Pressure* to [decrease] by [about the same] (Fig. 16.1). The underlined parts of that sentence are defined using drop-down menus, and the resulting relationship is also depicted by a graph showing a visual representation of this relationship. Defining relationships with words helps students overcome the mathematical obstacles typically associated with creating computational models, and allows them to focus on a conceptual understanding of the relationships between variables (Stephens & Ke, 2017). This feature of the software engages students in aspect 2, resulting in a model that represents an instantiation of the student's thinking about the workings of some phenomenon that can now be tested.

Once a system includes variables and relationships between those variables, the model can be run, generating tables and graphs that provide feedback on the behavior of the model. To simplify comparing model output with other data sources, SageModeler is embedded in CODAP, the Common Online Data Analysis Platform (Finzer & Damelin, 2016). CODAP is an intuitive graphing and data analysis platform that takes the outputs generated by the system model, as well as any other data source—from published data sets to results of computational models or student physical experiments—and combines them into a single analytic environment.

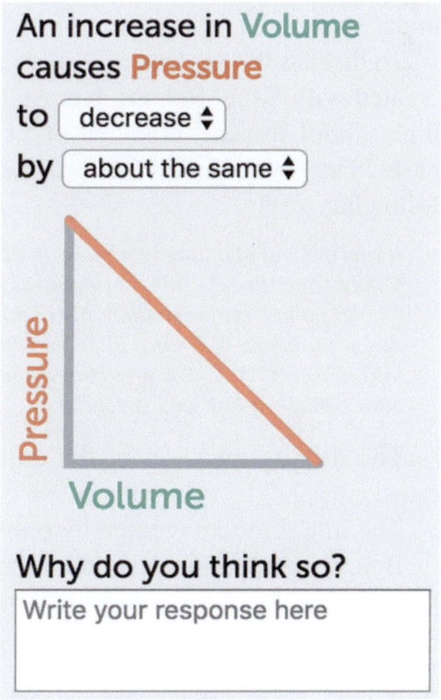

Fig. 16.1 Defining a relationship between two variables in SageModeler using the relationship inspector: an increase in volume causes pressure to [decrease] by [about the same]

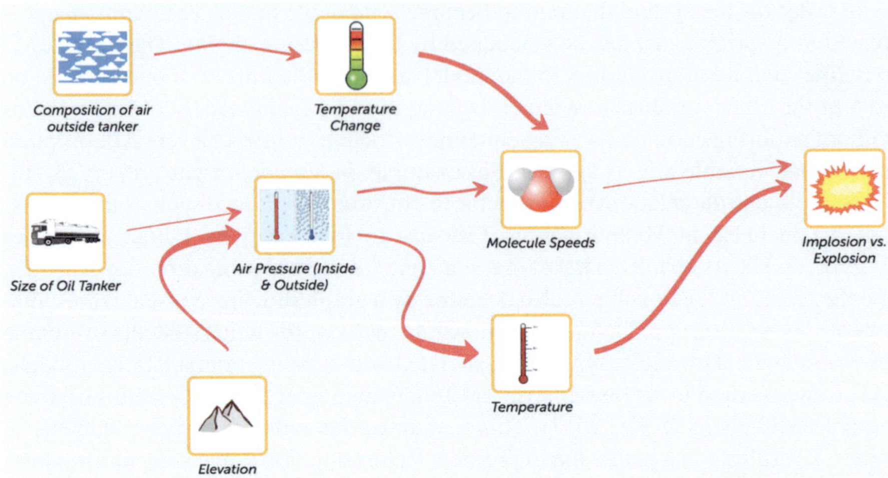

Fig. 16.2 Initial model created by one student pair. The red arrows indicate positive relationships (i.e., as one variable increases, so will the other). The thickness of the arrow represents how much of an effect one variable has on another: a thin arrow indicates a small effect, a thick arrow indicates a large effect, a gradually thickening arrow indicates a change by 'more and more'

Students use the feedback from the visualizations in CODAP to inform iterative cycles of creating, testing, and evaluating their models. Here students engage in aspects 3 and 4, using model output in comparison with an external data set to validate choices made about model components and relationships, resulting in a working model that can be used to explain and make predictions about the phenomenon under study.

To discuss the aspects of system modeling competence in the context of a model created with SageModeler, we use the initial and final models created by a pair of high school students engaged in a chemistry unit about the emergent properties of gases. The phenomenon that was the focus of the unit was framed for students in the following way:

> *It was the end of a long work day on the railway. It was a cooler day and a chilled rain was falling from the sky. A few of the workers were given the task of steam cleaning one of the 67,000 pound, half inch thick steel tankers. When they were done, they sealed up the tanker and went home. Not long after they left, disaster struck, and the steel-walled tanker collapsed in on itself. So, how can something that can't be seen crush a 67,000-pound oil tanker made of half inch steel?*[3]

The driving question of the unit was, 'How can something that can't be seen crush a 67,000 lb. oil tanker made of half inch steel?'

The initial model created by one pair of students is shown in Fig. 16.2.

Below we expand upon each aspect of system modeling competence, illustrating them with examples from student models in Figs. 16.2 and 16.3.

---

[3]Chemistry unit lead author Erin Cothran, a teacher at Hudson High School, in Hudson, MA.

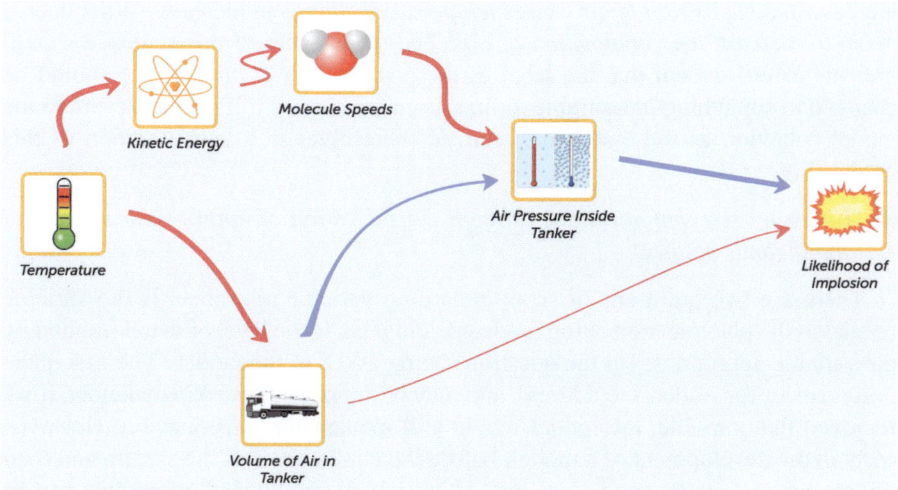

Fig. 16.3 Final model from the same student pair in Fig. 16.2, after 3 weeks of classroom activities and four model revisions. The red arrows indicate positive relationships (as one variable increases, so does the other), and the blue arrows indicate negative relationship (as one variable increases, the other decreases). The thickness of the arrow represents how much of an effect one variable has on another

16.4 Results

16.4.1 *Define Boundaries of a System by Choosing Components in the Model That Are Relevant to the Phenomenon Under Investigation*

When constructing a model of some phenomenon it is important that the model is expansive enough to include all the relevant components and relationships to produce appropriate behaviors, but not so expansive that it complicates the model, hindering understanding of the system. This aspect can manifest itself in two ways.

(a) *Distinguishing between objects and variables*

In order for a system model created using the modeling tool to generate data, each component of the model must be a variable that represents a measurable quantity or quality, something that can be defined on a low-to-high scale. Many students, especially in early iterations of their models, include objects or other components that have no inherent measurement scale. One scaffold that was designed in the modeling tool to support students in this aspect was the text in the relationship box used to define the effect between the variables (Fig. 16.1). When defining these relationships, students are asked to choose the appropriate semi-quantitative effect (about the same, a little, a lot, etc.). If students did not label the variable appropriately as a measurable variable, the sentence will not make sense (for example, ‘an increase in *composition of air* causes *temperature change* to increase.' What does it mean to increase the *composition of air*? The strangeness of this sentence should indicate to the student that the label of the *composition of air* variable should be changed to something measurable, or in this case removed if it is not relevant to the model behavior, as these students realized themselves in a later iteration of this model (Fig. 16.3).

(b) *Choosing relevant variables through consideration of appropriate scope and significance of effect*

There are two questions of scope regarding variable relevance: Is the variable related to the phenomenon being modeled, and if so, is the level of detail implied by the variable appropriate for the questions being asked of the model? The first question is easier for students to address, and may be supported by asking whether, if we removed that variable, the model would still explain the phenomenon. However, early in the development of a model, before the components of the system and their effects are well understood, decisions about which components to include can be challenging. The second question, regarding the level of detail a model should include, tends to be harder to define. It would be inappropriate for every model to drill down to the level of atomic or subatomic interactions, while some models do require this level of detail. Because of this, the scope of variables to include is related not only to the phenomenon being modeled, but also to the features of the phenomenon that are important to understand. In the case of emergent properties of gases, a molecular-level understanding, while not absolutely necessary, provides a richer and more widely applicable model.

Even when variables are all clearly within an appropriate boundary of the system being modeled and are at an appropriate level of detail, some variables will have a greater effect than others. Variables can be related to the phenomenon but have so little effect on the model output as to be insignificant. Including these variables only complicates the model and obstructs exploration of the salient features. In Fig. 16.2 the variable *Elevation* is included and linked to *Air pressure (Inside and Outside)*. While it is scientifically correct that elevation will affect air pressure, the effect will be insignificant on a model of this phenomenon, which occurs under typical atmospheric conditions.

16.4.2 Determine Appropriate Causal and Correlational Relationships Between Components in the Model

Defining the interactions and relationships between elements in the model is critical for developing a good scientific model. These relationships will determine the outcome of the model. When using modeling tools, this aspect can manifest itself in two ways:

(a) *Defining logically correct and scientifically accurate relationships to represent interactions between variables*
There are *several* ways a link between two variables could be incorrect:
(i) There may be no relationship between variable A and variable B. While A and B may covary, a change in one may not be the cause of a change in the other. While linking these two variables together in a model might produce expected outcomes, there would be no rationale for making a causal chain by linking these two variables together.
(ii) There is a relationship, but the way the relationship is defined doesn't match the real-world behavior of the interaction between variable A and variable B. For example, some students correctly predicted that an increase in (external) pressure would decrease the volume of a gas (assuming the container can change size), but they defined a directly proportional relationship rather than an inversely proportional one.
(iii) The direction of causality is reversed from the correct orientation. It was not uncommon for one or more relationships in a causal chain to be reversed.

An example of this aspect can be seen in how the pair of students reversed the linkage between air pressure and temperature from their initial model (Fig. 16.2) to their final model (Fig. 16.3). In the phenomenon explored in this unit *temperature* affects the *air pressure* rather than the other way around, and students recognized this during their model building and testing iterations. These students also changed the relationship between *air pressure* and *likelihood of implosion* from positive to negative, as they realized that an increase in pressure inside the tanker will decrease the chance of the tanker implosion.

(b) *Defining direct relationships between variables*
This is *one* of the most complicated tasks for students when constructing models. There are two considerations related to the directness of relationships:
(i) Large gaps in the causal chain. For example, linking *temperature* and *pressure* might be acceptable if a model is describing what happens at a macroscopic level. However, if the expectation is for a molecular-level explanation, one could argue that other variables should come between *temperature* and *pressure*. Perhaps *temperature* -> *molecular kinetic energy* -> *speed of molecules* -> *number and strength of molecule collisions* -> *pressure* might be more appropriate.
(ii) Inclusion of indirect relationships between variables. Often students will show one variable having an effect on two or more other variables in the model. It is not typical that one variable truly has a *direct* effect on many others; even more unlikely to be accurate is when a variable is connected to both the beginning of a causal chain and to later parts of the same chain of linked variables.

Identifying large gaps, b(i), overlaps with the issue of defining system boundaries, because defining a relationship in this way results in missing relevant variables. Identifying indirect relationships, b(ii), is the greater challenge for students.

In Fig. 16.3, the variable *Volume of Air in Tanker* is linked in two ways to the *Implosion* variable. There are several issues with this, one of which has to do with appropriate labeling of variables. During various model iterations the students attached different meaning to the *Volume of Air* variable. At one point it was meant to refer to the amount of air molecules, and at other points as referring to the volume of the tanker. Though they had not untangled all of these issues by the end of the unit, they did consider it problematic that *Volume* affected *Implosion* via two different causal chains in their model. One of them noted in an interview that the link that did not include *Air Pressure* was not necessary and that they planned to remove it.

16.4.3 Using Evidence and Reasoning to Build, Evaluate, and Revise Models

A well-designed model, which has explanatory and predictive power regarding real-world phenomena, should use evidence to justify which variables are included and how the relationships are defined. Evidence for the inclusion of individual model components can be in the form of collected empirical data or from external data sources, and should be supported by reasoning based on scientific principles.

For example, at the beginning of the properties of gases curricular unit, students were introduced to a phenomenon and driving question and asked to construct an initial model that they thought would help answer the driving question. Because they had not had much experience beyond observing the phenomenon, most of the relationships and some of the variables they used in their initial models were speculative, based on prior knowledge or intuitions about how to define the relationships. As the unit progressed, students conducted experiments, explored simulations, and discussed articles that gave them a foundation on which to defend their choices for specific relationships they had defined. We often observed students modifying those relationships and adding and deleting variables soon after engaging with these validating data sources.

16.4.4 Interpreting the Behavior of the Model to Determine Its Usefulness in Explaining and Generating Predictions About the Phenomena

One of the goals was to create a tool and associated curriculum that would support students in experiencing the modeling practice similar to the way in which scientists engage in modeling and system thinking. Thus, this aspect of system modeling competence is related to students' ability to use models in much the way scientists do. Achieving this aspect of competence means that not only have students created a testable model, but that they also understand how to run it and can make visualizations of the output, compare the model behavior with expected real-world behavior, make predictions related to similar events, and use it to answer the unit's driving question about the core phenomenon being modeled.

Although the students in the emergent properties of gas unit considered the validity of individual relationships between directly-connected pairs of variables in their models, they showed less competence in considering the behavior of their models as a whole as a predictor of the investigated phenomena. For example, the pair of students who produced the model shown in Fig. 16.3 did not appear to notice that their model gave two different predictions for the effect of temperature on implosion. Their final model showed both a positive relationship between temperature and pressure and a negative relationship between those same two variables. Situations like this could exist in which two different causal pathways exist, or a feedback loop might cause an oscillating effect. However, two different predictions for the effect of a single variable usually indicates a problem with the model. These students constructed graphs of the relationships between different pairs of variables in their model, but did not consider the overall behavior of the model as feedback that could have helped them detect, diagnose, and resolve problems, in this case, arising from inconsistency in the way they were thinking about the variable *Volume of Air in Tanker*.

16.5 Discussion

In the examples provided above, students encountered some challenges with causal reasoning similar to those described by Jonassen and Ionas (2008), Schauble (1996), and Koslowski and Masnick (2002). These challenges were most apparent when students were asked to provide evidence and justifications for the variables and relationships defined in their models and to explain how their model addressed the driving question in the unit. However, students' model-based explanations improved after each model revision, indicating that the technology-rich modeling environment and curricular materials offered support to students in developing their causal reasoning.

In line with the goals of scientific modeling practice (NRC, 2012; Schwarz et al., 2009), this chapter provided examples of how students utilized a modeling tool to construct, use, evaluate, and revise their own models. Iterative cycles of model testing and revision stand at the heart of the modeling practice (Lehrer & Schauble, 2006; Magnani, Nersessian, & Thagard, 1999). Students iteratively revised their models to explain the phenomenon under investigation by honing the boundaries of the system and improving relationships between the variables in their models. This process is recommended by Schwarz et al., (2009) and aligns with the modeling cycle presented in Chap. 2 of this book and described by Krell, Upmeier zu Belzen, and Krüger (2014), in which students move between the experimental world and the model world. Our evidence supports the notion that students require repeated appropriate opportunities to use validating data sources to develop, test, evaluate and revise their models in the pursuit of gains in this competence. Students were able to take the ideas they learned in their real-world investigations and incorporate them, with the support of the teacher and their peers, into their computational models. Although we saw improvements with each iteration, the examples in this chapter indicate that it was not easy for students to build towards all four aspects of system modeling competence discussed here. Most students made progress toward them, showing the greatest gains in aspects 1 and 2 related to model structure, but results suggest that progress toward aspects 3 and 4 would have benefited from more explicit connections to validating data sources and more explicit support to use the models to make predictions about specific phenomena.

The impetus for developing a modeling tool for students was to support them in engaging in developing, using, testing and revising models—key aspects of the modeling practice. The focus was primarily on the more general aspects (aspects 3 and 4) articulated in the framework for system modeling competence. This informed how we designed the tool to support sense-making with models through comparative data analysis and ease of model construction. We provided scaffolds to encourage student articulation of evidence for defining particular relationships and designed tools for sharing models and supporting peer review. These features scaffolded students in their growth toward aspect 3—using evidence and reasoning to build, evaluate, use, and revise models. We also built the tool to utilize an existing data analytic environment, which was designed for student visualization of data and facilitated the comparison of data across multiple data sources. This feature of the tool design supports student growth in aspect 4—interpreting the behavior of a model to determine its usefulness in explaining and making predictions about phenomena.

At the same time that the framework for system modeling competence was informing the design of the modeling tool, the experience with students during the tool development influenced the development of the framework for system modeling competence. Aspects 1 and 2 primarily grew out of challenges we observed students encountering when building system models. Some of these were anticipated, such as the need to include only measurable variables as model components, but other barriers toward student generation of useful models proved to be significant obstacles. In particular, the issues students encountered in causal reasoning and graph literacy spawned many discussions about how to address these issues both through software scaffolds, such as the design of the relationship inspector (Fig. 16.1), to pedagogical scaffolds, which included teacher materials to support targeted discussions around these issues and easy ways for the teacher to project student models for discussion.

The framework for modeling competence (FMC; Chap. 1) arose in a different context from ours, but connections can be drawn between the two approaches. Grosslight, Unger, Jay, and Smith (1991) defined three general levels of model understanding. In level I, students consider models as mere copies of reality. In level II, students still consider models as copies of reality, but start to understand that models can be used for different purposes and focus on certain features of the target reality. In level III, students perceive models as representational tools for developing and testing ideas, and understand that the role of the modeler and the models can change as the understanding develops. Based on these levels, Krell, Upmeier zu Belzen, and Krüger (2014) developed the FMC. According to their model, three levels of understanding were suggested for each of five modeling aspects (Chap. 1). Levels I and II in this framework consider models as descriptions of reality, while level III considers models as predictive research tools for testing different hypothesis (Krell et al., 2014; Krell & Krüger, 2016; Krell, Reinisch, & Krüger, 2015).

We suggest that students with level III modeling competence would perform at a high level on the four aspects of system modeling competence. This would be expected from students with level III competence in testing and changing models, because it is necessary for students to consider the four aspects described in this chapter—choosing the relevant components of the system, determining the appropriate causal relationships between components, using evidence to support the model, and interpreting the behavior of the model—to successfully test and change their models. When considering the aspect *purpose of models*, students who understand the predictive power of the model (level III) should be more competent in the aspect of interpreting the behavior of the model, as the output of the simulation should be used for predicting real-world behavior of the system being modeled. It is important to emphasize that the four system modeling aspects described in this chapter are not defined as levels, but as ideas about what students need to know and be able to do.

We hold views on the aspect nature of models somewhat different from those presented in Chap. 2. We believe that the explanatory power of a model is just as important as the predictive power of it. A good model that provides a complete and appropriate explanation of the investigated phenomena is just as meaningful for students' learning as a model that is used to test hypotheses and predict changes in the system.

16.6 Conclusions

Modeling and system thinking play an important role in supporting students in developing useable knowledge of science (Pellegrino & Hilton, 2012). Technologically advanced learning tools such as the modeling tool described in this chapter hold a potential to support students' engagement in the modeling practice and provide an environment for demonstrating competence in modeling, systems thinking, and causal reasoning (Fishman, Marx, Blumenfeld, Krajcik, & Soloway, 2004; Quintana et al., 2004; Wagh, Cook-Whitt, & Wilensky, 2017; Yoon et al., 2015). However, as others have found (e.g., Fretz et al., 2002), students face substantial challenges when learning to engage in modeling, systems thinking, and causal reasoning. We believe that modeling tools that engage students in the process of understanding phenomena through constructing, testing, evaluating, and revising models increase students' competence related to the four aspects of system modeling competence described in this chapter.

We have described the development of an online computational modeling tool, discussed four key aspects that are part of a the framework for system modeling competence, and illustrated them with examples of competence from students' models in a high school unit using the modeling tool. We believe that building toward these four important aspects of system modeling competence provides a way to evaluate whether students are productively engaged in the modeling practice. This articulation of the framework for system modeling competence can also provide guidance for designers considering the creation of tools, curricula, and teacher supports that will encourage student growth related to systems and system modeling.

References

Booth Sweeney, L., & Sterman, J. D. (2000). Bathtub dynamics: Initial results of a systems thinking inventory. *System Dynamics Review, 16*(4), 249–286.

Booth Sweeney, L., & Sterman, J. D. (2007). Thinking about systems: Student and teacher conceptions of natural and social systems. *System Dynamics Review, 23*(2–3), 285–311.

Chinn, C., & Brewer, W. (2001). Models of data: A theory of how people evaluate data. *Cognition and Instruction, 19*(3), 323–393.

Clement, J. (2000). Model based learning as a key research area for science education. *International Journal of Science Education, 22*(9), 1041–1053.

Clement, J. (2008). *Creative model construction in scientists and students*. Dordrecht, The Netherlands: Springer.

Damelin, D. (2017). *Technical and conceptual challenges for students designing systems models.* Paper presented at the Annual Meeting of the National Association for Research in Science Teaching (NARST), San Antonio, TX.

Damelin, D., Krajcik, J., McIntyre, C., & Bielik, T. (2017). Students making system models: An accessible approach. *Science Scope, 40*(5), 78–82.

Dörner, D. (1980). On the difficulties people have in dealing with complexity. *Simulations and Games, 11*(1), 87–106.

Dörner, D. (1996). *The logic of failure: Recognizing and avoiding error in complex situations* (R. Kimber & R. Kimber, Trans.). New York: Metropolitan Books.

Finzer, W., & Damelin, D. (2016). Design perspective on the common online data analysis platform. In C. E. Konold (Chair), *Student thinking, learning, and inquiry with the common online data analysis platform.* Symposium conducted at the meeting of the American Educational Research Association, Washington, DC.

Fishman, B., Marx, R. W., Blumenfeld, P., Krajcik, J., & Soloway, E. (2004). Creating a framework for research on systemic technology innovations. *The Journal of the Learning Sciences, 13*(1), 43–76.

Fretz, E. B., Wu, H. K., Zhang, B., Davis, E. A., Krajcik, J. S., & Soloway, E. (2002). An investigation of software scaffolds supporting modeling practices. *Research in Science Education, 32*(4), 567–589.

Grosslight, L., Unger, C., Jay, E., & Smith, C. L. (1991). Understanding models and their use in science: Conceptions of middle and high school students and experts. *Journal of Research in Science Teaching, 28*(9), 799–822.

Harrison, A. G., & Treagust, D. F. (2000). A typology of school science models. *International Journal of Science Education, 22*(9), 1011–1026.

Hmelo-Silver, C. E., & Pfeffer, M. G. (2004). Comparing expert and novice understanding of a complex system from the perspective of structures, behaviors, and functions. *Cognitive Science, 28*(1), 127–138.

Jacobson, M. J., & Wilensky, U. (2006). Complex systems in education: Scientific and educational importance and implications for the learning sciences. *The Journal of the Learning Sciences, 15*(1), 11–34.

Jonassen, D. H., & Ionas, I. G. (2008). Designing effective supports for causal reasoning. *Educational Technology Research and Development, 56*(3), 287–308.

Koslowski, B., & Masnick, A. (2002). The development of causal reasoning. In U. Goswami (Ed.), *Blackwell handbook of childhood cognitive development* (pp. 257–281). Oxford, UK: Blackwell Publishing, Ltd.

Krajcik, J., & Merritt, J. (2012). Engaging students in scientific practices: What does constructing and revising models look like in the science classroom? *The Science Teacher, 79*(3), 38.

Krell, M., & Krüger, D. (2016). Testing models: A key aspect to promote teaching-activities related to models and modelling in biology lessons? *Journal of Biological Education, 50*(2), 160–173.

Krell, M., Upmeier zu Belzen, A., & Krüger, D. (2014). Students' levels of understanding models and modelling in biology: Global or aspect-dependent? *Research in Science Education, 44*(1), 109–132.

Krell, M., Reinisch, B., & Krüger, D. (2015). Analyzing students' understanding of models and modeling referring to the disciplines biology, chemistry, and physics. *Research in Science Education, 45*(3), 367–393.

Lehrer, R., & Schauble, L. (2006). Cultivating model-based reasoning in science education. In R. K. Sawyer (Ed.), *The Cambridge handbook of the learning sciences* (pp. 371–387). New York: Cambridge University Press.

Lehrer, R., & Schauble, L. (2015). Development of scientific thinking. In R. Lerner, L. Liben, & U. Müller (Eds.), *Handbook of child psychology and developmental science: Vol. 2. Cognitive processes* (7th ed., pp. 671–714). Hoboken, NJ: Wiley.

Magnani, L., Nersessian, N., & Thagard, P. (Eds.). (1999). *Model-based reasoning in scientific discovery*. New York: Kluwer Academic/Plenum Publishers.

Metcalf-Jackson, S., Krajcik, J., & Soloway, E. (2000). Model-it: A design retrospective. In M. Jacobson & R. Kozma (Eds.), *Advanced designs for the technologies of learning: Innovations in science and mathematics education*. Hillsdale, NJ: Erlbaum.

National Research Council. (2012). *A framework for K-12 science education: Practices, crosscutting concepts, and core ideas*. Washington, DC: The National Academies Press.

NGSS Lead States. (2013). *Next generation science standards: For states, by states*. Washington, DC: The National Academies Press.

Passmore, C., Gouvea, J. S., & Giere, R. (2014). Models in science and in learning science: Focusing scientific practice on sense-making. In *International handbook of research in history, philosophy and science teaching* (pp. 1171–1202). Amsterdam, The Netherlands: Springer.

Pellegrino, J. W., & Hilton, M. L. (2012). *Education for life and work: Developing transferable knowledge and skills in the 21st century*. Washington, DC: The National Academies Press.

Quintana, C., Reiser, B. J., Davis, E. A., Krajcik, J., Fretz, E., Duncan, R. G., et al. (2004). A scaffolding design framework for software to support science inquiry. *The Journal of the Learning Sciences, 13*(3), 337–386.

Rutten, N., Van Joolingen, W. R., & Van Der Veen, J. T. (2012). The learning effects of computer simulations in science education. *Computers & Education, 58*(1), 136–153.

Schauble, L. (1996). The development of scientific reasoning in knowledge-rich contexts. *Developmental Psychology, 32*(1), 102–119.

Schwarz, C. V., Reiser, B. J., Davis, E. A., Kenyon, L., Achér, A., Fortus, D., et al. (2009). Developing a learning progression for scientific modeling: Making scientific modeling accessible and meaningful for learners. *Journal of Research in Science Teaching, 46*(6), 632–654.

Stephens, L., & Ke, L. (2017). *Explanations and relationships in students' mental and external models.* Paper presented at the Annual Meeting of the National Association for Research in Science Teaching (NARST), San Antonio, TX.

Wagh, A., Cook-Whitt, K., & Wilensky, U. (2017). Bridging inquiry-based science and constructionism: Exploring the alignment between students tinkering with code of computational models and goals of inquiry. *Journal of Research in Science Teaching, 54*(5), 615–641.

Yoon, S. A., Anderson, E., Koehler-Yom, J., Klopfer, E., Sheldon, J., Wendel, D., et al. (2015). Designing curriculum and instruction for computer-supported complex systems teaching and learning in high school science classrooms. *J-STEM: Journal of Research in STEM Education, 1*(1), 4–14.

Zimmerman, C. (2007). The development of scientific thinking skills in elementary and middle school. *Developmental Review, 27*(2), 172–223.

Chapter 17
Learning Abstraction as a Modeling Competence

Jonathan T. Shemwell and Daniel K. Capps

17.1 Introduction

A challenge for modeling instruction is that students tend to think of models as literal interpretations of their referents, what Grosslight, Unger, Jay, and Smith (1991) called "copies of reality" (p. 817). Literal interpretation runs counter to authentic modeling and thus stands in the way of modeling competence. The authors first encountered literal interpretation when piloting a novel approach to modeling featured in the present chapter, which is called synthesis modeling (Capps & Shemwell, 2016). The approach had two phases. In the initial phase, pre-synthesis, students used data and other information to construct and evaluate a pair of conceptual models, one for each of two ways that deserts form. In the second, synthesis phase, students drew on the models from the first phase to generate, or synthesize, an omnibus "general model" of desert formation. When this technique was piloted in classrooms, students thought very differently about their synthesized, general models than their pre-synthesis models. In particular, they viewed the pre-synthesis models as straightforward descriptions of "what happens," whereas they saw the general model as a transformation on these phenomena, conveying more essential information about them. In other words, they thought of their general model as a model, while their pre-synthesis models were closer to literal interpretations.

Our explanation for the contrast in student thinking just described began with the recognition that synthesis modeling was an abstraction process, which can be described as seeking the underlying structure from two scenarios that contain it

J. T. Shemwell (✉)
Department of Curriculum and Instruction, University of Alabama, Tuscaloosa, AL, USA
e-mail: jtshemwell@ua.edu

D. K. Capps
Department of Mathematics and Science Education, University of Georgia, Athens, GA, USA

A. Upmeier zu Belzen et al. (eds.), *Towards a Competence-Based View on Models and Modeling in Science Education*, Models and Modeling in Science Education 12, https://doi.org/10.1007/978-3-030-30255-9_17

(Catrambone & Holyoak, 1989; Gick & Holyoak, 1980, 1983). Such processes, which have been extensively studied in the field of analogical learning, enable learners to think of their resulting ideas as being abstracted or pulled away from their sources (Gentner, 2010; Gick & Holyoak, 1980). Putatively then, the students in our study had learned that their general models were abstractions, and it was on this basis that they thought of them as models and not literal interpretations. This explanation indicates a possible omission in the modeling literature. Namely, there is wide agreement that models are fundamentally abstractions (Crawford, 2016; Morrison & Morgan, 1999), yet no authors have explicitly proposed that students should learn this essential fact (e.g., National Research Council, 2007; Passmore, Gouvea, & Giere, 2014). In Capps and Shemwell (2016), the authors pointed out this omission in the context of a study of student learning through synthesis modeling.

Extending from Capps and Shemwell (2016), the thesis of the present chapter is that thinking of models as abstractions can be of service in assessment, specifically as an aspect of modeling competence. According to Klieme, Hartig, & Rouch (2008), three requirements define competence as a unique form of educational objective. First, and most importantly, competence underlies authentic performance in a domain (Klieme et al., 2008). Second, and related to this, competence depends on transferable knowledge, meaning knowledge that can be applied under transformation to novel scenarios. Third, competence is separable from general intellectual capacity, making it responsive to focused instruction. In what follows, we show that thinking of models as abstractions meets each of Kleime et al.'s requirements while also defining a key understanding within modeling practice. In conjunction, a modification is proposed to the framework for modeling competence (FMC, Chap. 1) such that thinking of models as abstractions occupies a more salient position.

17.2 Theoretical Background

17.2.1 Models as Abstractions

An abstraction is an idea that is removed, or pulled away, from its source (Merriam-Webster, 2015). While on its own this definition may not seem to say much, when applied to models it can be very informative. Nersessian (2008) succinctly defined models as organized units of representations that mimic the structure of their referents. Within her definition, mimicking structure translates to extracting key information. It also denotes a degree of transformation on that information such that a model's structure differs in appearance from that of its referents (i.e., mimicking; not reflecting). This transformational aspect of models is crucial: models are meant to represent essential information distinct from surface details (Schwarz et al., 2009). Under the Merriam-Webster definition, Nersessian's concept of mimicking structure becomes abstracting structure, which is to say pulling structure away from source phenomena. We prefer this phrasing because it renders the requisite transformation more active and explicit. Meanwhile, philosophers of science seem well settled on the point that models are abstractions in the sense presented here. For example, Morrison and Morgan (1999) defined models as "a kind of rendering—a partial representation that either abstracts from, or translates into another form" (p. 27).

Modeling researchers in science education have broadly acknowledged that models are abstractions (Gilbert & Justi, 2016; Gobert & Buckley, 2000; Schwarz et al., 2009; Treagust, Chittleborough, & Mamiala, 2002). However, these same researchers have not referred explicitly to abstraction when defining models for educational audiences. For example, Passmore et al. (2014) defined models as "entities that represent some aspects of a phenomenon to some degree" (p. 1176). Similarly, Justi and Gilbert (2003) defined a sophisticated notion of a model as being "a representation of a part of something" (p. 1375). And, writing for a broad audience, National Research Council (2007) described models as "deliberate simplifications" (p. 152). Taken as a body, these authors almost seem to avoid the term abstraction. Speculatively, this is because abstraction seems like a vague and enigmatic idea, less useful for practitioners than more familiar terms like simplification. By contrast, the view advanced here is that abstraction, or particularly, abstraction of structure, provides a more useful instructional-level definition of what models are than existing, colloquial definitions of models. Specifically, in the concept of pulling structure away from a phenomenon, abstraction succinctly defines how a model corresponds to its referent, namely with fidelity to key information, but under transformation. Existing verbiage like accurate, but not exact reflections, or deliberate simplifications are less precise and thus less informative in denoting this relationship. Further, the concept of abstraction also indicates the essential modeling process, which is to pull structure away from a source or sources and make this structure explicit.

17.2.2 Models as Literal Interpretations

As outlined in the introduction, students commonly generate model-like representations without thinking of them as models (Mahr, 2011). Instead, they think of them as veridical approximations of their referents, what the present chapter refers to as literal interpretations, what others call copies of reality (Grosslight et al., 1991), and still others describe as replications of the original (Krell, Upmeier zu Belzen, & Krüger, 2014; Upmeier zu Belzen & Krüger, 2010). To the authors of Grosslight et al. (1991) and their Piagetan-minded contemporaries, Ryan and Aikenhead (1992), literal interpretation was difficult to eradicate because it stemmed from a global disposition to think in literal terms. Later scholars have not necessarily taken this view, but collectively, they have showed literal interpretation to be a strong tendency among novice modelers, not only among learners (Gogolin & Krüger, 2017; Harrison & Treagust, 1996; Ingham & Gilbert, 1991; Krell, Reinisch, & Krüger, 2015), but also their teachers (Dogan & Abd-El-Khalick, 2008; Lin & Chen, 2002; van Driel & Verloop, 1999).

17.2.3 Literal Interpretation and Modeling Competence

Given that novices are prone to thinking of models as literal interpretations, it would be useful to frame an instructional objective, perhaps a competence, around learning a truer conception of what models are. Unfortunately, there is a vacancy of precise ideas in the modeling literature about what this truer conception would be. As already pointed out, when writing for educational audiences, researchers have largely resorted to vague terminology to define what students should think that models are (Justi & Gilbert, 2003; National Research Council, 2007; Passmore et al., 2014). Another trend has been to avoid (or perhaps neglect) attempting such definitions. This trend is evident in studies of innovations designed to counteract literal interpretation (Harrison & Treagust, 1996, 2000; Lin & Chen 2002; Sarri & Viiri, 2003). As a body, authors of these studies have pointed out many ways of teaching against literal interpretation, meaning what models are not. But, they have not explicitly defined or attempted to teach conceptions of what models are that would make them not literal interpretations. For example, Harrison and Treagust (2000) combatted literal interpretation by having students work with multiple models of the same phenomenon (see also Lin & Chen, 2002). They reasoned that if students could see that all of the models differed from each other, they would know that no one of them could be a literal interpretation. However, the instruction did not include the further step of helping students learn characteristics of their models that discriminated them from literal interpretations.

Despite the literature's reluctance to define what students should learn what models are, Upmeier zu Belzen and Krüger (2010), make good strides in outlining a competence that opposes literal interpretation. The authors present a framework with three levels of student thinking about how a model should relate to its referent. At the lowest level, students do not consider this relationship, so their thinking defaults to literal interpretation. At the middle level, students think of models as approximate or idealized representations of known phenomena. At the highest level, students think of models as conjectural tools with which to investigate unknown phenomena. Reasoning pragmatically, the authors' FMC comprises ideas about models that are both ontological (i.e., what models are) and epistemological (i.e., the purpose and validity of models). Considering first the ontological aspect of the framework, this seems to be limited by the underlying literature's vacancy of definition as described above. Specifically, the middle level provides scant information about the relationship between models and their referents because it reflects the literature's use of colloquial terminology to define what models are (Justi & Gilbert, 2003; National Research Council, 2007; Passmore et al., 2014). Meanwhile, the highest level does not have a strong ontological component. Instead, it emphasizes models' conjectural role in knowledge generation, an epistemological characteristic. On this latter point, additional information is available in the framework's most overtly ontological dimension, which the authors call nature of models (Fig. 1.3). Here the highest level of competence describes thinking that models are "theoretical constructions." This phrasing can be taken to have both ontological and epistemological meaning. Epistemologically, it includes the idea that theories are conjectural, as models must be. However, the term theory also refers to general ideas developed from particular instances. Thus, the authors also touch upon the ontological dimension of modeling competence described and exemplified herein, abstraction. This subject is discussed at the conclusion of the chapter.

17.2.4 Abstraction as a Competence

In response to the existing limitations on defining what students should learn that models are, which in turn limit the FMC, the proposal advanced here is that students should learn that models are abstractions. Extending from this starting point, the thesis of the present chapter is that thinking of models as abstractions is a useful dimension of modeling competence, potentially to be overlayed on the FMC. This thesis incorporates the proposal that thinking of models as abstractions can meet three distinct requirements of a competence as defined by Klieme et al. (2008). Considering these requirements in detail, the first, and most prominent is that competence, by definition, underlies authentic performance in a domain. For authentic performance in modeling, the obvious detractor is thinking that models should be literal interpretations. Accordingly, an important question relating to this chapter's thesis is how effectively thinking of models as abstractions counteracts thinking of them as literal interpretations, thus providing crucial support for authentic modeling practice. This question helps frame the first of three research questions to be addressed herein. It is stated as how *precisely* thinking of models as abstractions *runs counter* to thinking of them as of literal interpretations.

A second research question stems from the requirement that competence be domain-specific and thus learnable through focused instruction (Klieme et al., 2008). Necessary to the present chapter's thesis, then, is the possibility of making visible progress in thinking of models as abstractions over a few lessons, or perhaps a few weeks, on what should be called an instructional timescale. This requirement is no small hurdle. The capacity for abstract thinking has long been associated with fluid intelligence (Meer, Stein, & Geertsma, 1955). Thus it seems reasonable to suspect that learning to think of models as abstractions would progress slowly, following the development of general intellectual capacity through inculturation and schooling (Martinez, 2015). However, researchers in analogical learning have shown that people can learn to think abstractly about scenarios in a short period of time (Catrambone & Holyoak, 1989; Gentner, 2010; Gick & Holyoak, 1980, 1983; Loewenstein, 2017). And, of course, the synthesis modeling featured in the present chapter utilizes the analogical learning approach. Perhaps then, growth in thinking that models are abstractions is achievable on an instructional timescale using techniques like synthesis. Whether this is so is the second research question for the present chapter.

As a third question, Klieme et al. (2008) explained that competence is more than learned procedures or skills that can be performed when they are called for. Rather, it is the capacity to meet a range of real-world demands, akin to adaptive expertise

(Hatano & Inagaki, 1984). Consequently, it would be of limited value to ask whether students can think about models as abstractions within a single context. The question, rather, is whether their thinking can be transferable to novel modeling scenarios. This is the third research question for the present chapter.

In sum, the viability of the present chapter's thesis rests on the answers to three research questions:

1. How precisely does thinking of models as abstractions run counter to literal interpretation?
2. Can this thinking be learned on an instructional time scale?
3. Can it transfer to new modeling contexts?

To answer these questions, evidence is presented of students learning to think of models as abstractions in the context of synthesis modeling. It consists primarily of claims to student thinking about model abstraction as indicated by responses to assessment items before and after instruction. This information is supplemented with data from transcribed think-aloud interviews of students as they worked the items.

17.3 Method

As a brief overview, Capps and Shemwell (2016) used a pretest-posttest control-group study to investigate learning about model abstraction among early high school students. The study featured a synthesis approach to modeling desert formation as introduced at the outset of the present chapter and described below. Identical pre- and posttests included three items related to student knowledge of model abstraction, plus several more items addressing domain content. In the treatment condition, 175 students, taught by three teachers,[1] took the pretest, engaged in a unit that included synthesis modeling, and then took the posttest. In the control group, 49 students taught by three other teachers took the same pre- and posttests surrounding subject matter unrelated to the synthesis modeling unit. The control group served the limited but important purpose of checking for the possibility of increasing scores by taking the test twice (i.e., a testing effect). As an auxiliary component of the study, 12 students reworked items on the posttest while thinking aloud.

17.3.1 Synthesis Modeling

Figure 17.1 illustrates the synthesis modeling process, featuring a model constructed by a small group of students in the Capps and Shemwell (2016) study. The students were modeling the air and moisture patterns by which deserts form. They synthesized a general model of desert formation (right), drawn from two source models (left)

[1] These were the students' normal classroom teachers. However, they were particularly well prepared to teach via synthesis modeling, as they had helped to develop the approach.

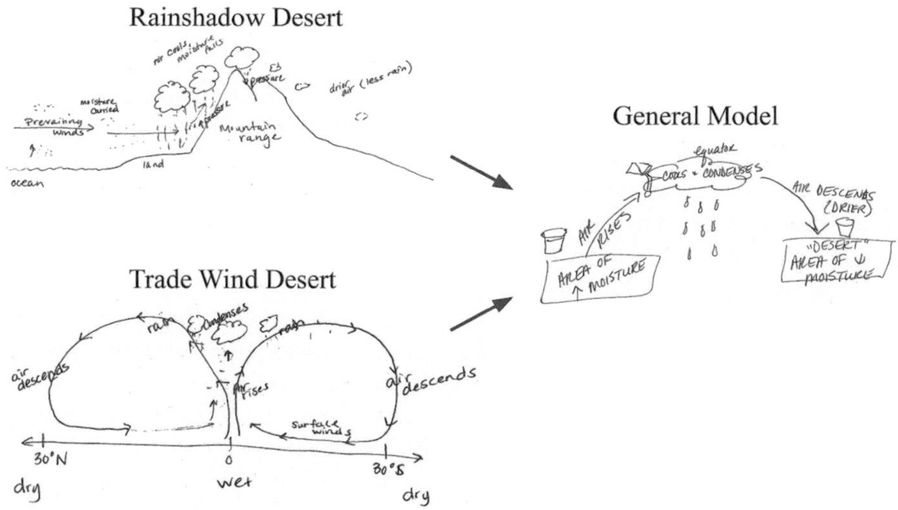

Fig. 17.1 Students drew on two component models (left) to synthesize a general model (right)

that they had developed over the previous several days of the unit. Zooming in on the figure, the source model at top left represents rain shadow deserts, in which air loses moisture as it rises over a mountain range before descending dry over a desert region. The source model at bottom left represents trade wind deserts, in which air loses moisture as it ascends in the equatorial region and moves poleward, eventually descending dry near the mid latitudes (30°N and 30°S). The synthesized general model at the right abstracts the underlying structure of both of these phenomena. It shows that wet air rises, moisture precipitates out, and the resulting dry air moves and descends upon a new location. Importantly, the general model does not reflect either source model directly; nor is it a simple intersection of their shared features. Instead, it provides transformation on the sources, extracting a structure that is transferable to either of them.

17.3.2 *Instructional Procedure*

Since one of the research questions is about the time scale of learning, it is important to know that students synthesized their general models in approximately one 50-min period of instruction. They did this in small groups using whiteboards. Overall, the period included two intensive periods of model generation of about 15 min each. These were followed by teacher-facilitated whole group presentations and questioning.

17.3.3 Test of Knowledge of Model Abstraction

Of the three modeling items featured on the pre- and posttests, two were designed to measure knowledge of model abstraction. The more elaborate of these is shown in Fig. 17.2. It is called the abstracting item, so named because it measured students' ability to abstract when constructing a model. Its prompt gave an image of an igloo and referred to it as a specific model for a house. It then asked students to use words and/or pictures to construct a general model for a house (i.e., dwelling) by thinking of other kinds of houses they knew of. Students earned a correct score when they generated a model that abstracted a dwelling away from an igloo and other typical instantiations of dwellings, most commonly wood frame houses (Fig. 17.2, bottom right and left). Additionally, the item had three features designed to discriminate thinking of models as abstractions compared to reiterating ideas or facts learned from synthesis instruction. First, as shown at the top of Fig. 17.2, an example accompanied the item. It depicted a "general model" of a sports ball that was constructed from several specific sports balls. Given this example, performance did not depend crucially on decoding the phrase "general model" in order to know abstraction was needed. Rather, students had to recognize the example as an instance of abstraction, namely representing essential structure apart from surface features. Second, the modeling context was novel (i.e., a dwelling rather than desert formation). This meant that students could not "get around" abstraction by recalling an abstract model they had learned during instruction. Rather, they had to transfer their notion of abstraction to this new context. Third, the students were not given two

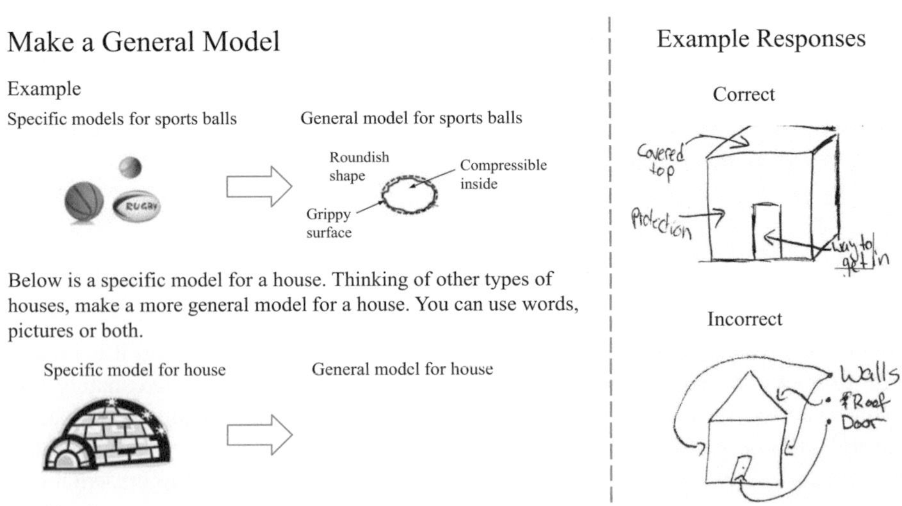

Fig. 17.2 Abstracting item from the pre- and posttests with example incorrect and correct responses

different sources from which to abstract, as in synthesis, but one source. Therefore, they could not answer the item by repeating the synthesis procedure.

The second modeling item was much simpler. In the present chapter it is called it the instantiation item.[2] It prompted students to explain the advantages of a general model over a specific model. A correct score had to meet two requirements. One was to state that a general model should represent a range of situations. The second was to indicate that these situations were diverse (i.e., not identical). As an example, a student giving a correct response said, "A general model would be more specific because it covers a wide variety of different examples instead of just specific ones." Thus, the item discriminated students who thought of general models as instantiable on situations without outward resemblance to each other, and by implication, without outward resemblance to the model.

The third modeling item, which is called the synthesis modeling item, did not measure knowledge of abstraction but rather how disposed students were to draw on the synthesis model (i.e., the model at the right on Fig. 17.1) as an explanation of desert formation. The item consisted of a simple, open-ended question, asking, in general, how deserts form. The item was scored categorically, according to whether students explained desert formation using the synthesis model or, instead, they explained it using either (or both) of the component models.

17.3.4 Think-Alouds

In the think-aloud component of the study, 12 students, chosen at convenience, reworked the abstracting and instantiation items while thinking aloud in the presence of a researcher. This occurred about 2 weeks after the students had completed the posttest for the main study. The purpose of the think-alouds was to understand how students thought about model abstraction as they worked the items. The protocol was structured so that students would "verbalize their thoughts rather than describe or explain them" (Ericsson & Simon, 1998, p. 181). To begin each interview, a researcher modeled the think-aloud process with an arithmetic problem and then asked the student to practice thinking aloud on a similar problem. Then, the student thought aloud while generating written answers to the two items. The researcher stood by taking notes and reminded students to think-aloud if they fell silent. Each think-aloud was recorded and transcribed. Those accompanying correct responses were analyzed for how the participant thought about model abstraction.

[2] In Capps and Shemwell (2016) this item is called the transferability item because it reflected students' understanding that models transfer to their referents (under transformation), as distinct from translating to them (without transformation).

17.4 Results

17.4.1 Test Scores

In the synthesis model group, scores on the three modeling items increased measurably from pretest to posttest. There was no increase in control group scores, making a testing effect unlikely. Concentrating on the synthesis group, for the abstracting item, pretest scores showed 58% of students answering incorrectly, meaning they did not abstract any of the features of a dwelling, compared to 42% of students who abstracted at least some features. Among these latter, only 6% abstracted all the features in their model, as in the example at the bottom right of Fig. 17.2. On the posttest, the distribution flipped, with 59% of students abstracting at least some features, and 41% of students not abstracting. Further, the proportion of students abstracting all features tripled, to 19%. For the instantiating item, the increase in scores was larger, with students going from 7% correct at pretest to 42% at posttest. As detailed by Capps and Shemwell (2016), the increases for both items were statistically significant.

Considering the consistency of responses between items, students who were correct on the instantiation item were about as likely as incorrect students to abstract at least some features on their dwelling model (48% versus 52%). However, they were nearly three times as likely to abstract all features (30% versus 11%). Correspondingly, students who were incorrect on the instantiation item were three times more likely to abstract no features on the abstracting item. Thus, a similar knowledge base seemed to support performance on both items.

Scores on the synthesis modeling item also went up. Unsurprisingly, no students explained desert formation in terms of the synthesis model at pretest. At posttest, 35% did so, signifying that they had internalized the synthesis model to a greater degree than the majority of students, whose explanations featured one or both of the pre-synthesis models. More important was the relationship between internalizing the synthesis model and learning about abstraction. For abstracting, students who internalized the synthesis model were more likely to have correct scores on the abstracting item, 72% versus 49%. Further, they were much more likely to abstract all features on the item, 32% versus 9%. For instantiation, students who internalized the synthesis model were twice as likely to have correct scores on the instantiation item, 64% to 29%. Taken together, these results suggest that learning the synthesis model, which presumably happened when students engaged in synthesis to construct this model, was the source of students' learning to think of models as abstractions.

17.4.2 Think-Aloud Interviews

For the think-aloud analysis, at issue was how students thought about model abstraction when answering either the abstracting or instantiating item correctly. For the abstracting item, 7 of 12 students had correct written responses. Their verbal outputs indicated two related patterns of thinking about abstraction. One, evident in three students' utterances, centered on reduction to essentials:

Student 1. If you really strip it all down, what you have got.
Student 2. A house is something that you can live in comfortably with a roof and an entrance and sides, and a good model for that would be something pretty generic for a house.
Student 3. A general model can't show like, specific features or else it becomes a specific model, and it's describing a specific, like a situation or event so, I guess we should just keep it basic.

Parenthetically, it is notable that student 3 added the additional idea of avoiding specific features, which also surfaced for the instantiating item. Meanwhile, a common characteristic of these three utterances was a descriptive notion of what it means for a model to be abstract (e.g., generic). Such ideas were absent in the remaining four students' interviews. Their thinking about abstraction resided at the level of particular model features, revolving around the need for versatility when representing characteristics like the shape of a house, or key elements such as walls, roof, and door:

Student 4. It [the house] could be any shape.
Student 5. I can do the general shape of the house and, [drawing] that's kind of general.
Student 6. There is a solid shell on the outside of the house, to protect from weather, the elements, like rain or anything.
Student 7. A house has to have a place to enter, which is the same for every house.

As all four utterances show, these students thought in terms of the need for model features to cover a range of situations. However, unlike the first three students, who verbalized a quality of models that provided for versatility, namely stripped down, basic, and generic, students in the second group left their ideas implicit.

The think-aloud results for the instantiation item were less informative, partly due to the misfortune of having only 4 of 12 students give correct responses to the item. Compounding this problem was a procedural error for two of the four students in which the researcher did not ensure that the students generated the written responses necessary to the think-aloud process. This left only two students who both answered the item correctly and had valid protocols. However, the verbalizations for both students showed thinking that was consistent with that of the abstracting item. Specifically, both students stressed that general models must represent classes of phenomena. The first spoke of this as genre: "It's not as, particular. It can explain a certain genre of situations instead of just one." The second spoke of variety: "It covers a wide variety of things, rather than just having one specific model." Thus, both emphasized that general models cannot be constrained by any one situation, echoing student 3's idea that a general model can't show specific features. More generally, their thinking was commensurate with the idea that general models must be abstracted away from their sources.

17.5 Discussion

17.5.1 Abstraction Versus Literal Interpretation

The first research question, how precisely thinking of models as abstractions runs counter to literal interpretation, is most directly answered using the think-aloud results. For the abstracting item, the more articulate students verbalized reducing the given instance, an igloo, to its essentials, which they conceived in terms of stripping it down or keeping it generic or basic. Less articulate students indicated that model features needed to be generic, but they did not express this idea descriptively. Both results pushed precisely in the opposite direction of literal interpretation, which would have students focus on a given instance and try to capture as many of its particular features possible (Grosslight et al., 1991; Tasquier, Levrini, & Dillon, 2016). And, while results from the complementary instantiation item were thin, those that were available also directly opposed literal interpretation, stressing that a model cannot be confined by specific instances that it might represent.

The quantitative test results for the abstracting and instantiation items also show that student thinking about models was opposite that of literal interpretation, particularly in light of how these items were coded. To score well on the abstracting item, students had to avoid the inclusion of details that would limit the model to particular instances. At posttest, 19% of students did this for all of the features of their model. Thus, they purposefully constructed a model that would not reflect its source at a surface level. In the same vein, another 41% abstracted at least some features in their model. For the instantiating item at posttest, a similar percentage of students, 42%, indicated that a general model must apply to a diverse set of referents, implying that no one referent would resemble the model on the surface. Again, this way of thinking is antithetical to literal interpretation, wherein a model is meant to correspond to a given referent in all of its particulars (Grosslight et al., 1991).

17.5.2 Timescale of Learning

The reported gains from pre- to posttest also bear on the second research question, which was whether model abstraction could be learned on a timescale of a few lessons, meeting the learnability requirement for competence (Klieme et al., 2008). This requirement was, indeed met, as students made measurable gains in thinking of models as abstractions, apparently as a result of a single 50-min period of instruction. As a possible objection to this claim, the pretest was given before all instruction began, and it took students several lessons to construct the two component models of desert formation prior to synthesis. Thus, students could have learned about model abstraction steadily through the course of the unit and not during the synthesis period. However, there was no reasonable opportunity for them to do so.

Further, there was a clear relationship between learning the synthesis model and performing well on the two abstracting items, suggesting that the synthesis lesson was the source of learning about abstraction. Moreover, even if significant learning did occur prior to synthesis, the whole unit took fewer than ten lessons to complete, which is still very much an instructional time scale. Withal, the evidence shows that learning to think of models as abstractions does not depend exclusively on the slow process of building the general capacity for abstract thinking (Martinez, 2015).

17.5.3 Transfer

The third research question was whether students could transfer thinking of models as abstractions to new contexts. The answer is yes, as shown by performance on the abstracting item. Students who synthesized a "general" desert formation model transferred their idea of abstraction to a novel modeling situation, that of a dwelling. Further, the item did not call for model synthesis by combining two or more sources, which would have been transfer of a learned procedure (Bassok & Holyoak, 1989). Rather, it required students to abstract from a single instance of a phenomenon by thinking of other similar instances, thus transferring the underlying structure of synthesis.

Transfer may be prompted or spontaneous (Schwartz, Varma, & Martin, 2008). Quite obviously, the abstracting item measured prompted transfer, since the item told students that they should abstract, which it framed as making a "more general" model (Fig. 17.2). This fact limits the interpretation of model abstraction as a competence. Authentic performance depends on recognizing when knowledge is applicable, meaning that students should not need strong cues to realize that they should abstract when constructing models. On the other hand, the ability to abstract on cue is definite progress toward competence. For instance, if students who had attained this ability were to drift into literal interpretation during modeling instruction, a cue or signal to abstract could get them back on track. In the present study, one cue to abstract was the term "general model," which was used in the synthesis process and maintained in the item prompts. This scenario raises the question of whether a "general" model should set apart as a distinct type. In answer, while the so-called general model would be patently abstract, and manifestly intended to represent a phenomena under transformation, all models are abstractions to one degree or another (Morrison & Morgan, 1999). Thus, while the term "general" might be useful as a scaffold to cue abstraction, or perhaps to signal a high degree of abstraction, its use should fade as students come to realize that any model can be classified as general.

As a final thought on transferability, the abstracting item did not measure a robust type of transfer. Indeed, the item was designed to be maximally sensitive to learning, so it made transfer as easy as possible. Obviously, most modeling scenarios would not do this. Further, the instantiation of a model in the item, a dwelling, was a very limited one. A dwelling is essentially an object, while many scientific models represent processes (Bokulich, 2011). Thus, the item did not show transfer for anything like a broad range of modeling situations. Acknowledging these limitations, transfer was nevertheless observed, and as the result of a very short period of modeling instruction. Thus, it seems reasonable to hope for other forms of transfer that would be needed for authentic modeling practice, especially given a longer program of teaching.

17.6 Conclusion

To conclude, thinking of models as abstractions is a good candidate for inclusion under the aspect, nature of models, in FMC (Chap. 1). There are four reasons why. First, abstraction neatly denotes functional awareness of perhaps the most fundamental characteristic of models, namely that they are representations that pull structure away from that of their referents. Second, and concomitantly, thinking of models as abstractions runs directly counter to the ubiquitous, mistaken idea that models should be literal interpretations. Third, this way of thinking can transfer to novel modeling scenarios. Fourth, and finally, it is a way of thinking that is readily learnable via focused instruction.

Applied to current views of competence, our proposal suggests a modification to the initial aspect of the FMC. This aspect, which is described as nature of models (Fig. 1.3), sorts students' conceptions of models into three levels: copies (level I); idealized representations (level II); and theoretical constructions (level III). The suggested modification would put these levels on a unipolar scale indicating the degree to which students think of models as abstractions. Speculatively, the middle level of this scale could be denoted by thinking implicitly that models are abstractions, indicating a preference for abstraction but without the ability to describe or justify this idea. Such a level would fit the finding that some students in the present chapter readily thought about how to abstract features of a model but without having descriptive ideas of what it meant to abstract. By contrast, higher levels could be defined by the presence of descriptive notions of abstraction. Meanwhile, the lowest level would indicate the absence of thinking in terms of abstraction.

The switch to an abstraction scale would improve the nature of models aspect of the FMC in two ways. First, the term abstraction would carry more and better information about the relationship between a model and its referent than does the existing terminology. Here the reference is not just to the terminology found in the FMC (Chap. 1), but also to the broader modeling literature (Justi & Gilbert, 2003; National Research Council, 2007; Passmore et al., 2014). Second, thinking of models as abstractions would provide for a more clearly unidimensional assessment construct. By contrast, in the FMC, it is not clear whether the upper two levels, idealized representations and theoretical constructions, lie along a single dimension of measurement. To explain, while "idealized representations" is clearly ontological, "theoretical constructions" seems to incorporate the epistemological idea that models are conjectural. In the suggested formulation, this epistemological component would be shifted to another dimension of the framework.

At a more general level, the present chapter argues for an ontological dimension of competence separate from epistemological considerations. Admittedly, such a separation may or may not rest on a secure philosophical foundation (Mahr, 2011). Nevertheless, pragmatically speaking, it would allow assessment frameworks to more squarely address the crucial issue of what students should think that models are. Moreover, it seems possible to define at least one aspect of the nature of models, namely abstraction, that is useful enough to warrant a separate ontological dimension. Indeed, this dimension may be exactly what is needed if students are to surpass the literal interpretation hurdle and thus progress toward authentic modeling practice.

References

Bassok, M., & Holyoak, K. J. (1989). Interdomain transfer between isomorphic topics in algebra and physics. *Journal of Experimental Psychology: Learning, Memory, and Cognition, 15*(1), 153–166.

Bokulich, A. (2011). How scientific models can explain. *Synthese, 180*(1), 33–45.

Capps, D. K. & Shemwell, J. T. (2016). *Synthesis modeling as a way of learning through model revision*. Paper presented at the annual meeting of the National Association for Research in Science Teaching, Baltimore, MD.

Catrambone, R., & Holyoak, K. J. (1989). Overcoming contextual limitations on problem-solving transfer. *Journal of Experimental Psychology: Learning, Memory, and Cognition, 15*(6), 1147–1156.

Crawford, B. A. (2016). *Transforming teachers' thinking about engaging students in scientific modeling in school classrooms*. Presentation at the annual meeting of the National Association for Research in Science Teaching, Baltimore, MD.

Dogan, N., & Abd-El-Khalick, F. (2008). Turkish grade 10 students' and science teachers' conceptions of nature of science: A national study. *Journal of Research in Science Teaching, 45*(10), 1083–1112.

Ericsson, K. A., & Simon, H. A. (1998). How to study thinking in everyday life: Contrasting think-aloud protocols with descriptions and explanations of thinking. *Mind, Culture, and Activity, 5*(3), 178–186.

Gentner, D. (2010). Bootstrapping the mind: Analogical processes and symbol systems. *Cognitive Science, 34*(5), 752–775.

Gick, M. L., & Holyoak, K. J. (1980). Analogical problem solving. *Cognitive Psychology, 12*(3), 306–355.

Gick, M. L., & Holyoak, K. J. (1983). Schema induction and analogical transfer. *Cognitive Psychology, 15*(1), 1–38.

Gilbert, J., & Justi, R. (2016). *Modelling-based teaching in science education*. Basel, Switzerland: Springer.

Gobert, J. D., & Buckley, B. C. (2000). Introduction to model-based teaching and learning in science education. *International Journal of Science Education, 22*(9), 891–894.

Gogolin, S., & Krüger, D. (2017). Diagnosing students' understanding of the nature of models. *Research in Science Education, 47*(5), 1127–1149.

Grosslight, L., Unger, C., Jay, E., & Smith, C. L. (1991). Understanding models and their use in science: Conceptions of middle and high school students and experts. *Journal of Research in Science Teaching, 28*(9), 799–822.

Harrison, A. G., & Treagust, D. F. (1996). Secondary students' mental models of atoms and molecules: Implications for teaching chemistry. *Science Education, 80*(5), 509–534.

Harrison, A. G., & Treagust, D. F. (2000). A typology of school science models. *International Journal of Science Education, 22*(9), 1011–1026.

Hatano, G., & Inagaki, K. (1984). Two courses of expertise. Research and Clinical Center for Child Development annual report, 6 (pp. 27–36).

Ingham, A. M., & Gilbert, J. K. (1991). The use of analogue models by students of chemistry at higher education level. *International Journal of Science Education, 13*(2), 193–202.

Justi, R., & Gilbert, J. (2003). Teachers' views on the nature of models. *International Journal of Science Education, 25*(11), 1369–1386.

Klieme, E., Hartig, J., & Rauch, D. (2008). The concept of competence in educational contexts. In J. Hartig, E. Klieme, & D. Leutner (Eds.), *Assessment of competencies in educational contexts* (pp. 3–22). Göttingen, Germany: Hogrefe & Huber.

Krell, M., Reinisch, B., & Krüger, D. (2015). Analyzing students' understanding of models and modeling referring to the disciplines biology, chemistry, and physics. *Research in Science Education, 45*(3), 367–393.

Krell, M., Upmeier zu Belzen, A., & Krüger, D. (2014). Students' levels of understanding models and modelling in biology: Global or aspect-dependent? *Research in Science Education, 44*(1), 109–132.

Lin, H. S., & Chen, C. C. (2002). Promoting preservice chemistry teachers' understanding about the nature of science through history. *Journal of Research in Science Teaching, 39*(9), 773–792.

Loewenstein, J. (2017). Structure mapping and vocabularies for thinking. *Topics in Cognitive Science*, 1–17.

Mahr, B. (2011). On the epistemology of models. In G. Abel & J. Conant (Eds.), *Rethinking epistemology* (Vol. 1, pp. 301–352). Berlin, Germany: De Gruyter.

Martinez, M. E. (2015). *Education as the cultivation of intelligence*. Mahwah, NJ: Routledge.

Merriam-Webster. (2015). Retrieved from https://www.merriam-webster.com/dictionary/abstract.

Meer, B., Stein, M. I., & Geertsma, R. (1955). An analysis of the miller analogies test for a scientific population. *American Psychologist, 10*(1), 33–34.

Morrison, M., & Morgan, M. S. (1999). Models as mediating instruments. *Ideas in Context, 52*, 10–37.

National Research Council. (2007). In R. A. Duschl, H. A. Schweingruber, & A. W. Shouse (Eds.), *Taking science to school: Learning and teaching science in grades K–8*. Washington, DC: The National Academies Press.

Nersessian, N. J. (2008). *Creating scientific concepts*. Cambridge, MA: MIT Press.

Passmore, C., Gouvea, J. S., & Giere, R. (2014). Models in science and in learning science: Focusing scientific practice on sense-making. In *International handbook of research in history, philosophy and science teaching* (pp. 1171–1202). Amsterdam, The Netherlands: Springer.

Ryan, A. G., & Aikenhead, G. S. (1992). Students' preconceptions about the epistemology of science. *Science Education, 76*(6), 559–580.

Saari, H., & Viiri, J. (2003). A research-based teaching sequence for teaching the concept of modelling to seventh-grade students. *International Journal of Science Education, 25*(11), 1333–1352.

Schwartz, D. L., Varma, S., & Martin, L. (2008). Dynamic transfer and innovation. In S. Vosniadou (Ed.), *Handbook of conceptual change* (pp. 479–508). Mahwah, NJ: Erlbaum.

Schwarz, C. V., Reiser, B. J., Davis, E. A., Kenyon, L., Achér, A., Fortus, D., et al. (2009). Developing a learning progression for scientific modeling: Making scientific modeling accessible and meaningful for learners. *Journal of Research in Science Teaching, 46*(6), 632–654.

Tasquier, G., Levrini, O., & Dillon, J. (2016). Exploring students' epistemological knowledge of models and modelling in science: Results from a teaching/learning experience on climate change. *International Journal of Science Education, 38*(4), 539–563.

Treagust, D. F., Chittleborough, G., & Mamiala, T. L. (2002). Students' understanding of the role of scientific models in learning science. *International Journal of Science Education, 24*(4), 357–368.

Upmeier zu Belzen, A., & Krüger, D. (2010). Modellkompetenz im Biologieunterricht. [Modeling Competence in Biology Classes]. *Zeitschrift für Didaktik der Naturwissenschaften, 16*, 41–57.

Van Driel, J. H., & Verloop, N. (1999). Teachers' knowledge of models and modelling in science. *International Journal of Science Education, 21*(11), 1141–1153.

Part V
Attainments and Challenges

Chapter 18
Attainments and Challenges for Research on Modeling Competence

Jan van Driel, Dirk Krüger, and Annette Upmeier zu Belzen

An understanding of the nature of models and model building is an integral component of science literacy.
S.W. Gilbert (1991, p. 78)

18.1 Introduction

Models rank among the main products of science, and models are constructed and applied in research in all scientific disciplines (Gilbert, Boulter, & Elmer, 2000; Gilbert & Justi, 2016; Harrison & Treagust, 2000). If we want students to "learn science in a way that reflects how science actually works" (NRC, 1996, p. 214), the role of models and modeling should be central in science education, from the early years into higher education.

To develop students' modeling competence, we need to distinguish between the major overarching goals of science education. Hodson (2014) argued that these include:

- *Learning science*—acquiring and developing conceptual and theoretical knowledge.
- *Learning about science*—developing an understanding of the characteristics of scientific inquiry, the role and status of the knowledge it generates (…)

J. van Driel (✉)
Department of Education, Melbourne University, Melbourne, Australia
e-mail: j.vandriel@unimelb.edu.au

D. Krüger
Department of Biology, Biology Education, Freie Universität Berlin, Berlin, Germany

A. Upmeier zu Belzen
Department of Biology, Biology Education, Humboldt-Universität zu Berlin, Berlin, Germany
e-mail: jannette.upmeier@biologie.hu-berlin.de

A. Upmeier zu Belzen et al. (eds.), *Towards a Competence-Based View on Models and Modeling in Science Education*, Models and Modeling in Science Education 12, https://doi.org/10.1007/978-3-030-30255-9_18

- *Doing science*—engaging in and developing expertise in scientific inquiry and problem-solving. (Hodson, 2014, p. 2537)

Referring to Chap. 1, in which the idea for the book is tied in with previous research, and the terms model and modeling are defined on the basis of a literature review, we discuss how these goals are related to modeling competence. Related to *learning science*, we emphasize the medial function of models in imparting and acquiring knowledge. We explain that models are themselves products of science activity and thus tools for teaching and reflecting on the nature of the natural sciences (*learning about science*). Finally, we argue that, related to *doing science,* it is important that students are engaged in the process of modeling: constructing and testing their own models, or applying and reconstructing existing models, to develop skills on the basis of authentic modeling experiences.

To achieve these goals, it is important that teachers, textbook authors, science education researchers, and curriculum developers are aware of how models are actually developed and used by present-day scientists in their research endeavors. However, textbooks for secondary science education often present models as static facts or as final versions of scientific knowledge. The possible limitations of a model or the way in which a particular model was developed are seldom addressed. Moreover, textbooks rarely include modeling assignments that engage students in actively constructing or testing models (Erduran, 2001). Research on teacher knowledge in the domain of models and modeling in science has indicated that many science teachers have limited knowledge about the nature of models and the act of modeling (e.g., Borrmann, Reinhardt, Krell, & Krüger, 2014; Krell & Krüger, 2016; Van Driel & Verloop, 2002).

The present book provides an overview of recent international research on the teaching and learning of models and modeling in science, with a focus on modeling competence. The different sections of the book address the abovementioned issues. Section A focuses on the nature of models and modeling in science and how teaching and learning in this domain can be aligned with authentic scientific practices. Section B describes five studies on ways to assess modeling competence, while teachers' knowledge and strategies are central in Section C. Finally, Section D presents the findings of five studies on the development of students' modeling competence. This final chapter reviews the major contributions of this book to the research-based literature on models and modeling in science education and addresses some of the challenges and implications for further research and development.

18.2 Attainments: What Have We Learned?

Central in this volume is the framework for modeling competence (FMC), which is introduced and explained in Chap. 1. Competences have been defined in various ways in the literature. In this volume, we have adopted the view that competences include knowledge and skills that manifest themselves during performance.

Competences are not personal qualities but are learnable, and their development is related to specific domains and contexts. The FMC includes five aspects, each of which can be performed at three levels of increasing (theoretical) sophistication. Only level III meets all the abovementioned goals because, at this level, models are seen and used as research tools, whereas levels I and II focus on the medial role of models. In subsequent chapters, authors have referred to this framework, used it, or applied it, and some have suggested amendments or further refinements.

On the basis of a philosophical analysis, Aduriz-Bravo (Chap. 2) concluded that the FMC implies that school science would be analogous to scientists' science and that it allows students to work in ways that mirror those of scientists. Schwartz (Chap. 4) related modeling competence to the nature of science (NOS), demonstrating that one cannot be seen in isolation of the other: Understanding how scientists generate and develop knowledge requires insights into the epistemic nature of models and modeling. In other words, an understanding of NOS is implied in the FMC, although it is not explicitly stated. The author argued that a "requisite for understanding NOS and exhibiting level III of modeling competence involves acknowledging that a scientific model is not an exact replica of reality but a representation that serves to explain features and relationships that the scientists find curious and significant with respect to their questions" (p. 19). The chapter concluded with implications for the design of education, enabling science learning that closely reflects "how science actually works." The authentic practices of scientists were France's (Chap. 5) focus, whereby she described two detailed cases that demonstrate how models and modeling are front and center in these practices. The author argued that students' modeling competence will benefit from experiencing the messiness, creativity, and complexity of authentic modeling processes. These chapters are consistent with the framework proposed by Van der Valk, Van Driel, and De Vos (2007), which consists of a set of common features of models as recognized by practicing scientists. The framework of these authors focuses on the nature, purpose, and functions of models and recognizes the role of creativity.

Some of the subsequent chapters focused on ways to assess the modeling competence of both students and teachers or others. Taken together, these chapters demonstrate the need to develop multi-method approaches to capture the abovementioned complexity of modeling practices. In combination with verbal data (e.g., collected through interviews, think-aloud protocols), visual data can provide insights into what gets noticed during the act of modeling. Ubben, Salisbury, and Daniel (Chap. 6) described the affordances of eye-tracking technology in the context of assessing and researching modeling competence. Mathesius and Krell (Chap. 7) reviewed more traditional, "closed-ended" writing tasks that are used to assess the understanding of models and modeling in science, concluding that most existing instruments fail to capture the multidimensional nature of modeling competence in a valid manner. Applying the FMC, Schouten, Van Joolingen, and Leenaars (Chap. 8) studied the effects of drawing-based modeling, supported by a digital tool, on students' modeling competence. They reported that, although students' gains were limited, the models and reasoning logs created by students provided valuable insight into students' development of modeling competence and are thus appropriate instruments for assessing students' modeling competence.

Crawford and Flanagan (Chap. 10) focused on teachers' modeling competence in the context of the broader goal of developing scientific literacy among their students. They concluded that programs for teacher learning are needed to provide sustained and meaningful experiences concerning all aspects of models and modeling. Such programs contribute to the development of teachers' modeling competence, which, in turn, will serve to foster their students' critical thinking skills and scientific literacy. In the next chapter (Chap. 11), Ke and Schwarz illustrated how teachers can be supported to meaningfully engage their students in modeling and, by addressing epistemic considerations, scaffold them in developing their modeling competence. Crucially, the authors argued, teachers need to understand and enact productive interactions with their students. Campbell, McKenna, An, and Rodiguez (Chap. 12) also focused on meaningful student-teacher interactions. These authors explored how teachers might use "redirection" (Lineback, 2015) as a responsive methodological construct in an MLF environment. The authors demonstrated that when appropriate curricular and pedagogical frameworks are connected, a consistent focus on modeling can help a teacher place students' ideas in the foreground, scrutinize students' ideas, and use different forms of responsive instruction. As a result of this, students' modeling competence might flourish.

Other chapters described promising strategies for developing students' modeling competence. Krell and Hergert (Chap. 9), for instance, showed how an instructional sequence involving a black box could facilitate modeling practices. The approach encourages students to make their modeling activities explicit, thus fostering their meta-modeling knowledge. Louca and Zacharia (Chap. 14) used the FMC as a reference for capturing the development of the modeling competence of elementary school students who participated in a modeling-based learning environment. The authors drew attention to the role of the teacher who might use the FMC as a guide for pushing students' thinking in a particular direction. Thus, the FMC can provide teachers with a productive tool for scaffolding students to reach level III of the aspects of the framework. Similarly, Forbes and Schubert-Lange (Chap. 15) discussed the FMC as a tool that teachers can use to capture, assess, and support elementary school students' modeling competence. They described a study that laid the foundation for continuing collaborative research, aimed at supporting 3rd grade students' modeling competence in a particular domain (i.e., water systems). One of their conclusions was that if teachers are going to create supportive learning environments to foster students' modeling competence, they cannot actually make this happen unless they are prepared and supported. Bielik, Stephens, Damelin, and Krajcik (Chap. 16) focused on system modeling competence. The authors reported on an online computational modeling tool that can support students' modeling practices and systems thinking in the context of a high school chemistry unit. They proposed a framework of four key aspects of system modeling competence that is consistent with the FMC and extends this by incorporating the notion of systems thinking.

18.3 Challenges: Proposed Adaptations or Extensions of the FMC

The FMC was hypothesized as an initial competence model in the field of models and modeling. In accordance with Koeppen, Hartig, Klieme, and Leutner (2008), it refers to the cognitive facet of competence, whereas motivational components are not included. The FMC is conceptualized as comprising the abilities needed to engage in modeling practices (*Doing science,* see above) as well as knowledge of models (*Learning science)* and the modeling process in science (*Learning about science).* Although the FMC does not refer *explicitly* to meta-modeling knowledge (Schwarz & White, 2005), the focus on the understanding of models and modeling processes is meant in the sense of meta-modeling knowledge (cf. Krell & Krüger, 2017). This was recognized by Constantinou, Nicolaou, and Papaevripidou (Chap. 3) who described the modeling-based learning framework (MLF). These authors argued that modeling competence includes the abilities to practice modeling and demonstrate meta-modeling knowledge, "which refers to developing an understanding of the nature of models and an appreciation of the purpose of scientific modeling" (p. 8) in addition to having metacognitive knowledge about the modeling process.

Building on their prior work, Ke and Schwarz (Chap. 11) used the Epistemologies-in-Practices (EIP) framework, which focuses on the nature and purpose of models. Comparing this framework with the FMC, the authors noted that both frameworks share a focus on using models as sense-making tools to explain and predict phenomena. However, they concluded that the two frameworks have different perspectives: While the EIP framework is situated in specific modeling settings and practices, "the FMC points to individuals' general understanding about models (e.g., nature of models and multiple models) and the process of modeling (e.g., testing and changing models) assuming they can be applied to different modeling contexts" (p. 19). This idea is consistent with the view that the FMC is content-free (Chap. 1). Ke and Schwarz (Chap. 11) explicitly used the term *explain* as a scientific activity in their EIP framework. By contrast, in the FMC, the term *explain* (level II) can be generally understood as a pedagogical as well as a research activity (Rocksén, 2016). On level II, explaining is used in a pedagogical sense, and it refers to making something understandable for somebody, for example, by describing or clarifying known relationships and correlations between the variables that are part of the phenomenon under investigation. In a scientific context, however, scientific explanations are put forth to answer three questions: What do we know? Why does it happen? How do we know? (Rocksén, 2016). First, if the term *explain* is used in a biological context, explanation refers to a proximate causation, which explicates a biological function in terms of physiological or environmental influences. In this case, explaining is related to how something happens. Second, an explanation ultimately refers to a cause, which explicates traits in terms of evolution. In this case, explaining is related to how something happened. Because these two scientific meanings of explanation share hypothetical components, they would be located on level III of the FMC.

Lehrer and Schauble (Chap. 13) have worked with elementary students for many years, inventing, testing, and revising models as important mechanisms that can support learning in ways that are largely consistent with the FMC. However, these authors prefer the term "representational competence" to describe their goals for young children. They see the relationship between representations and models as a fluid continuum, aiming to expand representational competence "toward more elaborated forms of modeling that acknowledge multiple models of the phenomena under investigation" (p. 24).

Capps and Shemwell (Chap. 17) focused on the notion of thinking of models as abstractions. They proposed a modification to the FMC such that thinking about models as abstractions is recognized as a key element under the aspect of the "nature of models" and is given a more prominent position in the framework. They argued that the term abstraction reflects the most fundamental characteristic of models, that is, "they are representations that pull structure away from that of their referents" (p. 11). Therefore, they explained that they prefer the term abstraction over the terminology in the FMC. The same authors also challenged the two upper levels of the aspect of the "nature of models" in the FMC, that is, idealized representations (level II) and theoretical reconstructions (level III), arguing that it is unclear whether these levels "lie along a single dimension of measurement. To explain, while 'idealized representations' is clearly ontological, 'theoretical constructions' seems to incorporate the epistemological idea that models are conjectural. In the suggested formulation, this epistemological component would be shifted to another dimension of the framework" (p. 12). However, level III entails epistemological perspectives across all aspects of the FMC.

18.4 Ideas for Future Research

This book focuses on a competence-based approach to models and modeling in science education, the underlying approach referring to competence models, which are prominent in German-speaking countries (Blömeke, Gustafsson, & Shavelson, 2015; Klieme, Hartig, & Rauch, 2008; Chap. 1). However, alternative approaches are well established (e.g., Chaps. 3 and 14), for example, based on the notion of learning progressions, which is prevalent in the United States (Lehrer & Schauble, 2015; Schwarz, Reiser, Achér, Kenyon, & Fortus, 2012; Chaps. 4 and 11) or the use of rubrics (Andrade, 2000; Burke, 2006). Rubrics entail indicators of a given competence in terms of descriptions of different states. While the competence-based approach was initially grounded in theory, learning progressions for scientific modeling (e.g., Schwarz et al., 2009) are also based on empirical findings on student thinking and learning. This leads to the question of which conceptual aspects of such models might be similar or different. The fact that different approaches have been developed under different conditions such as cultural backgrounds, national curricula, and standards as well as different school systems shows the need for international discussions.

Upmeier zu Belzen, Alonzo, Krell, and Krüger (2019) started working to address parts of this question by comparing learning progressions and competence models in the field of models and modeling by applying the criteria: kinds of models, model structure, application to teaching and learning, and evaluation through research. The authors concluded that differences in the origins and original purposes led to different emphases in research efforts. Learning progressions are grounded in empirical findings on student thinking and learning and have resulted in strong hypotheses on how to foster students' modeling competence. Competence models, often structured into aspects and levels, are theoretically grounded; and evaluation efforts tend to focus on the empirical investigation of the structure of the model. However, in order to reflect on the purposes of such research, it might be helpful to think about what researchers working with these different approaches can learn from each other.

As several options for the assessment of competences are executed in this book, we are able to describe student performance at single time points with regard to special aspects of theoretical models by descriptive data. To date, these studies have focused on demonstrating that each cell of the FMC describes a distinct separable part of students' modeling competence (Krell, Upmeier zu Belzen, & Krüger, 2016). But still, Mathesius and Krell (Chap. 7) concluded that, "clearly more research is necessary to develop and evaluate scales and questionnaires for the assessment of the different aspects of modeling competence" (p. 14). Future research should therefore focus on the development and use of instruments that are designed for assessing modeling competence, for example, in large-scale longitudinal studies and for diagnosing individual student learning (Gogolin & Krüger, 2018). In addition, causal explanations for the observations of students' modeling competence are still lacking, and thus, we recommend controlled experimental studies with pre- and post-tests on the conditions of competence development. This could be done with different assessment methods, also including assessments of performance (Shavelson, 2013). Performance-based assessment might also be a relevant source of evidence for validating performed competences.

In the context of assessment efforts, validation becomes more and more important in terms of the interpretation of test scores with regard to theory. To date, objectivity and reliability are considered in most studies using assessment instruments (Chap. 7), but different sources of validity evidence have not been investigated as much: test content, response processes, internal structure, relations-to-other-variables, and consequences of testing (AERA, APA, & NCME, 2014).

Constantinou, Nicolaou, and Papaevripidou (Chap. 3) contended that in the MLF, modeling practices and meta-knowledge are equally important, "while the FMC refers only to learners' cognitive reflections about models and modeling" (p. 13). They concluded that further research is needed to investigate and clarify the role of reflection and metacognition in modeling practices, in both the contexts of scientific work and science learning. Similarly, Krell and Hergert (Chap. 9) questioned the relationship between meta-modeling knowledge and modeling processes. The authors challenged the common assumption of a positive relationship, that is, that meta-modeling knowledge guides the practice of modeling (e.g. Schwarz et al., 2009), and they argued that future research is needed to provide empirical evidence to support or reject this assumption.

Based on approaches from a specialist perspective, the classification of models and modeling processes from a student perspective should also be investigated (Krell, Upmeier zu Belzen, & Krüger, 2014a; Meisert, 2008). This should avoid the development of plain ontological categories (cf. Ritchey, 2012) and include the classification of modeling processes besides the classification of models themselves (e.g. modeling pedagogies: Campbell et al., 2015; Campbell et al., 2013).

There are repeated indications that what students think about models and modeling depends on the respective model (Krell, 2013; Krell, Upmeier zu Belzen, & Krüger, 2014b) and the relevant discipline (e.g., biology, chemistry, or physics; Krell & Krüger, 2017; Krell, Reinisch, & Krüger, 2015). It is necessary to examine the extent to which the specific context can be systematically measured as a "difficulty-generating task characteristic" (Hartig & Frey, 2012; Prenzel, Häußler, Rost, & Senkbeil, 2002) and how this influences the completion of the tasks (Krell et al., 2014b). It is necessary to clarify how specialized knowledge, the knowledge of certain models, and the meta-knowledge of models are related to problems. Related to this, Forbes and Schubert-Lange (Chap. 15) called for future research to investigate the impact of students' modeling competence on their reasoning about familiar and unfamiliar scientific concepts. In conclusion, we recommend that future research focus on students' use of scientific models across different content domains and types of models as well as within different modeling practices and epistemic considerations.

Several chapters (i.e., Chaps. 11, 12, 14, and 15) demonstrated how the FMC can provide teachers with a tool for capturing and supporting their students' modeling competence. These chapters concluded that teachers need to be prepared and supported in order to be able to create supportive learning environments that actually foster students' modeling competence. Specifically, this requires teachers to understand and enact meaningful interactions with their students (Chaps. 11 and 12). Specific learning resources and programs are needed to assist teachers in developing this expertise. Research is necessary to investigate the efficacy of such programs. Ke and Schwarz (Chap. 11) concluded that there is still a lot "to be learned about how to meaningfully engage students in scientific modeling and how to support teachers in doing so. Understanding the interactions of teaching and learning along with supporting teachers to enact these productive interactions will be critical to advancing the field" (p. 24).

18.5 Conclusion

This book opened with a statement about the increased recognition of models and modeling as a core practice of scientific work and, consequently, of the importance of engaging students in an authentic modeling practice as a key element of their science education. The research presented in this book spanned a wide range in terms of theoretical and methodological perspectives as well as the contexts and settings of the authors. Despite this variety, there seemed to be a consensus about the idea that, ranging from students' early years to the university level, if science education in the twenty-first century aims to provide students with up-to-date scientific experiences, a central role should be played by learning scientific models, learning to model, and learning about models and modeling. This book brings together a number of approaches and perspectives that are geared toward providing such learning opportunities and reports on research evidence for the effectiveness of some of these approaches. The FMC, as shown in this book, can play a prominent role in this context. It can be used to design curriculum materials and teaching-learning sequences for students as well as (pre-service) teachers. In addition, it can be used as an analytical framework for the assessment of students' modeling competence and teachers at a particular time or to monitor the development of these competences over time. As became apparent in this book, the FMC can be used productively alongside other perspectives and approaches. Through its use in a variety of contexts, the FMC itself is in need of further development. Suggestions for future research that can inform this development include instruments and procedures for assessing modeling competence, the relationships between meta-modeling and meta-cognitive knowledge and modeling practices, and longitudinal research on the development of students' modeling competence over time.

References

Andrade, H. (2000). Using rubrics to promote thinking and learning. *Educational Leadership, 57*(5), 13–18.

AERA, APA, & NCME. (2014). *Standards for educational and psychological testing*. Washington, DC: American Educational Research Association.

Blömeke, S., Gustafsson, J.-E., & Shavelson, R. J. (2015). Beyond Dichotomies. Competence viewed as a continuum. *Zeitschrift für Psychologie, 223*, 3–13. Hogrefe Publishing.

Borrmann, J. R., Reinhardt, N., Krell, M., & Krüger, D. (2014). Perspektiven von Lehrkräften über Modelle in den Naturwissenschaften [Teachers' perspectives on models in science]. *Erkenntnisweg Biologiedidaktik, 13*, 57–72.

Burke, K. (2006). *From standards to rubrics in 6 steps*. Heatherton, Australia: Hawker Brownlow.

Campbell, T., Oh, P., & Neilson, D. (2013). Reification of five types of modeling pedagogies with model-based inquiry (MBI) modules for high school science classrooms. In M. Khine & I. Saleh (Eds.), *Approaches and strategies in next generation science learning* (pp. 106–126). Hershey, PA: Information Science Reference.

Campbell, T., Oh, P., Maughn, M., Kiriazis, N., & Zuwallack, R. (2015). A review of modeling pedagogies: Pedagogical functions, discursive acts, and technology in modeling instruction. *Eurasia Journal of Mathematics, Science &Technology Education, 11*, 159–176.

Erduran, S. (2001). Philosophy of chemistry: An emerging field with implications for chemistry education. *Science & Education, 10*, 581–593.

Gilbert, J. K., Boulter, C. J., & Elmer, R. (2000). Positioning models in science education and in design and technology education. In J. K. Gilbert & C. J. Boulter (Eds.), *Developing models in science education* (pp. 3–18). Dordrecht, The Netherlands: Kluwer Academic Publishers.

Gilbert, S. W. (1991). Model building and a definition of science. *Journal of Research in Science Teaching, 28*, 73–79.

Gilbert, J. K., & Justi, R. (2016). *Modelling-based teaching in science education*. Springer international publishing.

Gogolin, S., & Krüger, D. (2018). Students' understanding of the nature and purpose of models. *Journal of Research in Science Teaching, 55*(9), 1313–1338.
Harrison, A., & Treagust, D. F. (2000). A typology of school science models. *International Journal of Science Education, 22*, 1011–1027.
Hartig, J., & Frey, A. (2012). Konstruktvalidierung und Skalenbeschreibung in der Kompetenzdiagnostik durch die Vorhersage von Aufgabenschwierigkeiten. *Psychologische Rundschau (2012), 63*, 43–49.
Hodson, D. (2014). Learning science, learning about science, doing science: Different goals demand different learning methods. *International Journal of Science Education, 36*(15), 2534–2553.
Klieme, E., Hartig, J., & Rauch, D. (2008). The concept of competence in educational contexts. In J. Hartig, E. Klieme, & D. Leutner (Eds.), *Assessment of competencies in educational contexts* (pp. 3–23). Göttingen, Germany: Hogrefe.
Koeppen, K., Hartig, J., Klieme, E., & Leutner, D. (2008). Current issues in competence modeling and assessment. *Zeitschrift für Psychologie* [Journal of Psychology], *216*, 61–73.
Krell, M. (2013). *Wie Schülerinnen und Schüler biologische Modelle verstehen: Erfassung und Beschreibung des Modellverstehens von Schülerinnen und Schülern der Sekundarstufe.* I (Dissertation). Berlin: Logos.
Krell, M., Upmeier zu Belzen, A. & Krüger D. (2014a). How year 7 to year 10 student's categorise models: Moving towards a student-based typology of biological models. In D. Krüger & Ekborg, M. (Eds.), *Research in biological education* (pp. 117–131). Westermann.
Krell, M., Upmeier zu Belzen, A., & Krüger, D. (2014b). Context-specificities in students' understanding of models and modelling: An issue of critical importance for both assessment and teaching. In C. Constantinou, N. Papadouris, & A. Hadjigeorgiou (Eds.), *E-book proceedings of the ESERA 2013 conference. Science education research for evidence-based teaching and coherence in learning. Part 6. Nature of science: History, philosophy and sociology of science.* Nicosia, Cyprus: European Science Education Research Association.
Krell, M., Reinisch, B., & Krüger, D. (2015). Analyzing students' understanding of models and Modeling referring to the disciplines biology, chemistry, and physics. *Research in Science Education, 45*(3), 367–393.
Krell, M., & Krüger, D. (2016). Testing models: A key aspect to promote teaching-activities related to models and modelling in biology lessons? *Journal of Biological Education, 50*, 160–173.
Krell, M., Upmeier zu Belzen, A., & Krüger, D. (2016). Modellkompetenz im Biologieunterricht. In A. Sandmann & P. Schmiemann (Eds.), *Biologiedidaktische Forschung: Schwerpunkte und Forschungsstände*. Logos, Band 1 (pp. 83–102).
Krell, M., & Krüger, D. (2017). University students' meta-modelling knowledge. *Research in Science & Technological Education, 35*, 261–273.
Lehrer, R., & Schauble, L. (2015). The development of scientific thinking. In R. M. Lerner, L. S. Liben, & U. Mueller (Eds.), *Handbook of child psychology and developmental science* (pp. 671–714). Trenton, NJ: Wiley.
Lineback, J. E. (2015). The redirection: An Indicator of how teachers respond to student thinking. *The Journal of the Learning Sciences., 24*(3), 419–460.
Meisert, A. (2008). Vom Modellwissen zum Modellverständnis - Elemente einer umfassenden Modellkompetenz und deren Fundierung durch lernerseitige Kriterien zur Klassifikation von Modellen. *Zeitschrift für Didaktik der Naturwissenschaften, 12*, 243–261.
NRC (National Research council). (1996). *National science education standards.* Washington, DC: National Academic Press.
Prenzel, M., Häußler, P., Rost, J., & Senkbeil, M. (2002). Der PISA-Naturwissenschaftstest: Lassen sich die Aufgabenschwierigkeiten vorhersagen? *Unterrichtswissenschaft, 30*(1), 120–135.
Ritchey, T. (2012). *Outline for a morphology of modelling methods.* Contribution to a General Theory of Modelling. Acta Morphologica Generalis 1(1).
Rocksén, M. (2016). The many roles of "explanation" in science education: A case study. *Cultural Studies of Science Education, 11*, 837–868.

Schwarz, C. V., Reiser, B. J., Achér, A., Kenyon, L., & Fortus, D. (2012). MoDeLS. In A. C. Alonzo & A. W. Gotwals (Eds.), *Learning progressions in science: Current challenges and future directions* (pp. 101–137). Rotterdam, The Netherlands: Sense Publishers.
Schwarz, C. V., Reiser, B. J., Davis, E. A., Kenyon, L., Achér, A., Fortus, D., et al. (2009). Developing a learning progression for scientific modeling: Making scientific modeling accessible and meaningful for learners. *Journal of Research in Science Teaching, 46*(6), 632–654.
Schwarz, C. V., & White, B. (2005). Metamodeling knowledge: Developing students' understanding of scientific modeling. *Cognition and Instruction, 23*(2), 165–205.
Shavelson, R. J. (2013). On an approach to testing and modeling competence. *Educational Psychologist, 48*(2), 73–86.
Upmeier zu Belzen, A., Alonzo, A. C., Krell, M., & Krüger, D. (2019). Learning progressions and competence models – A comparative illustration through models and modeling in science education. In E. McLoughlin, O. Finlayson, S. Erduran, & P. Childs (Eds.), *Bridging research and practice in science education*. Selected Papers from the ESERA 2017 Conference. Chapter 16. Springer.
Van der Valk, T., Van Driel, J. H., & De Vos, W. (2007). Common characteristics of models in present-day scientific practice. *Research in Science Education, 37*, 469–488.
Van Driel, J. H., & Verloop, N. (2002). Experienced teachers' knowledge of teaching and learning of models and modelling in science education. *International Journal of Science Education, 24*(12), 1255–1272.

MIX
Papier aus verantwortungsvollen Quellen
Paper from responsible sources
FSC® C105338

If you have any concerns about our products,
you can contact us on
ProductSafety@springernature.com

In case Publisher is established outside the EU,
the EU authorized representative is:
Springer Nature Customer Service Center GmbH
Europaplatz 3, 69115 Heidelberg, Germany

Printed by Libri Plureos GmbH
in Hamburg, Germany